Horizons in World Physics, Volume 260

# INSTABILITIES OF RELATIVISTIC ELECTRON BEAM IN PLASMA

# Horizons in World Physics Series

Horizons in World Physics, Volume 260

# Instabilities of Relativistic Electron Beam in Plasma

## Valery B. Krasovitskiy

Nova Science Publishers, Inc.
*New York*

For permission to use material from this book please contact us:
Telephone 631-231-7269; Fax 631-231-8175
Web Site: http://www.novapublishers.com

**NOTICE TO THE READER**
The Publisher has taken reasonable care in the preparation of this book, but makes no expressed or implied warranty of any kind and assumes no responsibility for any errors or omissions. No liability is assumed for incidental or consequential damages in connection with or arising out of information contained in this book. The Publisher shall not be liable for any special, consequential, or exemplary damages resulting, in whole or in part, from the readers' use of, or reliance upon, this material. Any parts of this book based on government reports are so indicated and copyright is claimed for those parts to the extent applicable to compilations of such works.

Independent verification should be sought for any data, advice or recommendations contained in this book. In addition, no responsibility is assumed by the publisher for any injury and/or damage to persons or property arising from any methods, products, instructions, ideas or otherwise contained in this publication.

This publication is designed to provide accurate and authoritative information with regard to the subject matter covered herein. It is sold with the clear understanding that the Publisher is not engaged in rendering legal or any other professional services. If legal or any other expert assistance is required, the services of a competent person should be sought. FROM A DECLARATION OF PARTICIPANTS JOINTLY ADOPTED BY A COMMITTEE OF THE AMERICAN BAR ASSOCIATION AND A COMMITTEE OF PUBLISHERS.

LIBRARY OF CONGRESS CATALOGING-IN-PUBLICATION DATA
*Available upon request*

ISBN 1-60021-515-7

Published by Nova Science Publishers, Inc. ✦ New York

# Contents

# PREFACE

This book is devoted to the nonlinear theory of the collective interaction between a modulated beam of relativistic charge particles and narrow electromagnetic and Langmuir wave packets in plasma or gas slow-wave systems. Regular oscillations excited by a relativistic electron beam under the conditions of Cherenkov resonance and the anomalous Doppler effect can be used to generate coherent microwave radiation and accelerate charge particles in plasma.

The idea of self-focusing of an intense relativistic electron beam due to the beam-plasma instability is discussed. The electromagnetic radiation carries away the pulse, and the beam surface electrons are subject to the action of radiation drag. In this way, the actual possibility of radial beam focusing by coherent radiation is confirmed in principle.

The beam instability is frequently encountered in laboratory, geophysical and astrophysical plasma studies. Therefore, the book will be of interest to experts in plasma physics and relative sciences, as well as to all those who are interested in the problems of modern physics.

To solve the applied problems of plasma physics, the general methods of macroscopic electrodynamics are used. Most of the chapters are supplemented by appendices that contain detailed mathematical derivations and may be useful for lecturers and students.

# INTRODUCTION

Among diverse plasma instabilities one can single out the beam plasma instability discovered by Akhiezer and Fainberg [1] and Bohm and Gross [2] in 1949. Despite a lot of research carried out in this area since, the attention of physicists theorists and experimentalists in this problem remains unflagging. This great interest in the phenomenon of beam instability is explained by the practical value of this phenomenon which fathered appearance of a new direction in plasma physics, namely that of plasma electronics [3,4]. On the other hand, the beam plasma instability (reversal of the Landau damping) is the simplest model of the collective resonance particle-beam interactions the studies on which, accumulated as experience, can be used for tackling more sophisticated plasma physics challenges.

The initial theoretical studies of various beam instability types demonstrated that the majority of them could be reduced to the Cerenkov effect and anomalous or normal Doppler effect [4]. The linear theory allowed for conditions of instability appearance and increments [5-8] and stimulated the running of experiments in electron beam-plasma interactions.

The very first successes in advancing the nonlinear beam plasma theory are associated with using charged particle beams for turbulent heating of fusion plasmas [9]. The quasi-linear theory allowed one to analyze the process of the beam and plasma distribution non-linear relaxation function (production of plateaus) during beam interactions across a broad oscillation spectrum with low amplitudes when about 30% of the beam energy is transformed into the plasma oscillation energy [10-14].

An alternative research direction to the quasi-linear theory is associated with using plasmas as a slow-wave waveguide for charged particle acceleration [15-18]. The needed high electrostatic field strengths come into being at relatively low plasma densities when the waveguide is transparent for accelerated particles. An effective method of longitudinal wave excitation in plasmas is the beam instability. However, a non-modulated beam excites oscillations across a broad spectrum in the plasma and a considerable portion of energy goes to beam and plasma heating. That is why the pre-requisute for an effective acceleration is an oscillation spectrum narrowing, that is to say the beam excitation of the regular wave in the plasma. This problem was solved by preliminary beam modulation at the plasma input that prevented the instability transition into the kinetic regime [16].

Violation of the natural progression of the beam instability in the external modulating field causes all beam and plasma electrons to couple with a narrow wave packet which is a plane wave with a slow changing amplitude and phase. The original research [19] conducted in this experimental layout indicated that the exponential growth of low perturbations is stabilized by the beam non-linearity on account of resonance particles being captured by wave or waveguide property variation of the slow-wave medium. In the both cases, there occur field amplitide nonlinear oscillations with the period on the order of the reverse of the linear theory.

A monovelocity electron beam at the nonlinear stage of the beam plasma instability breaks up into bunches that move periodically from the field slow-wave phases into the accelerating ones. To solve this problem the numeric simulation has to be done [20,21]. The Langmuir oscillation excitation with such a beam that has been preliminarily broken up into separate electron-ion bunches, with the spatial modulation period being equal to the plasma wavelength, was considered in the paper [22]. This model describes longitudinal non-linear oscillations in a plasma, excited by a low-density beam, and has an analytical solution that qualitatively corresponds to the non-modulated beam nonlinear stage. Besides, relativistic electron bunch interactions with longitudinal plasma oscillations [23] is to be considered separately for collective charged particle acceleration by charge density waves [24,25].

This monograph is written on the problems of the nonlinear theory of interaction of the charged particle monovelocity beam with regular longitudinal and electromagnetic waves in the slow-wave plasma/gas waveguides. The focus of attention is paid to elaboration of simple mathematical models, allowing, as a rule, to derive such analytical solutions that permit the explicit physical interpretation. As distinct from the quasi-linear approach, where the amplitudes of harmonics remain (linear), with the field back action on the beam being determined by a broad spectrum over the wave numbers, in a wave, in our case, is excited with a fixed wave number and frequency. For low-density beams, when the unstable perturbation increments are small relative to eigen oscillation frequencies of the slowing medium, the method of slow amplitudes and phases [26] is employed to solve the sets of nonlinear equations in their full derivatives vs. time or cooordinate.

The goal of the Chapter 1 is to carry out an analytical and numerical study of the instability of the relativistic electron beam with allowance for both the beam and plasma nonlinearities. In the case of a weakly relativistic beam, the beam-wave resonance detuning is caused by both the large spread in the electron velocities and the bunching of a fraction of the electrons in the decelerating/accelerating phases of the wave. At the same time, a nonlinear saturation of the field amplitude due to a change in the waveguide properties of the plasma in a strong electric field (change in the phase velocity of the wave) can be disregarded. But in the transitional region of the intermediate relativistic energies the competition between the two nonlinear effects appear. The plasma heating that cannot be described in the hydrodynamic approach because of the asynchronous nature electron oscillations is taken into account in numerical experiments.

Particle-in-cell simulations of the plasma and beam showed that the nonlinear relaxation of a monoenergetic relativistic electron beam in a dense plasma occurs in three stages. In the first stage, the exponential growth of a small perturbation with the highest hydrodynamic rate results in the trapping of resonant electrons by the wave and in the saturation of an unstable mode at the first maximum of the field amplitude. The nonlinear phase oscillations of the trapped-electron bunches that arise in the second stage are accompanied by oscillations of the field amplitude and the onset of an oscillatory (modulational) instability, during which the short-wavelength perturbations (of order the Debye length) damp via Landau damping and the wave field energy is converted into the energy of plasma electrons. In the final stage of the instability, energy exchange between the beam and the plasma is almoscompletely absent. Numerical investigation of the plasma-beam interaction show that the nonlinear relaxation of a monoenergetic relativistic electron beam in a dense plasma is in a good agreement with the theoretical results.

It is supposed in Chapter 2 that the inhomogeneity of particle distribution in the transverse cross-section for the finite radius beam causes a "hook-up" of potential and electromagnetic oscillation branches and an energy transformation of the beam-excited Langmuir wave into electromagnetic energy on the density gradient.That is why the instability increment is determined not only by the energy, yielded per unit volume of beam and plasma, but it also depends on the Pointing vector flow across the beam boundary into plasma. Under the conditions of the beam instability a slow Langmuir wave is excited in the plasma that is transformed on the plasma-beam boundary into electromagnetic wave propagating beyond of plasma in a dense dielectric.

The electromagnetic radiation carries away the pulse, and the beam surface electrons are subject to the action of radiation drag. An equation has derive for the radiation force (which is quadratic in the field amplitude) acting on the surface of a rectilinear beam of arbitrary cross section moving in a plasma-filled waveguide with dielectric walls. This is the radiative-damping force, which is proportional to the beam density, since the radiation is coherent. Since the field amplitude increases exponentially in time, the process of radial beam focusing by coherent radiation pressure is highly transient. This effect may be of importance in a scheme for the release of controlled thermonuclear energy where an intense relativistic electron beam is shot onto a dense thermonuclear target to ignite a sequence of thermonuclear micro-explosions. The concentration of the beam energy into a small volume is necessary and this might be achieved with the described electromagnetic beam self-focusing.

Chapter 3 deal with excitation of nonlinear quasi-transverse Lanmuir waves in a plasma with a relativistic low-density electron beam.It is assumed that the beam radius is small relative to the wavelength, with the oscillations being polarized across the beam velocity. We discuss here some nonlinear mechanisms which act to suppress the transverse Cherenkov and cyclotron instabilities in the "thin-beam" model. Our investigation shows that the nonlinear evolution of the plasma–beam system depends on the growth rate of instability which is determined by the modulation frequency. This theory expands the results of one-dimensional nonlinear theory which are represented in Chap. 1 and Chap. 2.

Chapter 4 elaborates on the nonlinear theory of regular oscillation enhancement in a semi-finite plasma, using a modulated electron beam. In the case of longitudinal Langmuir perturbations the plasma and beam nonlinearity is taken into account so that the papers [27] and [28] are the limiting cases of the common problem. Beam interactions with the transverse Langmuir oscillations (radial focusing or defocusing) are considered in the thin beam approximation.

The electromagnetic instability of "super speed-of-light" charged particle beam in the longitudinal magnetic field in slow-wave structures is studied in Chapter 5. The nonlinear theory of charged beam interactions with a slow regular electromagnetic wave under the conditions of the anomalous and normal Doppler effects is including the effects of coherent radiation excitation by beam and charged particle acceleration by radiation pressure.

An effective physical mechanism of charge acceleration in the magnetic field by radiation pressure is autoresonance acceleration. The resonance remains during relativistic increasing of the electron mass, and in the absence of the radiation drag force and Coulomb collisions, the energy of the particle in applied field grows unlimitedly. In the case of the finite density beam the problem outgrows the limits of the single particle approach, and the accelerated particle energy gets limited by the beam spatial charge.

Final Chapter 6 studies the nonlinear Langmuir and electromagnetic oscillations in two-level and three-level media with electron beam. Consideration is given to the instability of a two-level atomic beam moving at the super-speed of light under the conditions, in which the collective pumping mechanism of laser system is realized.

The reverse effect of energy transformation of the three-level system into the plasma wave energy under the conditions of the normal Doppler effect is considered in the paper. In the meanwhile, unstable becomes the medium with inverse level population, with the beam acting as a waveguide for the Langmuir wave. Conversion of polarization and Cerenkov

5

losses of energy of the modulated beam in a medium with inverse level population creates a principal possibility for charged particle beam acceleration in an unstable gaseous medium.

This monograph presents results of theoretical research, conducted with the author's participation at Kharkov Institute of Physics & Technology, Rostov State University and Keldysh Institute of Applied Mathematics during 1965-2000. The original research [19,22] was carried out jointly with the late Prof. V.I. Kurilko, being the propulsion of Academician Ya.B. Fainberg's ideas in the area of novel charged particle acceleration methods in plasma and gaseous media as well as charged particle acceleration, using the light.

A great contribution to elaboration of theory and numeric methods was made by researchers and post-graduate students of the Chair of Theoretical and Nuclear Physics of Rostov State University: V.G. Dorofeenko, G.V. Fomin and S.I. Osmolovskii. The numeric simulation of plasmas and the beam, using the particle method, was made by Yu.A. Volkov at the Keldysh Institute.

# CHAPTER 1

## TWO-STREAM INSTABILITY OF A RELATIVISTIC ELECTRON BEAM IN A PLASMA

The instability of a monoenergetic electron beam in a dense plasma, $\nu^{1/3} \ll 1$ ($\nu = n_b/n_p$, where $n_b$ and $n_p$ are the beam and plasma density, respectively) is known to be accompanied by the conversion of the beam energy of Langmuir oscillations occurring with the characteristic time $t_{NR} \simeq \left(\nu^{1/3}\omega_p\right)^{-1}$. For $t > t_{NR}$ the trapping of the resonant electrons by the field of the wave suppresses the instability if the field of the wave $E \simeq \nu^{1/3}\omega_p v_0/e$ (where $v_0$ is the initial beam velocity and $\omega_p$ is the electron plasma frequency) is weak. As a result, only a small part of the energy of a nonrelativistic beam is converted into the energy of oscillations and the plasma electron motion can be described by the linearized equations [1-4].

The relativistic increase in the mass of the electrons causes a decrease in the growth rate of the instability and increases the time over which the energy exchange between the beam and the wave is efficient $t_R \simeq \gamma_0 t_{NR}$ ($\gamma_0 = (1 - \beta_0^2)^{-1/2}$, $\beta_0 = v_0/c$), i.e., the time during which the energy of the electron plasma oscillations $E^2 \simeq \alpha 8\pi n_b mc^2 \gamma_0$ increases at the rate proportional to $\alpha = \gamma_0 \nu^{1/3}$ [5-11].

For larger beam energies $\alpha \simeq 1$, a larger number of relativistic electrons travels in the accelerating phases of the wave. This is related to the fact that the masses of these electrons increase during acceleration while their velocities remain close to the speed of light; as a result, no substantial slippage of these electrons with respect to the wave occurs. In the interaction, part of the energy the decelerated electrons is transferred to the accelerated electrons, and the efficiency of conversion of the beam energy into the plasma electron oscillations decreases for large beam energies [10]. Consequently, as the parameter $E(\alpha)$ increases, the curve $E(\alpha)$ reaches a maximum and then decreases [7,10,11].

For relativistic beam energies $\alpha \simeq 1$, the increase in the relativistic mass of the electrons is so large that the frequency of the phase oscillations is small in comparison with the instability growth rate, and, consequently, we can describe the beam electron dynamics by the linearized equations [12]. Besides, the nonlinear shift in the frequency of the Langmuir oscillations is accompanied by a decrease in the phase velocity of the wave which causes detuning between the beam and the excited wave [13]. The qualitative estimates in [5,6] and [13] lead to the following formula for the maximum field energy density:

$$\frac{E^2}{8\pi} \simeq n_b mc^2 \gamma_0 \begin{cases} \alpha, & \alpha \ll 1, \\ \alpha^{-2}, & \alpha \gg 1. \end{cases} \tag{1.1}$$

The goal of the present chapter is to carry out an analytical and numerical study of the instability of the relativistic electron beam with allowance for both the beam and plasma nonlinearities. The asymptotic dependence $E \sim \sqrt{\alpha}$ corresponds to a weakly relativistic beam. In this case, the beam-wave resonance detuning is caused by both the large spread in the electron velocities and the bunching of a fraction of the electrons in the decelerating (accelerating) phases of the wave. The asymptotic dependence $E \sim 1/\alpha$ agrees well with the numerical results in the range of ultrarelativistic energies. In the transitional region of the intermediate energies, the dependence $E(\alpha)$ is more complicated due to the competition between the two nonlinear effects [14].

At ultrarelativistic energies, the discrete set of modulation frequencies $\omega_n$ for which there exists a solitary-pulse solutions and for which energy is returned to the beam is found below by numerical and analytic methods. The hydrodynamics solutions of this type exist for a discrete set $\epsilon_n = 1 - \omega_p^2/\omega_n^2$ of parameters value at which the frequency of the phase oscillations is much less than the instability growth rate [15,16].

The plasma heating that cannot be described in the hydrodynamic approach because of the asynchronous nature electron oscillations must be taken into account in numerical experiments. Particle-in-cell simulations of the plasma and beam showed that the nonlinear relaxation of a monoenergetic relativistic electron beam in a dense plasma occurs in three stages [17,18]. In the first stage, the exponential growth of a small perturbation with the highest hydrodynamic rate results in the trapping of resonant electrons by the wave and in the saturation of the unstable mode at the first maximum of the field amplitude. The nonlinear phase oscillations of the trapped-electron bunches that arise in the second stage, are accompanied by oscillations of the field amplitude and the onset of an oscillatory (modulational) instability, during which the short-wavelength perturbation ($k\lambda_d \leq 1$, where $\lambda_d$ is the Debye length) are damped via Landau damping and the wave field energy is converted into the energy of plasma electrons (ions). In the final stage of the instability, energy exchange between the beam and the plasma is almost completely absent.

## §1.1. Kinetic theory of the instability of a monoenergetic electron beam.

We use the equations of relativistic hydrodynamics for plasma electrons that include the equation of motion for the momentum $P$ [velocity $V = P/\Gamma$, $\Gamma = (1-V^2/c^2)^{-1/2}$] and the continuity equation (the plasma density is denoted as $N$) and the Poisson equation

$$\frac{\partial P}{\partial t} + V\frac{\partial P}{\partial x} = eE, \quad \frac{\partial N}{\partial t} + \frac{\partial}{\partial x}NV = 0,$$

$$\frac{\partial E}{\partial x} = 4\pi e\left[N - N_{\mathrm{i}} + \sum_{s=-\infty}^{\infty}\delta[x - sl - x_s(t)]\right]. \tag{1.1.1}$$

Here, $N_{\mathrm{i}}$ is the density of the immobile ions. These equations, together with the equation of motion for the beam electrons

$$\frac{d}{dt}\frac{mc\beta_s}{\sqrt{(1-\beta_s^2)}} = eE(t, x_s), \quad \beta_s = \frac{1}{c}\frac{dx_s}{dt} \tag{1.1.2}$$

form a self-consistent set for determining the electric field $E$ in a plasma.

We use a model representation of the electron beam in the form of negatively charged (with a surface charge density $e\sigma$) sheets $x = x_s(t)$ whose velocities at the initial time t=0 are equal to $v_0$ [19].

From equations (1.1.1) and (1.1.2), the energy and momentum equations follows:

$$\frac{\partial}{\partial t}\left(N\mathcal{E} + \frac{E^2}{8\pi}\right) + \frac{\partial}{\partial x}(VN\mathcal{E}) = -e\sigma E \sum_{s=-\infty}^{\infty} \dot{x}_s(t)\delta[x - sl - x_s(t)],$$

$$\frac{\partial}{\partial t}(NP) + \frac{\partial}{\partial x}\left(VNP - \frac{E^2}{8\pi}\right) = -e\sigma E \sum_{s=-\infty}^{\infty} \delta[x - sl - x_s(t)],$$

(1.1.3)

where $\mathcal{E} = mc^2\Gamma$ is the energy of a plasma electron.

Averaging over the wavelength $\lambda$, we find the energy and momentum integrals

$$\left\langle N\mathcal{E} + \frac{E^2}{8\pi}\right\rangle + \frac{\sigma}{l}\sum_{s=1}^{S}\frac{mc^2}{\sqrt{1-\beta_s^2}} = C_1, \quad \langle NP\rangle + \frac{\sigma}{l}\sum_{s=1}^{S}\frac{mc\beta_s}{\sqrt{1-\beta_s^2}} = C_2,$$

(1.1.4)

where $S = \lambda/l$ is the number of electron bunches per wavelength,

$$\langle ...\rangle = \frac{1}{l}\int_{-\lambda2}^{\lambda/2}\langle ...\rangle \, dx.$$

From the Poisson equation (1.1.1) follows that plasma-beam system satisfy the neutrality condition

$$n_p + \frac{\sigma}{l} = N_{\text{i}}, \quad n_p = \langle N\rangle.$$

## 1. Set of averaged equations.

To obtain the set of equations for the amplitude and phase of the field, we pass to the new variables

$$t, x \to t' = t, \ x' = x - v_0 t.$$

For a beam of low density

$$\frac{\nu^{1/3}}{\gamma_0} \ll 1, \quad \nu = \frac{\sigma}{n_p l},$$

(1.1.5)

the temporal derivatives are small compared to the spatial ones

$$\frac{\partial}{\partial t} \ll v_0 \frac{\partial}{\partial x'}$$

and we need to take them into account only in the equation for the field. Under this simplifying assumption, from formulas (1.1.1), we obtain the equation

$$\frac{\partial^2}{\partial t\partial x'}(\mathcal{E} - 2v_0 P) + v_0\frac{\partial^2}{\partial x'^2}(v_0 P - \mathcal{E}) + 4\pi e^2 NV = -4\pi e\sigma$$

$$\times \sum_{s=-\infty}^{\infty}\dot{x}_s(t)\delta[x' - sl - x_s(t) + v_0 t],$$

(1.1.6)

containing the functions $\mathcal{E}, P, N$ and $V$, which can be expressed through the potential $\varphi$ by using the following integrals:

$$\mathcal{E} - v_0 P = 1 - \frac{e\varphi}{mc^2}, \quad \frac{N}{n_p} = \left(1 - \frac{V}{v_0}\right)^{-1},$$

(1.1.7)

9

where $E = -d\varphi/dx'$.

Introducing the expansion of (1.1.6) and (1.1.7) in powers of the dimensionless potential $\Phi$, we arrive at the following equation:

$$-\frac{2}{v_0}\frac{\partial^2\Phi}{\partial t \partial x'} + \frac{\partial^2\Phi}{\partial x'^2} + \frac{4\pi e^2}{mv_0^3}NV = \frac{4\pi e^2 \sigma}{mv_0^3}\sum_{s=-\infty}^{\infty}\dot{x}_s(t)\delta[x' - sl - x_s(t) + v_0 t], \qquad (1.1.8)$$

where

$$NV = n_p v_0 \Phi[1 + (2 - \beta_0^2/2)\Phi + (5 - 3\beta_0^2)\Phi^2],$$

$$\Phi = \frac{e\varphi}{mv_0^2}, \quad \beta_0 = \frac{v_0}{c}.$$

In the absence of the beam $\sigma = 0$, the equation for small amplitude (linear) plasma oscillations has the form

$$\Phi''_L(x') + k^2\Phi_L(x') = 0,$$

$$k = \omega_p/v_0, \quad \omega_p^2 = 4\pi e^2 n_p/m,$$

and the solution to this equation has the form

$$\Phi_L(x') = a\cos(kx' + \vartheta), \qquad (1.1.9)$$

where $a$ and $\vartheta$ are integration constants.

In the plasma with low-density beam, the solution has the form similar to (1.1.9), but in this case, the amplitude and phase of the wave are slowly varying functions of time: $|a'| \ll a$ $|\vartheta'| \ll \vartheta$. In addition, the nonlinear terms in (1.1.6) give rise to high-frequency harmonics of the fundamental frequency:

$$\Phi_{NL}(t, x') = A_0(t) + \mathrm{Re}\big[A_1(t)e^{ikx'} + A_2(t)e^{2ikx'}\big], \qquad (1.1.10)$$

where $A_1 = a\exp(i\vartheta)$.

After substituting (1.1.10) into (1.1.6) and spatially averaging over the period $\lambda = 2\pi/k$, we obtain the following set of nonlinear equations:

$$-\frac{i}{\omega_p}\dot{A}_1 + \left(2 - \frac{\beta_0^2}{2}\right)\left(A_0 A_1 + \frac{1}{2}A_2 A_1^*\right) + \frac{3}{8}(5 - 3\beta_0^2)|A_1|^2 A_1 = \frac{\nu}{v_0 S}\sum_{s=1}^{S}\dot{x}_s(t)\exp(-i\psi_s),$$

$$A_0 = -\left(1 - \frac{\beta_0^2}{4}\right)|A_1|^2, \quad A_2 = \frac{1}{6}(2 - \beta_0^2)A_1^2,$$

$$\qquad (1.1.11)$$

where $\psi_s = k[x_s(t) - v_0 t]$.

Since the plasma nonlinearity is pronounced for ultrarelativistic energies, the equation (1.1.11) can be simplified by setting $\beta_0 \simeq 1$ in the terms containing the plasma nonlinearity. The resulting closed set of equations, taking into account the nonlinearity of the both plasma and beam, can be introduced in the form

$$\frac{1}{\omega_p}\frac{dA}{dt} - \frac{3i}{16}|A|^2 A = -\frac{\nu}{v_0 S}\sum_{s=1}^{S}\dot{x}_s(t)\exp(-i\psi_s),$$

$$\frac{1}{\omega_p}\frac{d}{dt}\beta_s\gamma_s = \beta_0\mathrm{Re}A\exp(i\psi_s), \qquad (1.1.12)$$

$$\frac{1}{\omega_p}\frac{d\psi_s}{dt} = \frac{\dot{x}_s(t)}{v_0} - 1, \quad \beta_s = \frac{\dot{x}_s}{c}, \quad \gamma_s = \frac{1}{\sqrt{1 - \beta_s^2}}.$$

10

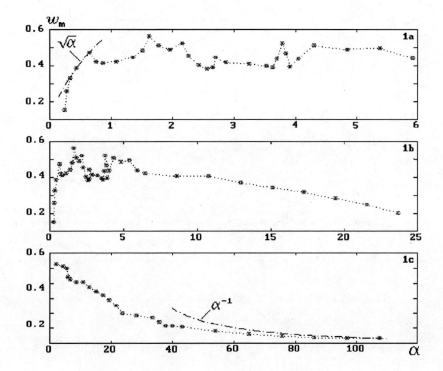

Fig. 1.1.1. Dependence of the maximum amplitude of the field $w_{\mathrm{m}}$ on the parameter $\alpha = \nu^{1/3}\gamma_0$. Dashed lines show the asymptotic dependencies for small, $\sqrt{\alpha}$, and large, $\alpha^{-1}$, beam energies.

In deriving equations (1.1.12) from formulas (1.1.2), (1.1.10) and (1.1.11), we assumed $A = -iA_1$ and omitted the small terms of the second order (proportional to $|A|^2$) on the right-hand side of the equation of motion for the beam electrons. These terms describe the oscillations at the harmonics of the plasma frequency.

Integrating (1.1.12) yields the conservation of the energy and momentum:

$$\frac{\beta_0^2}{2\nu}|A|^2 + \frac{1}{S}\sum_{s=1}^{S}\gamma_s = \gamma_0,$$

$$\left| A + i\frac{\nu}{S}\sum_{s=1}^{S}\exp(-i\psi_s)\right|^2 + \frac{3}{32}|A|^4 + \frac{2\nu}{\beta_0}\frac{1}{S}\sum_{s=1}^{S}\beta_s\gamma_s = 2\nu\gamma_0, \qquad (1.1.13)$$

$$\dot{x}_s(0) = c\beta_0, \quad \gamma_0 = (1 - \beta_0^2)^{-1/2}$$

which can be treated as a specific case with respect to the general formulas (1.1.4).

We can introduce the amplitude of the electric field $|\varphi| = k|E|$ and rewrite (1.1.12). Using the first formula (1.1.13), we introduce

$$w^2 = \frac{\beta_0^2}{2\nu\gamma_0}|A|^2 = \frac{|E|^2}{8\pi n_{\mathrm{b}}mc^2\gamma_0},$$

where $n_{\mathrm{b}} = \sigma/l$ is the average beam density, and pass to the "slow" time variable

$$\tau = \left(\frac{\nu}{2\gamma_0}\right)^{1/2}\omega_p t, \quad Q = \left(\frac{\nu\gamma_0^3}{2}\right)^{1/2}.$$

11

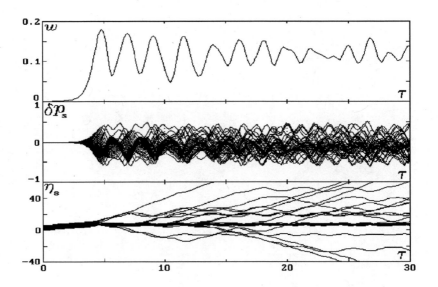

Fig. 1.1.2a. Evolution of the parameters of a nonrelativistic beam $\gamma_0 = 1.1$:

$$w = |E|/\sqrt{8\pi n_{\mathrm{b}} mc^2}, \quad \delta p_s = p_s/p_0 - 1, \text{ and } \eta_s = \psi_s + \vartheta.$$

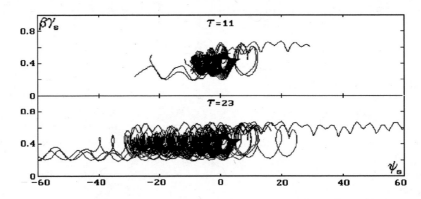

Fig. 1.1.2b. Spreading of the electron trajectories starting from the region $0 \leq \psi_s \leq 2\pi$.

From (1.1.12), we obtain the resulting set in the form

$$W' - \frac{3iQ}{4}|W|^2 W = -\frac{1}{S}\sum_{s=1}^{S}\beta_s \exp(-i\psi_s),$$

$$\beta'_s = \frac{2\gamma_0}{\gamma_s^3}\operatorname{Re} W \exp(i\psi_s), \quad \psi'_s = \frac{\gamma_0^2}{Q}\left(\frac{\beta_s}{\beta_0} - 1\right).$$

(1.1.14)

Because this is the phase of the beam electron with respect to the wave that is important for a physical interpretation, we will use the representation $W = w \exp(i\vartheta)$ and separate the real and imaginary parts of the field equation:

$$w' = -\frac{1}{S}\sum_{s=1}^{S}\beta_s \cos\eta_s, \quad \beta'_s = \frac{2\gamma_0}{\gamma_s^3}w\cos\eta_s,$$

$$\eta' = \frac{3Q}{4}w^2 + \frac{\gamma_0^2}{Q}\left(\frac{\beta_s}{\beta_0} - 1\right) + \frac{1}{S}\sum_{s=1}^{S}\beta_s \cos\eta_s,$$

(1.1.15)

12

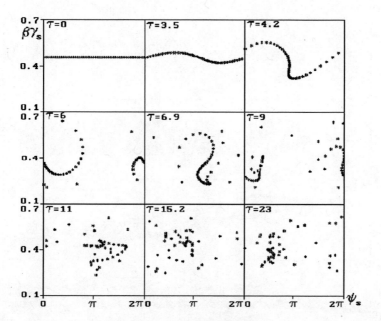

Fig. 1.1.2c. Phase plane, $\gamma_0 = 1.1$.

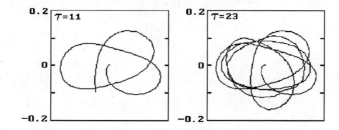

Fig. 1.1.2d. Field amplitude $w$ as a function of the phase $\vartheta$ in the polar coordinate system.

where $\eta_s = \vartheta + \psi_s$.

In the new variables, the formulas for conservation of the energy and momentum (1.1.13) have the form

$$\gamma_0 w^2 + \frac{1}{S} \sum_{s=1}^{S} \gamma_s = \gamma_0,$$

$$\frac{1}{S} \sum_{s=1}^{S} [(1 - \beta_0 \beta_s) \gamma_0 \gamma_s - 2Q\beta_0 w \sin \eta_s] = 1 + \frac{3}{8\beta_0^2} Q^2 w^4 \qquad (1.1.16)$$

$$+ \left(\frac{Q\beta_0}{\gamma_0 S}\right)^2 \sum_{s=1}^{S} \sum_{s'=1}^{S} \cos(\eta_s - \eta_{s'}),$$

and the nonlinear phase velocity of the wave $v_{\mathrm{NL}} = c\beta_{\mathrm{NL}}$ is determined by the formula

$$\beta_{\mathrm{NL}} = \beta_0 \left[1 - \left(\frac{\nu}{2\gamma_0}\right)^{1/2} \vartheta'\right]. \qquad (1.1.17)$$

13

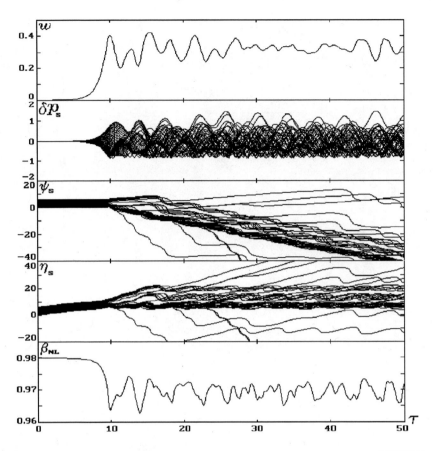

Fig. 1.1.3a. Evolution of the parameters of a relativistic beam $\gamma_0 = 5$: $w = |E|/\sqrt{8\pi n_b mc^2}$, $\delta p_s = p_s/p_0 - 1$, $\eta_s = \psi_s + \vartheta$, and $\beta_{\mathrm{NL}}$

## 2. Numerical integration.

For $\nu = 10^{-2}$ and $S = 50$, Figs. 1.1.1–1.1.9 show the results of the numerical integration of the set (1.1.14) for a wide range of beam energies $1.1 \leq \gamma_0 \leq 500$.

Fig. 1.1.1 shows the function $w_m(\alpha)$ (where $\alpha = 0.215\gamma_0$) which is obtained by means of numerical integration. The asymptotic dependence $w_m \sim \sqrt{\alpha}$ corresponds to a weakly relativistic beam. In this case, the beam-wave resonance detuning is caused by both the large spread in the electron velocities and the bunching of a fraction of the electrons in the decelerating (accelerating) phases of the wave [7]. The asymptotic dependence $w_m \sim \alpha^{-1}$ (see [13]) agrees well with the numerical results in the range of ultrarelativistic energies (Fig. 1.1.1c). In the transitional region of the intermediate energies (Fig. 1.1.1b), the dependence $w_m(\alpha)$ is more complicated due to the competition between the beam and plasma nonlinear effects [14]. Below, the analytical study clarifies the physical cause of the spikes that are present in Fig. 1.1.1.

The specific distinctive feature of the onset of the oscillations driven by the instability of a relativistic electron beam with $\gamma_0 = 1.1$ (Fig. 1.1.2a) is the scattering of the beam by the arising Langmuir oscillations. This scattering leads to a substantial spread in the beam electron momenta $\delta p_s = p_s/p_0 - 1$ in the stage of the transition to the nonlinear quasisteady state. In this case, the plasma response is still linear, and the nonlinear saturation of the growth of oscillation amplitude is caused by the slippage of the electrons (changing their phases $\eta_s$ with respect to the phase of the wave) (see Fig. 1.1.2b). The nonlinear oscillations

14

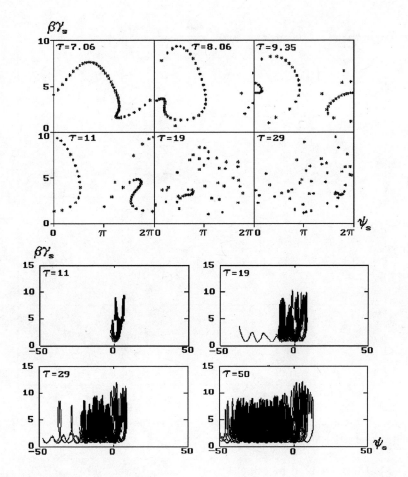

Fig. 1.1.3b. Phase plane and spreading of the electron trajectories starting from the region $0 \leq \psi_s \leq 2\pi$ for $\gamma_0 = 5$.

arise as a result of the interaction of the trapped electron bunches with the wave [1-4].

Fig. 1.1.2b shows the evolution of the electron trajectories starting from the region $0 \leq \psi_s \leq 2\pi$ in the phase space at the time $\tau = 0$. In Fig. 1.1.2c the phase plane is shown for the space interval $0 \leq \psi_s \leq 2\pi$. For the beam that was initially monoenergetic ($\gamma_0 = 1.1$), this figure demonstrates the breaking of the beam regularity and the arising multi-stream structure. Fig. 1.1.2d demonstrates regular nonlinear Langmuir oscillations excited by an initially monoenergetic nonrelativistic electron beam in a plasma.

The numerical integration shows that the increase in the initial energy of the pulse does not cause any qualitative modification of the instability features if $Q \leq 1$ (Fig. 1.1.3). However, the linear growth rate $\delta$ decreases (the linear stage of instability becomes more lengthy, $\tau_{\mathrm{L}} \cong \delta^{-1}$) and the field energy density increases as the relativistic factor increases [7]

$$W = W_0 \exp(\delta\tau), \quad \delta = \sqrt{3}/2Q^{1/3}, \quad w_m^2 \cong Q^{2/3}. \tag{1.1.18}$$

In the intermediate range of energies, $\gamma_0 = 10 \div 50$ the energy density of the field is already so high that the plasma nonlinearity, which results in an additional (compared to the linear theory) variation of the phase of the field, becomes essential. This is illustrated by Fig. 1.1.4 which depicts the functions $w(\tau)$ and $\vartheta(\tau)$ for $\gamma_0 = 20$ and $\gamma_0 = 50$ corresponding to the cases of nonlinear (solid curve) and linear plasma response.

In a plasma with a weakly relativistic electron beam, the saturation of the growth of the

15

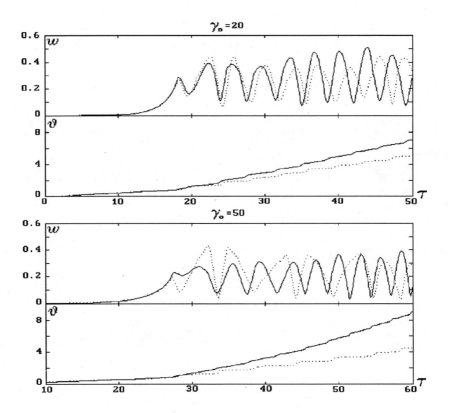

Fig. 1.1.4. Evolution of the amplitude $w$ and phase $\vartheta$ of the field for the nonlinear and linear (dots) interactions.

oscillation amplitude is caused by the electron diffusion in velocity space (Fig. 1.1.5). This diffusion results in detuning between the beam and the wave (i.e., destroys the matching condition $v_{\mathrm{ph}} = \omega/k = v_0$). The passage to the relativistic beam energies slows down this process and, according to formula (1.1.17), this is accompanied by a nonlinear decrease in the phase velocity of the wave. As a result, a part of the beam electrons is decelerated together with the wave, and the resonance interaction extends to the nonlinear stage and causes the maximum amplitude of the field to be higher than in the case of a linear plasma response.

The nonlinear oscillations driven by the ultrarelativistic beam, $\gamma_0 = 100$, are qualitative different. This is related to the fact that, for a small beam-electron spread in energies, the beam-plasma interaction is reversible (Fig. 1.1.6a). The saturation of the field amplitude is determined by the shift in the wave phase and is accompanied by the simultaneous detuning between the whole beam and the wave, so that all the electrons that have traveled in the decelerating phases get into the accelerating phases, and the field energy is transferred back to the beam. For even greater energies $\gamma_0 = 200$, this stage of the instability, which allows reversible solutions, becomes longer (Fig. 1.1.6b).

### 3. Hydrodynamic approximation.

The numerical calculations show that, as the beam energies increase to the values corresponding to $Q \gg 1$, the part of the beam energy transferred to the plasma increases substantially (compare Figs. 1.1.6-1.1.8); the beam spread in momenta (beam heating) decreases and the oscillations of the field amplitude are more regular. Figure 1.1.7 shows

16

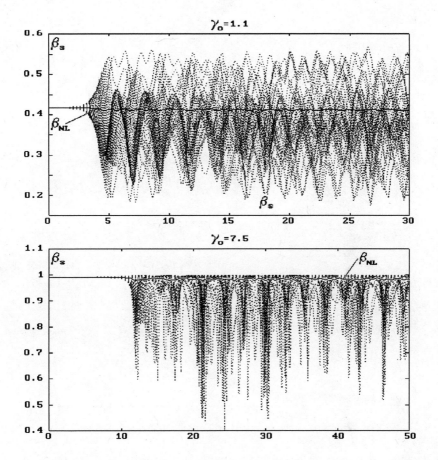

Fig. 1.1.5. Spread of the velocities for nonrelativistic ($\gamma_0$=1.1) and relativistic ($\gamma_0$=7.5) electron beam ($\beta_{\mathrm{NL}}$ is the nonlinear phase velocity of the wave).

the phase space for an ultrarelativistic beam. The onset of the instability does not cause the formation of multistream structures and can be described by the hydrodynamic approach [12,13].

The transition to the hydrodynamic description of the beam can be introduced by the expansion of the exponentials in Eqs. (1.1.14) in the small deviations of the electron trajectories from the initial trajectories

$$\psi_s = \psi_{so} + \tilde{\psi}_s, \quad \beta_s = \beta_0 + \tilde{\beta}_s,$$

$$|\tilde{\psi}_s| \ll 1, \quad |\tilde{\beta}_s| \ll \beta_0, \tag{1.1.19}$$

(with only the terms that are linear in the perturbation on the right hand parts of the equations) retained

$$W' - \frac{3iQ}{4}|W|^2 W = -\frac{\beta_0}{S}\sum_{s=1}^{S}(1 - i\tilde{\psi}_s)\exp(-i\psi_{so}),$$

$$\tilde{\psi}_s'' = \frac{1}{Q\beta_0}[W\exp(i\psi_{so}) + W^*\exp(-i\psi_{so})]. \tag{1.1.20}$$

In the above equation, we dropped the term $\sim\tilde{\beta}_s$, which is small according to (1.1.5).

17

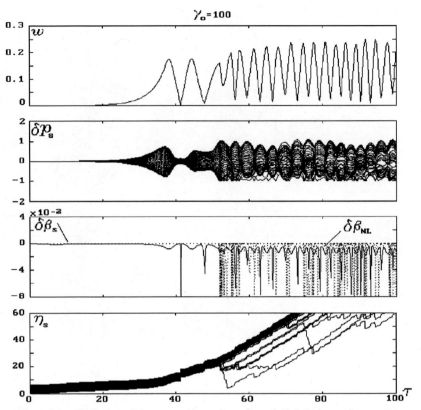

Fig. 1.1.6a. Evolution of the parameters of an ultrarelativistic beam for $\gamma_0 = 100$.

If the equalities

$$\sum_{s=1}^{S} \exp(-i\psi_{so}) = \sum_{s=1}^{S} \exp(-2i\psi_{so}) = 0 \qquad (1.1.21)$$

are satisfied, system (1.1.20) can be simplified to

$$W' - \frac{3iQ}{4}|W|^2 W = \rho, \quad \rho'' = \frac{i}{Q}W, \qquad (1.1.22)$$

and is identical to the set obtained in [13] from the hydrodynamic equation. The function

$$\rho = \frac{\tilde{n}_{\rm b}}{n_{\rm bo}} = \frac{\beta_0}{S} \sum_{s=1}^{S} i\tilde{\psi}_s \exp(-i\psi_{so})$$

can be interpreted here as a perturbation of the beam density, which satisfies the continuity equation. The conditions (1.1.21) are satisfied if the number of electron bunches (charged sheets) per period of the wave is sufficiently large, so that, in the limit $S \to \infty$, the summation over s ($\psi_{so} = 2\pi s/S$) can be changed to the integration

$$\lim_{S \to \infty} \int_1^S \exp(-2\pi i s/S)ds = (1/2\pi) \int_0^{2\pi} \exp(-i\psi)\,d\psi = 0.$$

From (1.1.14) and (1.1.18), we obtain the estimate

$$\tilde{\psi}_s'' \sim |\delta|^2 \tilde{\psi}_s \sim |W|/Q, \quad |\tilde{\psi}_s| \sim Q^{-1/3} \ll 1. \qquad (1.1.23)$$

18

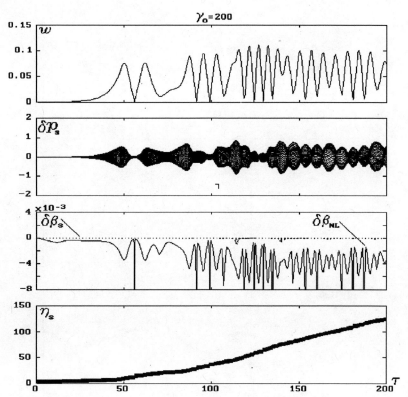

Fig. 1.1.6b. Evolution of the parameters of an ultrarelativistic beam for $\gamma_0 = 200$.

Because the oscillation amplitude is limited by the inequality $|W| \leq 1$, the expansion of formula (1.1.19) and relations (1.1.22) are valid for

$$Q^{1/3} = (\nu \gamma_0^3 / 2)^{1/6} \gg 1, \qquad (1.1.24)$$

i.e., for when a low-density beam can be described by the linear theory in the nonlinear stage of the instability.

In Fig. 1.1.8, we show the asymptotic solutions of equations (1.1.22) by solid curves; the solution of set (1.1.14) is shown by dots. For large relativistic factors $\gamma_0$, the asymptotic solution is virtually identical to the exact one for several first nonlinear pulsations of the field amplitude. In the later stages, the beam is heated and the curves show fairly different behavior. In this case, for $\gamma_0 \geq 200$, the perturbation of the beam density is small, $\rho \sim 10^{-2}$.

In the range of energies, $\gamma_0 \geq 100$, the electron diffusion in velocity space, $\psi_s$, is suppressed by the relativistic increase in the mass of the electrons. Therefore, the evolution of relative phase $\eta_s$ is determined principally by the evolution of the field phase $\vartheta(\tau)$ (see Fig.1.1.6a and Fig.1.1.6a). Function $w(\vartheta)$ for an ultrarelativistic beam, $\gamma = 100$ in the polar coordinate system is represented in Fig. 1.1.9. As the envelope of this curve is closed to circle, the nonlinear Langmuir oscillations are asymptotically closed to monochromatic one.

### 4. A single bunch per wavelength.

In the limiting case of a deep beam modulation, a single bunch is present in the region equal to the wavelength, $\beta_s = \beta$ and $\gamma_s = \gamma$ and system (1.1.15) and (1.1.16) can be

19

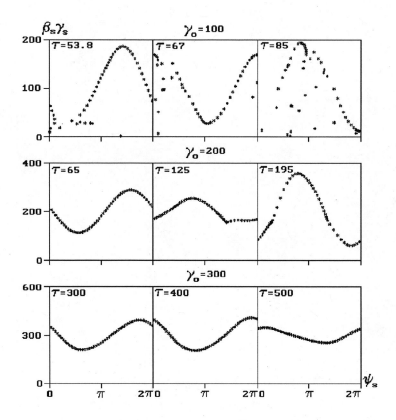

Fig. 1.1.7. Phase plane of an ultrarelativistic beam for $\gamma_0=100$, $\gamma_0=200$ and $\gamma_0=300$.

reduced to the form

$$w' = -\beta \cos \eta,$$

$$-2Q\beta_0 w \sin \eta = 1 + \frac{3}{8\beta_0^2}Q^2 w^4 - (1 - \beta_0\beta)\gamma_0\gamma. \qquad (1.1.25)$$

$$\gamma = \gamma_0(1 - w^2), \quad \beta = (1 - \gamma^{-2})^{-1/2},$$

These equations can be analyzed analytically in order to explain the nonmonotonic behavior of the function $w_m(\alpha)$ (see Fig. 1.1.1).

As follows from Eq. (1.1.25), for $|\eta_0| = \pi/2$, the onset of the instability is suppressed by the nonlinearity causing the detuning between the wave and bunches. Setting $\sin \eta_m = \pm 1$, we obtain the equation for the field amplitude $w_m$:

$$y_{\pm} = \pm 2Q\beta_0 w_m + 1 + \frac{3}{8\beta_0^2}Q^2 w_m^4 - (1 - \beta_0\beta_m)\gamma_0\gamma_m = 0, \qquad (1.1.26)$$

where $\beta_m = \beta(w_m)$ and $\gamma_m = \gamma(w_m)$.

Fig. 1.1.10 illustrates the changes in $w_m(\gamma_0)$ with the increase in the beam energy. For the small energies $\gamma_0 < 15$, the field amplitude is defined by the equation $y_+(w_{m1}) = 0$; for the large energies $\gamma_0 \geq 23$, by the equation $y_-(w_{m2}) = 0$. In the energy range $15 \leq \gamma_0 \leq 23$, equation (1.1.25) is not valid, and equation (1.1.26) has no roots $w_m \leq 1$ (Fig. 1.1.11).

In (1.1.25), we use the following expansion:

$$(1 - \beta_0\beta)\gamma\gamma_0 = 1 + \frac{w^4}{2\beta_0^2} + \frac{w^6}{2\beta_0^4}, \quad w \ll 1 \qquad (1.1.27)$$

20

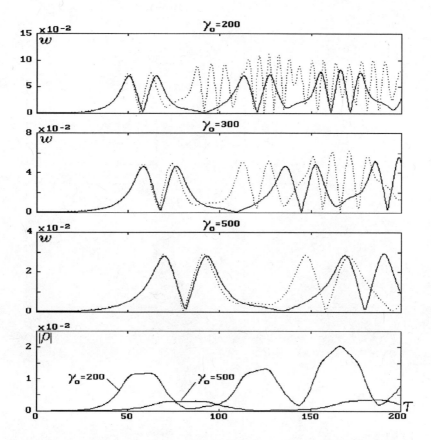

Fig. 1.1.8. The field amplitude $w$ and the modulation of the beam density $\rho$ correspond to the hydrodynamic equation (1.1.22). The dots show the kinetic solution of set (1.1.12) (Fig. 1.1.6d).

and we find the dependence of the phase on the field amplitude in the explicit form

$$4Q \sin \eta = \left(1 - \frac{3}{4}Q^2\right) \frac{w^3}{\beta_0^3} + \frac{w^5}{\beta_0^5}. \tag{1.1.28}$$

By setting $\sin \eta_{\mathrm{m}} = \pm 1$, we find the asymptotics for the curves in Fig. 1.1.11

$$w_{\mathrm{m}} \approx \begin{cases} (4Q)^{1/3}, & Q \ll 1 \\ (16/3Q)^{1/3}, & Q \gg 1, \end{cases} \tag{1.1.29}$$

which coincides with Eq. (1.1.1).

### 5. Basic results.

We have simulated the onset of the instability of a low density $\nu = 10^{-2}$ electron beam that propagates in a plasma for the beam-energy range $1.1 \leq \gamma_0 \leq 500$ and have determined the dependence of the maximum field amplitude on the parameter $\alpha = \nu^{1/3}\gamma_0 \leq 1$.

For parameters $\alpha \leq 1$, the detuning of the phase (Cherenkov) resonance between the beam electrons and the wave is caused by either the deceleration of individual electrons by

21

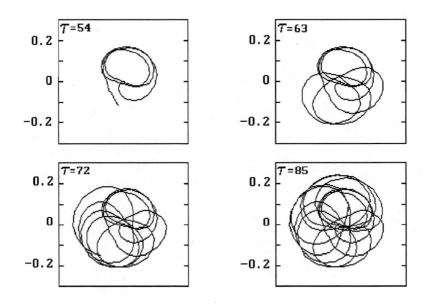

Fig. 1.1.9. Field amplitude $w$ as a function of the phase $\vartheta$ in the polar coordinate system for an ultrarelativistic beam $\gamma=100$.

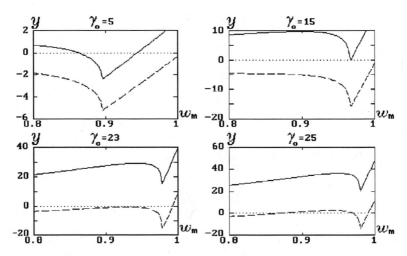

Fig. 1.1.10. Evolutions of the algebraic roots of the equation $y_\pm(w_m)=0$ for different beam energies.

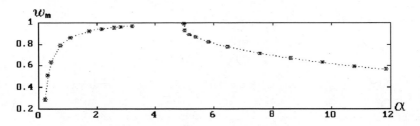

Fig. 1.1.11. Dependence of the maximum field amplitude $w_m$ on the parameter $\alpha=\nu^{1/3}\gamma_0$.

the field of the unstable mode. This is accompanied by the spreading of the beam not only in momenta but also in velocities.

The other mechanism for the suppression of the instability occurs for the ultrarelativistic beam energies of $\alpha^{1/2} \gg 1$. In this case, the spread of electrons in velocities $(\delta v \approx \nu^{1/3} v_0/\gamma_0)$ is negligible, the beam moves as a whole, and all its electrons go out of resonance with the wave simultaneously because of the nonlinear decrease in the phase velocity of the wave. The transition to the hydrodynamic description of the beam can be introduced by the expansion of the kinetic equations in the small deviations of the electron trajectories.

In the range of intermediate energies $\alpha \approx 1$, in which both these mechanisms produce the effects of the same order, the dependence of the maximum amplitude of the field on the energy is no longer monotonic and is defined by the interaction of the formed electron bunches with a wave having a varying phase velocity. The nonlinear resonance, which can occur in this case, corresponds to the conditions under which the deceleration of the relativistic electrons is compensated by the reduction in the phase velocity of the wave and is accompanied by an increase in the energy density of the oscillations.

### §1.2. Ultrarelativistic electron beam in a nonlinear plasma.

The equation of motion for the relativistic beam electron in a plasma wave $E(t, z) = E(t) \sin \omega_p (t - z/v_0)$ can be represented in the form

$$\ddot{\psi} + \Omega_{\text{ph}}^2 \sin \psi = 0,$$

$$\Omega_{\text{ph}}^2 = eE\omega_p/mv_0\gamma^3, \quad \psi = \omega_p \left[ t - z(t)/v_0 \right], \tag{1.2.1}$$

where $x(t)$ is the coordinate of electron. As the energy density of the oscillations is restricted by the inequality $E^2 < 8\pi n_b mc^2\gamma_0$, the frequency of the phase oscillations $\Omega_{\text{ph}}$ does not exceed the following value:

$$\Omega_m = (\nu/\gamma^5)^{1/4}\omega_p, \quad \nu = n_b/n_p.$$

In the linear stage, the exponential growth of a small perturbation with the highest hydrodynamic rate, $\delta \approx \nu^{1/3}\omega_p/\gamma_0$, results in the trapping of resonant electrons by the wave and in the arising of the nonlinear phase oscillations of the trapped-electron bunches with a character time of order $\Omega_{\text{ph}}^{-1}$. At ultrarelativistic beam energies, $\alpha^{1/2} \gg 1$ ($\alpha = \nu^{1/3}\gamma_0$), the frequency of phase oscillations $\Omega_{\text{ph}}$ is smaller than the growth rate $\delta$,

$$\Omega_m^2/\delta^2 \approx \alpha^{-1/2}, \tag{1.2.2}$$

and hence, a low-density beam, $\nu^{1/3} \ll 1$, can be described by the linear theory in the nonlinear stage of the instability.

We use the self-consistent equations of relativistic hydrodynamics for beam electrons together with the Poisson equation in a plasma

$$\frac{\partial \tilde{n}_b}{\partial t} + v\frac{\partial \tilde{n}_b}{\partial z} = -n_b\frac{\partial \tilde{v}_b}{\partial z}, \quad \frac{\partial \tilde{v}_b}{\partial t} + v\frac{\partial \tilde{v}_b}{\partial z} = \frac{e}{m\gamma^3}E,$$

$$\frac{\partial^2 E}{\partial t^2} + \omega_L^2 E = -4\pi e\frac{\partial}{\partial t}(v\tilde{n}_b + n_b\tilde{v}_b), \tag{1.2.3}$$

where $\tilde{v}_b$ and $\tilde{n}_b$ are the small perturbations of beam velocity and density, $v = (1-\gamma^{-2})^{1/2}$; the averaging beam density, $n_b$, is an integral of motion, and the relativistic factor $\gamma(t)$ is

23

determined by the energy conservation law (1.1.4) The dependence of plasma frequency, $\omega_L$, on the wave field amplitude is taken into account below.

To proceed, we seek solution of Eq. (1.2.3) in the form

$$n(t,x) = \operatorname{Re} \tilde{N}(t) \exp[i\Phi(t,x)], \quad E(t,x) = \operatorname{Re} E(t) \exp[i\Phi(t,x)],$$

$$\Phi(t) = k\left(\int_0^t v(t')dt' - x\right), \quad \dot{E} \ll \dot{\Phi}E.$$

As a result, we find the set of equations for the slowing varying complex functions [12]

$$\frac{d^2\tilde{N}}{dt^2} = \frac{ekn_b}{m\gamma^3}E, \quad \gamma = \gamma_0 - \frac{|E|^2}{8\pi n_b mc^2},$$

$$-k\left(2v\frac{dE}{dt} + E\frac{dv}{dt}\right) + i\left[(\omega_L^2 - k^2v^2)E\right] = 4\pi ekv^2\tilde{N}. \tag{1.2.4}$$

Note, that the first equation (1.2.4) coincides with Eq. (1.1.22) that found above from the asymptotic case of kinetic theory (see also **Appendix 1.1**).

The field amplitude is governed by a system of nonlinear hydrodynamic equations of motion of a plasma whose solution for an arbitrary electron beam density should be sought in the self-consistent approximation in the form of the wave (1.2.3). However, in the case of a low-density beam when the increment is small compared with the plasma frequency, the problem simplifies greatly because the beam makes contribution only to the increment and the waveguide properties of a beam-plasma system are governed mainly by the plasma. This makes it possible to ignore the presence of the beam in the determination of the nonlinear correction to the plasma frequency and then use Eq. (1.2.4) replacing in it the plasma frequency $\omega_L^2$ with its field-dependent value $\omega_L^2(E)$. In the case of waves of the $E(t-x/c)$ type the system of equations of motion of the plasma electrons and the Poisson equation lead to the following nonlinear equation

$$\Psi'' + \frac{\omega_p^2}{2}\left[\frac{1}{(1-\Psi)^2} - 1\right] = 0, \tag{1.2.5}$$

where $\psi = eE/mc\omega_p$, $\omega_p = \sqrt{4\pi e^2 n_p/m}$, and a prime denotes a derivative with respect to the total argument.

Introducing the expansion of (1.2.5) in powers of the dimensionless potential $\Psi \ll 1$, we arrive at the equation

$$\Psi'' + \omega_p^2\left(1 + \frac{3}{2}\Psi + 2\Psi^2\right)\Psi = 0. \tag{1.2.5'}$$

The nonlinear oscillations described by Eq. (1.2.5') are nearly harmonic and their frequency is [21]

$$\omega_L^2 = \omega_p^2\left(1 - \frac{3}{8}|\Psi|^2\right). \tag{1.2.6}$$

It should to be noted that the above expression is identical with the analogous formula (1.1.12).

Using Eq. (1.2.6) and expressing in Eq. (1.2.4) the beam velocity in terms of the field amplitude, $v^2 = v_0^2 - |E|^2/4\pi n_b m\gamma_0^3$, we obtain a nonlinear equation

$$\frac{d^2\tilde{N}}{dt^2} = i\frac{ekn_b}{m\gamma_0^3}E,$$

$$-2kv_0\frac{dE}{dt} + i\left[\omega_p^2 - k^2v_0^2 - \omega_p^2\frac{|E|^2}{E_p^2}\left(1 - \frac{8}{3\alpha}\right)\right]E = 4\pi ekv_0^2\tilde{N}, \tag{1.2.7}$$

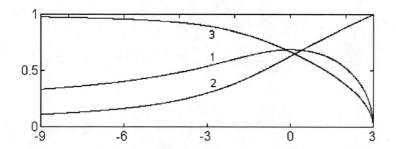

Fig. 1.2.1. Linear theory. Plots of several properties versus the parameter $\Delta$. (1) -
The growth rate; (2) - the initial ratio of the field amplitude $a_0$ to the beam density
modulation $N_0$; (3) - the initial phase of the beam.

where $E_p^2 = (32\pi/3)n_p mc^2$. In the ultrarelativistic limit $\alpha \gg 1$ considered here the contribution of the electron beam to the nonlinear term of Eq. (1.2.7) represents a small correction
to unity and the corresponding term can be omitted.

The system of equations (1.2.7) describing the oscillations of nonlinear plasma with an
ultrarelativistic beam can be transformed through the change of dimensionless variables

$$E = i\, E_m a(\tau) \exp(-i\vartheta), \quad \tilde{N} = n_b \sqrt{3/8\alpha}\, N(\tau) \exp(-i\varphi),$$

$$E_m^2 = \delta_0 E_p^2, \quad \delta_0 = \frac{\nu^{1/3}}{\gamma_0}, \quad \Delta = \frac{\epsilon}{\delta_0}, \quad \epsilon = 1 - \frac{\omega_p^2}{k^2 v_0^2}, \quad \tau = \delta_0 \omega_p t$$

to a system of equations for the "slow" amplitudes and phases of the field, $a$, $\theta$, and the
beam density modulation, $N$, $\varphi$:

$$a' = -\frac{N}{2}\sin(\vartheta - \varphi), \quad N'' - \varphi'^2 = -a\cos(\vartheta - \varphi),$$

$$\vartheta' = \frac{1}{2}(\Delta^2 + a^2) - \frac{N}{2a}\cos(\vartheta - \varphi), \quad (N^2\varphi')' = -aN\sin(\vartheta - \varphi). \tag{1.2.8}$$

From Eqs. (1.2.8) we find the integrals of motion

$$H = \frac{N'^2}{4} - \frac{a^2}{4}\left(\Delta + \frac{a^2}{2}\right) + \frac{1}{2}aN\cos(\vartheta - \varphi) + \frac{(P - a^2)^2}{4N^2},$$

$$P = a^2 - N^2\varphi', \tag{1.2.9}$$

which reflect energy and momentum conservation in a plasma with a beam. If there are no
waves in the plasma ($H = P = 0$), we can lower the order of Eqs. (1.2.8) by transforming
to the difference phase $\eta = \vartheta - \varphi$ and using (1.2.9):

$$a' = -\frac{N}{2}\sin\eta, \quad \eta' = \frac{1}{2}(\Delta + a^2) - \frac{N}{2a}\cos\eta - \frac{a^2}{N^2},$$

$$N' = \left[a^2\left(\Delta + \frac{a^2}{2}\right) - 2aN\cos\eta - \frac{a^4}{N^2}\right]^{1/2}. \tag{1.2.10}$$

Working in the linear approximation ($|\Delta| \gg a^2$), and omitting the terms $a^2/2$ and $a^4/2$
(which reflect the nonlinear dependence of the plasma permittivity on the field amplitude)
from the second and third equations of system (1.2.10), we find the following results for an

exponentially growing solution, $\sim \exp(\delta\tau)$, with $\delta > 0$:

$$\delta = \sqrt{3}\left(h_+^2 - h_-^2\right), \quad \cos\eta_0 = \frac{1}{2}\left[\Delta\left(h_+ + h_-\right) - 1\right],$$

$$\frac{a_0}{N_0} = h_+ + h_-, \quad h_\pm = \frac{1}{2}\left[1 \pm \sqrt{1 - \left(\frac{\Delta}{3}\right)^3}\right]^{1/3}. \tag{1.2.11}$$

(see **Appendix 1.2**).

These functions of the parameter $\Delta$ are shown in Fig. 1.2.1. The instability occurs after a threshold is reached, and it occurs under the condition $\Delta < 3$. The growth rate reaches a maximum of $\delta_m = \sqrt{3}/2^{4/3}$ at $\Delta = 0$ and decreases, $\delta \simeq (-\Delta)^{-1/2}$, as $\Delta \to -\infty$.

## §1.3. Soliton solutions.

The equation for the slowing varying complex field amplitude in a plasma with an ultrarelativistic electron beam (1.2.7) describes undamped nonlinear waves which grown in time in the linear stage of the instability, with a growth rate which depends on the parameter $\Delta$, for an arbitrary relation between the plasma frequency $\omega_p$ and the modulation frequency $\omega$ (see Fig. 1.2.1).

The existence of such waves in a plasma should clearly be accompanied by trapping of resonant electrons by a wave over a time on the order of $\Omega_{\mathrm{ph}}^{-1}$ and by significant deviations from the original values of the beam properties. It is thus worthwhile to study beam modulation regimes under condition such that field energy is returned to the beam (see Fig. 1.1.8), and the properties of the beam are unable to undergo irreversible changes as a result of phase mixing. In other words, we seek solutions of Eqs. (1.2.7) which satisfy the condition $d^n E/dt^n = 0$ (n=1,2...) at $t = \mp\infty$.

In contrast with the linear boundary-value problems of quantum mechanics, the problem at hand is nonlinear; by solving it one finds eigenvalues and eigenfunctions of a nonlinear operator. The reason is that the change of scales

$$y = E/E_p\sqrt{|\Delta|}, \quad x = |\Delta|\tau, \quad \Lambda = |\Delta|^{-3}$$

puts Eqs. (1.2.7) in the form

$$\hat{F}(y)y = \Lambda y,$$

$$\hat{F}(y) = 2i\frac{d^3}{dx^3} - \frac{d^2}{dx^2}\left(\mathrm{sign}\,\Delta + |y|^2\right). \tag{1.3.1}$$

On the practical side, this study leads to the conclusion that there exists a discrete set $\omega_n$, $\Delta(\omega_n) = \Delta_n$, of modulation frequencies for which an ultrarelativistic beam has nonlinear stability in a plasma [15].

### 1. Numerical simulation.

For numerical calculations it is convenient to put Eqs. (1.2.10) in the form of canonical equations

$$p' = \frac{\partial H}{\partial q}, \quad q' = -\frac{\partial H}{\partial p}, \quad p_N' = \frac{\partial H}{\partial N}, \quad N' = -\frac{\partial H}{\partial p_N} \tag{1.3.2}$$

26

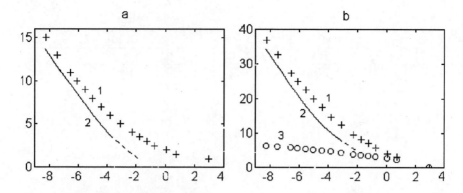

Fig. 1.3.1. a: 1—Ratio of the period of the nonlinear density oscillations, $T_N$, to the period of the nonlinear field oscillations, $T_E$, versus the parameter $\Delta$; 2—curve calculated in the adiabatic approximation.

b: $\Delta$ dependence of (1,2) the density maximum $N_m$ found through a numerical solution and in the adiabatic approximation, respectively, and (3) the largest of the field amplitude maxima $_{m,m}$.

for two pairs of canonical variables $p = a\sin\eta$, $q = a\cos\eta$ and $p_N = N'/2$, $N$ with the Hamiltonian

$$H(p,q,p_N,N) = p_N^2 - \frac{a^2}{4}\left(\Delta + \frac{a^2}{2} - \frac{a^2}{N^2}\right) + \frac{1}{2}qN. \qquad (1.3.3)$$

Since the function $H(p,q,p_N,N)$ is an even function of the variables $p$ and $p_N$, Eqs. (1.3.2) are symmetric under the transformations

$$p \to -p, \ p_N \to -p_N, \ \tau \to -\tau.$$

If a phase trajectory which develops in a 3D region $H(p,q,p_N,N) = 0$ of the phase intersects the curve

$$H(p = 0, q; \ p_N = 0, N) = 0,$$

at some time $\tau_0$, its evolution will therefore be symmetric with respect to this event. Accordingly, a trajectory of this sort, which emerges at $\tau \to \infty$ from a point of an unstable equilibrium position $p = q = p_N = N = 0$, returns to it as $\tau \to \infty$. Numerical calculations show that symmetric solutions of this sort do indeed exist.

Equations (1.3.2) with initial conditions (1.2.11) were integrated over a wide range of values of the parameter $\Delta$. The points on the curves in Fig. 1.3.1 show symmetric solutions for the discrete set of values $\Delta_n$ (see Table 1). These points are seen to conform to smooth curves. Each value of $\Delta$ on curve 1 (Fig. 1.3.1) corresponds to nonlinear oscillations with a given ratio of the oscillation periods of the field, $T_E$, and the density, $T_N$. If this ratio is the rational number $n_E/n_N$, Eqs. (1.3.2) have a symmetric solution, with $n_E$ field maxima and $n_N$ density maxima, which corresponds to a return regime of the beam-plasma instability (see Fig. 1.3.2 and Fig. 1.3.3).

### 2. Soliton solution near the instability threshold.

Since the instability is stopped by a nonlinearity at a low field amplitude near the threshold growth rate, $1 - \Delta/3 \ll 1$ (Fig. 1.2.1), one can expand Eq. (1.2.10) in powers of

27

**Table 1**

| $\Delta$ | 0.09298 | -0.7362 | -0.7307 | -1.2716 | -1.7771 |
|---|---|---|---|---|---|
| $n_E/n_N$ | 2/1 | 3/2 | 5/2 | 3/1 | 7/2 |
| $\Delta$ | -2.2661 | -3.08056 | -3.7901 | -4.4267 | -5.0115 |
| $n_E/n_N$ | 4/1 | 5/1 | 6/1 | 7/1 | 8/1 |
| $\Delta$ | -5.5571 | -6.0711 | -6.543 | -7.478 | -8.2115 |
| $n_E/n_N$ | 9/1 | 10/1 | 11/1 | 13/1 | 15/1 |

$|\eta| \ll 1$,

$$a' = -\frac{N}{2}\eta, \quad \eta' = \frac{1}{2}(\Delta + a^2) - \frac{N}{2a}\left(1 - \frac{\eta^2}{2}\right) - \frac{a^2}{N^2},$$

$$N' = \left[a^2\left(\Delta + \frac{a^2}{2}\right) - 2aN\left(1 - \frac{\eta^2}{2}\right) - \frac{a^4}{N^2}\right]^{1/2} \tag{1.3.4}$$

and use the linear asymptotic behavior in (1.2.11) as an initial condition,

$$\eta_0 = -2\delta, \quad a_0/N_0 = 1 - \delta^2/3,$$

$$\delta = (1 - \Delta/3)^{1/2}. \tag{1.3.5}$$

After some simple calculations, the system (1.3.5) simplifies to

$$a' = a\left(\delta^2 - \frac{a^2}{6}\right)^{1/2}, \quad \eta = -2\left(\delta^2 - \frac{a^2}{6}\right)^{1/2} \tag{1.3.6}$$

and has the soliton solution

$$a = \sqrt{6}\delta \operatorname{sech}\psi, \quad \eta = 2\delta \tanh\psi,$$

$$\psi = \delta\tau - \operatorname{arccosh}(\sqrt{6}\delta/a_0), \tag{1.3.7}$$

where $a_0$ is the initial perturbation, and the function $N(\tau)$ is found from the last of Eq. (1.3.4), $\Delta = \epsilon/\delta_0$, and $\delta_0 = \nu^{1/3}/\gamma_0$.

From (1.3.7) we find the maximum field energy density and the maximum amplitude of the beam density modulation:

$$\left\langle \frac{E^2}{4\pi} \right\rangle_{\max} = 8\delta^2\delta_0 n_p mc^2, \quad |\tilde{n}_b|_{\max} = \frac{4\delta}{\sqrt{\alpha}}n_b, \tag{1.3.8}$$

where $\alpha = \nu^{1/3}\gamma_0 \gg 1$.

A mechanical analogy is useful in explaining the appearance of a threshold stability of a beam-plasma system and in explaining the nonlinear stabilization of the growth of the wave amplitude. Above the instability threshold,

$$\omega > \omega_c = \omega_p(1 - 3\delta_0)^{-1/2}, \quad \Delta(\omega_c) = 3$$

the two oscillators - the plasma and the beam - are coupled only weakly, since their resonant frequencies are quite far apart. As the parameter $\Delta$ decreases, these frequencies move closer together, and the system becomes unstable when the plasma frequency in the frame of reference of the beam,

$$\Omega_p = \gamma_0(kv_0 - \omega_p) \approx \frac{1}{2}\gamma_0\omega\epsilon$$

28

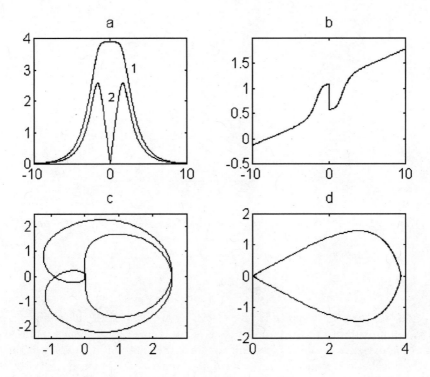

Fig. 1.3.2. Symmetric solution in the immediate vicinity of the resonance, $\Delta=0.09298$. a: 1—oscillations of the density, $N$; and 2—oscillations of the field, $a$; b: Difference phase, $\eta$; Projections of the phase trajectory onto the (c) $N,N'$ plane and (d) the $q,p$ plane.

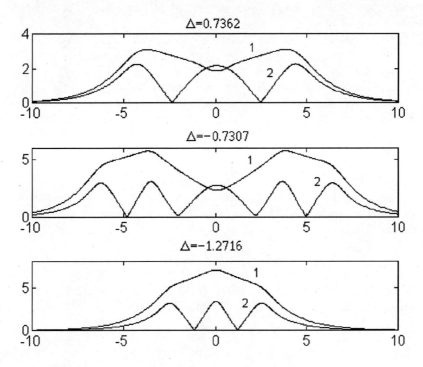

Fig. 1.3.3. Symmetric solutions for various values of parameter $\Delta$: 1—oscillations of the density; 2—oscillations of the field.

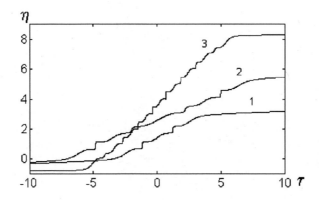

Fig. 1.3.4. Difference phase, $\eta$ for various values of $n_E/n_N=3/1$ (1), $5/2$ (2), $9/1$ (3).

approaches the frequency of the plasma waves of the beam in the plasma,

$$\Omega_b = \gamma_0\omega_b/\epsilon^{1/2}, \quad \omega_b = \delta_0^{3/2}\omega_p.$$

We thus find, in order of magnitude, $\Delta_c \approx (\Omega_p/\Omega_b)^{2/3} \approx 1$. In the nonlinear stage of instability, the growth of the field amplitude is accompanied by a change in the dielectric constant of the plasma $\epsilon_{NL} = \epsilon + \delta_0 a^2$, and a shift of the threshold frequency of the system. For a fixed perturbation frequency $\omega$, the instability thus saturates when the field amplitude reaches the value $a_m = \sqrt{6}\delta$ and we have $\Delta_{NL}(a_m) = 3$. Since the phase of the oscillations shifts simultaneously, the beam goes into a region of retarding phases of the field at the time $\tau_m = \delta^{-1}\text{arcch}(a_m/a_0)$, and the wave growth gives way to damping. As $\tau \to \infty$, the beam-plasma system returns to its original (unperturbed) state, as the phase approaches its asymptotic value $\eta_\infty = 2\delta$.

As $\Delta$ decreases as $\omega \to \omega_p$, the growth rate increases to its maximum value $\delta_m = \sqrt{3}/2^{4/3}$, and the necessary condition for the use of the approximation of a weak nonlinearity, $\delta \ll 1$, which we used above, is violated.

Numerical integration of Eq. (1.2.10) for arbitrary values $|\Delta| \sim 1$ shows that the phase $\eta$ increases without bound as time elapses, and complex nonlinear undamped waves characterized by two scales arise in a plasma with a beam. Asymptotic straight lines running parallel to the abscissa correspond to symmetric solutions (solitary pulses) as $\tau \to \infty$ (Fig. 1.3.4).

### 3. Adiabatic approximation.

As we move away from the resonance into the frequency region $\Delta < 0$, and the damping rate in (1.2.11) decreases to

$$\delta = |\Delta|^{-1/2}, \quad |\Delta| \gg 1 \tag{1.3.9}$$

a change occurs in the nature of the nonlinear oscillations (Fig. 1.3.4), The reason that a large number of periods of the field amplitude, $T_E \approx |\Delta|^{-1}$, fit into one period of the modulation of the beam density, $T_N \approx \delta^{-1}$:

$$T_N/T_E \approx |\Delta|^{3/2} \gg 1.$$

The problem can be solved in the adiabatic approximation, in which the density is assumed to remain constant over the field period [15].

30

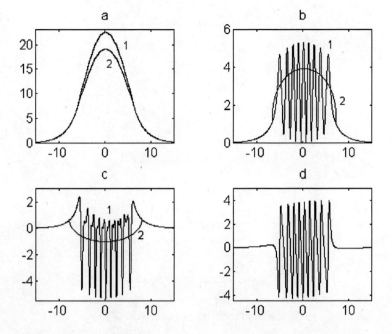

Fig. 1.3.5. Solitary pulse with $\Delta = -5.5571$ ($\mu = 0.15267$). a: The density amplitudes $N$; b: The field amplitudes $a$; c: The function $q$; and d: The function $p$. Curves 1 correspond to numerical solutions, and Curves 2 are analytical solutions.

We renormalize system (1.2.10),

$$A = \frac{|E|}{|\epsilon| E_p}, \quad \rho = \frac{4\pi e v_0}{\omega_p \epsilon^2 E_p} \tilde{n}_b$$

and write it in the form

$$\mu A' = -\rho \sin \eta, \quad \mu \eta' = A^2 - 1 - \frac{\rho}{A} \cos \eta,$$

$$\mu \rho' = \sqrt{(A^2/2 - 1)A^2 - 2A\rho \cos \eta}, \tag{1.3.10}$$

where the prime means the derivative with respect to $x = \delta t$, and where we have identified the small parameter of the problem:

$$\mu = \frac{2\delta}{\omega_p |\epsilon|} = \frac{2}{|\Delta|^{3/2}} \ll 1.$$

With $\mu = 0$, we find the following equation from (1.3.10):

$$A' = \frac{(1 - 3A^2/2)^{1/2}}{1 - 3A^2}, \tag{1.3.11}$$

$$\rho = A(1 - A^2), \quad \eta = 0.$$

A solution of this equation is

$$-\text{arccosh} \sqrt{2/3A^2} + 2\sqrt{1 - 3A^2/2} = \pm(x - x_0) + \sqrt{2} - \ln(1 + \sqrt{2}). \tag{1.3.12}$$

The ($\pm$) correspond to the wings of a cusp solution which has a singularity in its derivative at the point $x_0$, $A(x_0) = 3^{-1/2}$. At $\mu > 0$, this singularity disappears, but a numerical integration (Fig. 1.3.5) reveals that Eq. (1.3.12) describes only the tails of a solitary field pulse, since fast oscillations with a period $T_E \ll T_N$ arise when the singularity is crossed.

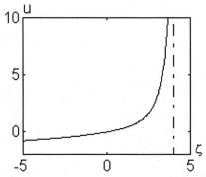

Fig. 1.3.6. Solution near the singular point of the field amplitude.

To study the solution near the singularity $x_0$ we introduce the new variables $A_1 = A - 3^{-1/2}$, $|A_1| \ll 3^{-1/2}$ and $\xi = x - x_0$, $|\xi| \ll x_0$. In the limit $\mu\eta' \ll 1$, one find from (1.3.10)

$$\mu\eta' = 3A_1^2 - \frac{\eta^2}{3} + \frac{\xi}{\sqrt{2}}, \quad \mu A_1' = \frac{2\eta}{3^{3/2}}. \tag{1.3.13}$$

To the left of the singularity, the asymptotic form of (1.3.13),

$$A_1 = \left(-\frac{\xi}{3\sqrt{2}}\right)^{1/2}, \quad -\xi \gg \mu^{4/5} \tag{1.3.14}$$

is the same as the asymptotic behavior of solution (1.3.12).

In the region $|\xi| \le \mu^{4/5}$ it is convenient to introduce the new variables $u = \mu^{-2/5}A_1$ and $\zeta = \mu^{-4/5}\xi$:

$$\frac{d^2 u}{d\zeta^2} = \frac{2}{\sqrt{3}}u^2 + \sqrt{\frac{2}{27}}\zeta. \tag{1.3.15}$$

The small term $\eta^2 \approx \mu^{6/5}$ was discarded in the transformation from (1.3.13) to (1.3.15).

A numerical solution of (1.3.15) reveals (Fig. 1.3.6) that the function $u(\zeta)$ increases monotonically near $\zeta = 0$, coincides with (1.3.14) at $-\zeta \approx 1$, and has a singularity at $\zeta_\infty \approx 4.3$:

$$u(\zeta) = \frac{3^{3/2}}{(\zeta_\infty - \zeta)^2}, \quad \eta(\zeta) = \left(\frac{3\mu^{1/5}}{\zeta_\infty - \zeta}\right)^3. \tag{1.3.16}$$

It follows from (1.3.16) that the width of the transition region is, in order of magnitude, $\Delta\zeta \approx 1$ and $\Delta\xi \approx \mu^{4/5}$.

At finite value $\mu > 0$, the singularity in the derivative at the point $x_0$ thus disappears. When we move into the region $x > x_0$, however, the singularity in (1.3.16) appears; near it we have the approximation $A_1 \ll 1$, used in the derivation of Eq. (1.3.13), is not valid. Furthermore, the nature of the solution changes to the right of the transition region, $x - x_0 \ge \mu^{4/5}$, according to the numerical calculation (Fig. 1.3.5).

The appearance of fast oscillations in the field amplitude, while amplitude of the beam density varies monotonically, makes it possible to integrate Eqs. (1.3.10) at constant values of $\rho$ and $\rho'$. Eliminating the phase $\eta$ from the first equation with the help of the third, we find

$$\mu^2 w'^2 + U(w) = 0, \quad w = A^2 - 1,$$

$$U(w) = \frac{1}{4}(w^2 - 1 - 2\rho'^2)^2 - 4\rho^2(1 + w). \tag{1.3.17}$$

Integrating (1.3.17), we find

$$w = \frac{w_1 + w_2}{2} - \frac{w_1 - w_2}{2} \frac{\lambda - \mathrm{cn}(\sqrt{fg}\,\xi/2\mu, k)}{1 - \lambda\,\mathrm{cn}(\sqrt{fg}\,\xi/2\mu, k)},$$

$$\lambda = \frac{f - g}{f + g}, \quad f, g = \sqrt{(w_{1,2} - b)^2 + d^2}, \tag{1.3.18}$$

$$k^2 = \frac{1}{2}\left[1 - \frac{(w_1 - b)(w_2 - b) + d^2}{fg}\right],$$

where $\mathrm{cn}(x, k)$ is the elliptic cosine, $\xi = x - x_0$, and $w_1 > w_2 \, and \, b + id$ are the roots of the fourth-degree equation $U(w) = 0$ (**Appendix 1.3**). The period of the nonlinear oscillations of the field amplitude is

$$T_E = \frac{8\mu}{\sqrt{fg}} K(k) \tag{1.3.19}$$

where $K(k)$ is the complete elliptic integral of the first kind [22].

In the central part of the pulse we have $f \approx g \approx 1$, and the oscillation period is $T_E \approx \mu$. As we approach the singularity $\rho_c = 2/\sqrt{27}$ and $\rho_c' = 1/\sqrt{6}$, however, the oscillation period increases (Appendix 1.4)

$$T_E \approx \mu \xi^{-1/4}, \quad \xi \to 0. \tag{1.3.20}$$

The condition for adiabatic behavior, $T_E' \ll 1$, is violated. At the limit of the range of validity of the approximation, $T_E' \leq 1$, at which expression (1.3.19) is still valid qualitatively, we find $T_E \approx \mu^{4/5}$, which agrees in order of magnitude with the width of the transition region in (1.3.16).

Near the singularity, the asymptotic form of (1.3.18) is bell-shaped,

$$A = \sqrt{1 + w} = \sqrt{3\,\frac{1 + (2\xi/9\mu)^2}{1 + (2\xi/3\mu)^2}} \tag{1.3.21}$$

and it describes an extreme peak in the field amplitude (Fig. 1.3.5). For $-\xi \gg \mu$, the asymptotic behavior in (1.3.21) is the same as in (1.3.16) (to within $\Delta x \approx \mu^{4/5}$).

Solutions which are averaged over the fast oscillations and which describe the monotonic variation of the beam density are of interest:

$$\langle\rho\rangle = T_E^{-1} \int_\xi^{T_E + \xi} \rho(\xi')\,\xi'.$$

For the calculations it is convenient to use the equation $\rho'' = -A\cos\eta$, which follows from (1.3.10). Averaging it over the nonlinear period (**Appendix 1.5**), we find

$$\rho'' = -\langle A\cos\eta\rangle = \frac{1}{4\rho}\left[1 + 2\rho'^2 + \frac{w_1^2 g - w_2^2 f}{f - g} - fg\,\frac{E(k)}{K(k)}\right], \tag{1.3.22}$$

where $E(k)$ and $K(k)$ are the complete elliptic integrals, and $\rho = \langle\rho\rangle$.

Equation (1.3.22) can be integrated by quadratures. Because of the complexity of the final expressions, however, we have carried out numerical integration. The results are shown in Fig. 1.3.5a (curve 3); they agree well with the results of a numerical solution of the original system of equations, (1.3.10).

Also plotted in Fig. 1.3.5 are the quantities

$$\langle q\rangle = \sqrt{-\Delta}\,\langle A\cos\eta\rangle, \quad \langle a^2\rangle = \sqrt{-\Delta}\,\langle A^2\rangle,$$

33

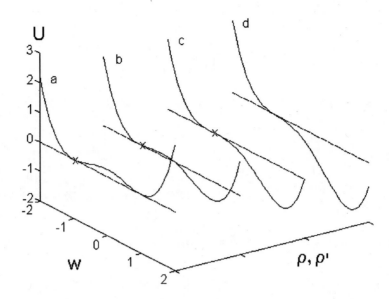

Fig. 1.3.7. Plot of the potential energy $U(w)$. (a) $\rho=0$, $\rho'=0$; (b) $\rho=0.273$, $\rho'=0.279$; (c) $\rho=\rho_c$, $\rho'=\rho'_c$; (d) $\rho=0.451$, $\rho'=0.475$. The crosses on the curves a-c shows the position of the particle.

where $\langle A \cos \eta \rangle$ is given by (1.3.22), and $\langle A^2 \rangle$ is calculated in a corresponding way:

$$
\begin{aligned}
\langle A^2 \rangle = {} & 1 + \frac{w_2 f + w_1 g}{f + g} + \frac{\pi \sqrt{fg}}{2K(k)} \\
& - \frac{fg(f - g)}{(w_1 - w_2)(f + g)} \left[ 1 + \frac{(w_1 - b)(w_2 - b) + d^2}{fg} \right] \frac{\Pi(n, k)}{K(k)},
\end{aligned}
\tag{1.3.23}
$$

where $\Pi(n, k)$ is the complete elliptic integral of the third kind [22],

$$
n = k^2 \frac{(f + g)^2}{(w_1 - w_2)^2}.
$$

The other quantities are defined in (1.3.18).

By analogy with mechanics, Eq. (1.3.17) can be interpreted as the motion of a particle with a mass of $2\mu^2$ in the potential well $U(w)$. The characteristics of the well evolve slowly in time (Fig. 1.3.7) as the functions $\rho$ and $\rho'$ described Eq. (1.3.22) vary. At the initial time, with $\rho = \rho' = 0$, the particle is at the bottom of the well, at the point $w = -1$ (Fig. 1.3.7a). With increasing $\rho$, the depth of the left-hand minimum decreases (Fig. 1.3.7b), at the singularity, with $\rho_c = 2/\sqrt{27}$ and $\rho'_c = 1/\sqrt{6}$, the curve has only a single minimum. The particle thus "rolls down" into the well on the right and becomes a nonlinear oscillator (Fig. 1.3.7, and d). After the density $\rho(x)$ has reached its maximum, the well evolves in the opposite direction, and the system returns to its original state as $x \to \infty$.

Finally, we can derive the asymptotic behavior of the spectrum of values $\mu_n$ corresponding to soliton solutions with $n \gg 1$ maxima of the field amplitude and a single density maximum (Fig. 1.3.5). For this purpose we note that a solitary pulse corresponds to an integral number of periods of the field amplitude oscillation, $\Delta x_n = n T_E$, between the singular points on the $\rho(x)$ curve ($\rho_c = 2/\sqrt{27}$, $\rho'_c = 1/\sqrt{6}$), as can easily be verified with the help of the mechanical model introduced above. When the solution returns to the slow branch, the position of the particle on the potential curve should be the same as that at the

34

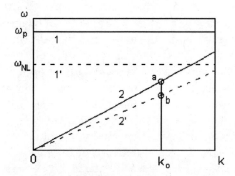

Fig. 1.3.8. Qualitative nature of an approach (a→b) of the Langmuir frequency $\omega_L(t)$ (1,1') to the instability frequency $\omega(t)=k_0 v(t)$ (2,2').

time at which the fast oscillations appeared (Fig. 1.3.7). This situation corresponds to the condition specified. A more rigorous perturbation theory [23] leads to an expression which differs from (1.3.18) by the replacement

$$x - x_0 = \xi \rightarrow T_E \int_0^\xi \frac{d\xi'}{T_E}$$

in the argument of the elliptic cosine. Using this expression and (1.3.19), we find the following from the relation $\Delta x_n = nT_E$:

$$\mu_n = \frac{G}{n}, \quad G = \frac{1}{8} \int_0^{\xi_c} \frac{\sqrt{fg}}{K(k)} \, d\xi \approx 1.17, \tag{1.3.24}$$

where the points 0 and $\xi_c$ correspond to the critical values

$$\rho_c = 2/\sqrt{27}, \quad \rho_c' = \pm 1/\sqrt{6}.$$

The integration has been carried out numerically. The corresponding spectrum of values of $\Delta = (2/\mu)^{2/3}$,

$$\Delta_n = \left(\frac{2n}{G}\right)^{2/3} = -1.43n^{2/3}, \tag{1.3.25}$$

is shown by curve 2 in Fig. 1.3.1, for comparison with the results of the numerical calculations.

Let us examine the mechanism for the nonlinear saturation of the instability at frequencies $\omega < \omega_p$; this mechanism is particularly obvious far from the plasma resonance, $|\epsilon| \gg \nu^{1/3}/\gamma_0$. In this case the presence of an external modulation singles out a beam mode $\omega = kv_0$, and the plasma mode is not excited in the initial stage of the instability. With increasing field amplitude, however, the frequency of the natural waves of the plasma,

$$\omega_{NL}^2 = \omega_p^2 \left(1 - |E|^2/E_p^2\right), \quad E_p^2 = (32\pi/3)n_p mc^2$$

decreases, and the beam wave branch and the plasma wave branch come closer together (Fig. 1.3.8). At a field amplitude $E_c^2 = |\epsilon|E_p^2/3$, the beam mode and the plasma mode couple nonlinearly.

It is important to note that in this stage of the instability the dielectric constant is $\epsilon_c = 2\epsilon/3 < 0$, the plasma remains opaque to the growing waves, and energy continues to build up in the beam mode. Later on, however, the behavior of the system changes radically,

35

and the rapid growth of the wave amplitude is accompanied by an abrupt transition (abrupt on the scale of the small parameter $\mu$) of the plasma to state with $\epsilon(|E|) > 0$. Beyond this point, the nonlinear interaction of the beam mode and the plasma mode is accompanied by a nonlinear process in which the beam and the wave go out of phase, and energy is returned to the beam because the two oscillators - the plasma and the beam - go out of phase.

An important distinction from the threshold (resonant) instability, however, is that waves appear at the frequency $\omega_p - \omega$ (which is considerably higher than rate $\delta = \omega_b/\sqrt{\epsilon}$, Fig. 1.3.1). These waves are accompanied by small-scale jumps in the wave phase velocity, like those described in Ref. [13]. In contrast with a cusp soliton [5], formed by growing and decaying beam wave branches (1.3.11), the solutions found above support the assertion that energy is returned to the beam only during the excitation of the plasma mode.

## 4. Main results.

In a plasma with an ultrarelativistic electron beam $\nu^{1/3}\gamma_0 \gg 1$, the nonlinear saturation of the instability stems from the dependence dielectric constant of the plasma on the field amplitude: $\epsilon_{NL} = \epsilon + |E|^2/E_p^2$. This dependence leads to a nonlinear change in the phase velocity $v_{NL} = v\left(1 - |E|^2/E_p^2\right)$ with respect to the beam velocity $v$; a further consequence is a deviation from the phase resonance of the beam with the wave. The electrons go into accelerating phases of the field, and energy is returned to the beam. Since the period of the nonlinear oscillations (which is on the order of the reciprocal of the growth rate) is small in comparison with the time scale for trapping of electrons by the wave (in comparison with the reciprocal of the phase oscillation frequency), the beam as a whole is displaced with respect to the wave, without breaking up into bunches.

Near the instability threshold, where the growth rate vanishes $\delta \to 0$, and the relaxation time of the phase velocity is quite long $t \approx T_E \approx \delta^{-1} \to \infty$, the beam electrons undergo a smooth transition into a region of accelerating phases of the field, and the solution assumes a soliton form. With increasing growth rate, the relaxation time of the phase velocity decreases. Since the process is inertial, the beam slips through the region of accelerating phases, enters a region of retarding phases, etc. The number of these nonlinear cycles is finite for the values of the parameter $\Delta_n$ (Fig. 1.3.1).

At modulation frequency $\omega < \omega_p$, far from resonance, the nature of the instability changes, since the plasma is initially opaque to the perturbations which arise in the beam, and the role of the plasma electrons reduces to one of screening these perturbations. As the field amplitude grows, however, the dielectric constant increases and the plasma is bleached when $\epsilon_{NL}$ changes sign. The beam subsequently interacts with the plasma wave, and the nonlinear process in which the phase matching is lost occurs as described above, over the large number of nonlinear cycles. For modulation frequencies close to the resonant frequency, $|\Delta| \ll 1$, the sign of $\epsilon$ is essentially irrelevant (Fig. 1.3.1), since the plasma bleaching time $T_N \approx \delta^{-1}$ is comparable in magnitude to the time scale of the nonlinear process in which the plasma matching is lost, $T_E \approx (|\epsilon|\omega_p)^{-1}$.

A necessary condition for the transport of the beam over a large distance in the plasma is a modulation at frequencies close to the resonance (Fig. 1.3.2); this modulation results in a rapid return of wave energy to the beam (the return regime of instability). The optimum instability regime for the excitation of waves with a high energy densities found at small growth rates, at $\omega < \omega_p$, where prolongation of the linear stage of the instability makes possible the buildup of a significant amount of energy in the beam mode (Fig. 1.3.1).

Furthermore, the appearance of a low-velocity electron beam as a result of the appearance of a return current $j/e = n_p u \approx n_b c$ is accompanied by the excitation of ion acoustic waves, with a growth rate [24]

$$\delta_i = \left(\frac{\pi}{8}\frac{m}{M}\right)^{1/2}\frac{k}{Q}\left(u - \frac{v_s}{Q}\right), \quad Q = \left(1 + k^2\lambda_D^2\right)^{1/2},$$

$$v_s = (T/M)^{1/2}, \quad \lambda_D = \left(T_e/m\omega_p^2\right)^{1/2},$$

(1.3.26)

if $u > v_s$. Consequently, the condition under which our approximation is valid, $\delta > \delta_i$, [with $\delta \approx \nu^{1/3}\omega_p/\gamma$ and $k \approx \lambda_D^{-1}$], is the inequality

$$\nu^{1/3} \gg \frac{m\gamma}{M}\frac{u}{v_s}.$$

(1.3.27)

## §1.4. Evolution of soliton solutions in a plasma with a modulated kinetic beam.

In §1.3, the hydrodynamic approach was taken to determine the spectrum of the modulation frequencies $\omega_n$, which are close the Langmuir frequencies $\omega_p$. For the spectrum the solution of the self-consistent set of equations describes a solitary pulse if the ratio between the periods of the nonlinear oscillations of the beam $T_N$ and field $T_E$ densities is a rational number (see Fig. 1.3.1).

Here, we analyze the asymmetry of the kinetic solutions (in comparison with the symmetric solutions considered above) that is introduced by asynchronous oscillations of the beam electrons in the wave field. Numerical integration carried out for the values of $\omega_n$ corresponding to hydrodynamic solitons shows that, in the kinetic model of the beam, the solutions are always asymmetric because of the irreversible heating of beam electrons under the action of the field of the unstable mode [16]. Solutions that are asymptotically close to those describing the solitons can exist only in the ultrarelativistic energy range, in which the velocity spread of the beam electrons is almost completely suppressed by the relativistic increase in the electron mass.

### 1. Continuous beam.

When the modulation frequency $\omega_n$ of the beam differs from the Langmuir frequency $\omega_p$, the set of nonlinear equations for a plasma with a modulated relativistic equation beam can be derived from the corresponding formulas (1.1.12):

$$\frac{1}{\omega_0}\frac{dA}{dt} - \frac{i}{2}\left(\epsilon + \frac{3}{8}|A|^2\right)A = -\frac{\nu}{v_0 S}\sum_{s=1}^{S}\dot{x}_s(t)\exp(-i\psi_s),$$

$$\frac{1}{\omega_0}\frac{d}{dt}(\beta_s\gamma_s) = \beta_0\mathrm{Re}\left[A\exp(i\psi_s)\right],$$

(1.4.1)

$$\frac{1}{\omega_0}\frac{d\psi_s}{dt} = \frac{\dot{x}_s(t)}{v_0} - 1, \quad \beta_s = \frac{\dot{x}_s}{c}, \quad \gamma_s = \frac{1}{\sqrt{1-\beta_s^2}}.$$

These equations contain the additional term $i\epsilon A$ at which $\epsilon = 1 - \omega_p^2/\omega^2$ is the permittivity of plasma.

37

Equations (1.4.1) have the energy and momentum integrals

$$\frac{\beta_0^2}{2\nu}|A|^2 + \frac{1}{S}\sum_{s=1}^{S}\gamma_s = \gamma_o,$$

$$\left|A + i\frac{\nu}{S}\sum_{s=1}^{S}\exp(-i\psi_s)\right|^2 + \frac{|A|^2}{2}\left(\epsilon + \frac{3}{16}|A|^2\right) + \frac{2\nu}{\beta_0}\frac{1}{S}\sum_{s=1}^{S}\beta_s\gamma_s = 2\nu\gamma_0,$$

$$x_s(0) = c\beta_0, \quad \gamma_0 = (1-\beta_0^2)^{-1/2}.$$

(1.4.2)

We introduce the dimensionless variables

$$y = \frac{A}{E_m}, \quad E_m^2 = \frac{32\pi}{3}\delta_0 n_p mc^2, \quad \delta_0 = \frac{\nu^{1/3}}{\gamma_0}, \quad \Delta = \frac{\epsilon}{\delta_0}$$

and transform equations (1.4.1) to

$$2y' - i(\Delta + |y|^2)y = -\sqrt{\frac{3\alpha^3}{2}}\frac{1}{S}\sum_{s=1}^{S}\beta_s\exp(-i\psi_s),$$

(1.4.3)

$$\beta_s' = \sqrt{\frac{8}{3\delta_0}\frac{1}{\gamma_s^3}}\mathrm{Re}[y\exp(i\psi_s)], \quad \frac{\beta_s}{\beta_0} = \delta_0\psi_s' + 1$$

where the prime means the derivative with respect to $\tau = \delta_0\omega_p t$.

To pass over to hydrodynamic description of the beam, we must expand the exponential functions from (1.4.3) in the small deviation of perturbed beam electron trajectories from the unperturbed one (in the parameter $\alpha^{-1/2} \ll 1$)),

$$\psi_s = \psi_{so} + \tilde{\psi}_s, \quad \beta_s = \beta_0 + \tilde{\beta}_s,$$
$$|\tilde{\psi}_s| \ll 1, \quad |\tilde{\beta}_s| \ll |\beta_0|.$$

(1.4.4)

and keep only terms that are linear in the perturbation $\tilde{\psi}_s$ on the right-hand sides of Eq. (1.4.3):

$$2y' - i(\Delta + |y|^2)y = -\sqrt{\frac{3\alpha^3}{2}}\frac{\beta_0}{S}\sum_{s=1}^{S}(1 - i\tilde{\psi}_s)\exp(-i\psi_{so}),$$

(1.4.5)

$$\tilde{\psi}_s'' = \sqrt{\frac{2}{3\alpha^3}}\frac{1}{\beta_0}[y\exp(i\psi_{so}) + y^*\exp(-i\psi_{so})].$$

When the equalities

$$\sum_{s=1}^{S}\exp(-i\psi_{so}) = \sum_{s=1}^{S}\exp(-2i\psi_{so}) = 0,$$

(1.4.6)

hold, equations (1.4.5) go over to the hydrodynamic equations describing a linear beam in a nonlinear plasma

$$2y' - i(\Delta + |y|^2)y = \rho, \quad \rho'' = iy,$$

(1.4.7)

when the function

$$\rho = \sqrt{\frac{3\alpha^3}{2}}\frac{\beta_0}{S}\sum_{s=1}^{S}i\tilde{\psi}_s\exp(-i\psi_{so})$$

(1.4.8)

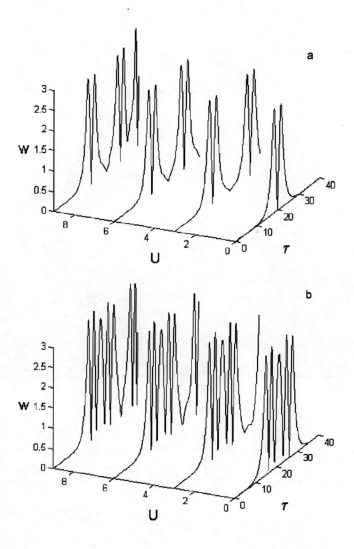

characterizes the perturbation of the beam density satisfying the continuity equation.

In the kinetic model of the beam, the asynchronous nature of the electron oscillations in the wave field manifests itself in that, for a fixed ratio $\nu = n_b/n_p$ of the beam to plasma densities, equations (1.4.3), along with the parameters $\Delta$, depend on the beam energy $\gamma_0$. Fig. 1.4.1 shows the electric-field amplitude $w = |y(\tau, \mathrm{U})|$ (where $\mathrm{U} = 10^3/\alpha$) obtained by integrating equations (1.4.3) numerically for $\nu = 10^{-2}$ and for different values of the parameter $\alpha$. The curves $w(0, \alpha)$ correspond to the solution in the limit $\mathrm{U} \to 0$ and can be obtained from equations (1.2.10) (see §1.3). The phase mixing which results from the displacement of the beam electrons with respect to the wave, leads to the momentum spread of the beam (Fig. 1.4.2) and causes the kinetic solutions to deviate from the hydrodynamic one.

A decrease in the beam modulation frequency in comparison with the Langmuir frequency (which corresponds to the shift to the parameter range $\Delta < 0$) is accompanied by the elongation of the solitary pulses and by an increase in the number of electric field maximums. Accordingly, the beam electrons interact with the wave over a longer time interval, and the irreversible momentum spread of the beam is found to larger than that in the case of short resonant pulses (Fig. 1.4.2).

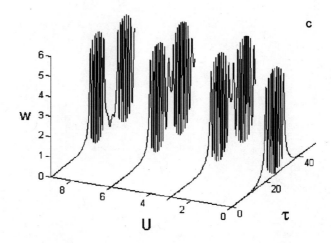

Fig. 1.4.1. Time evolution of the electric-field amplitude $w = |y(\tau,U)|$ with $U=10^3/\alpha$ for the beam modulation frequencies (a) $\Delta=0.09298$, (b) -0.7307 and (c) -5.55707 corresponding to hydrodynamic solutions.

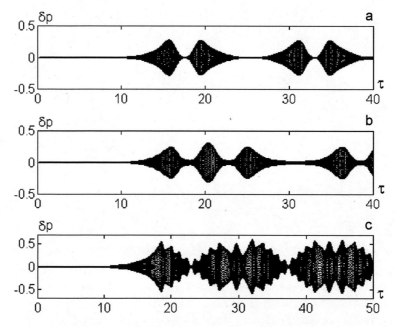

Fig. 1.4.2. Relative momentum spread $\delta p = p/p_0 - 1$ of the beam electrons for $\Delta$: (a) 0.09298, (b) -0.7307, and (c) -5.55707 corresponding to $U=6$.

## 2. Deep-modulated beam.

Let us consider the soliton solutions in a plasma with a highly modulated relativistic electron beam. In comparison with the continuous beam at which the density perturbation is small, we analyze an electron beam previously broken up into separate bunches.

In the limiting case of a deep beam modulation, only two bunches are present in the region equal to the wavelength, $\psi_{10} = 0$ and $\psi_{20} = \pi$, and it follows from (1.4.8):

$$\rho = \sqrt{\frac{3\alpha^3}{2}}\frac{\beta_0}{2}(\tilde{\psi}_1 - \tilde{\psi}_2), \quad \tilde{\psi}_2 = -\tilde{\psi}_1.$$

40

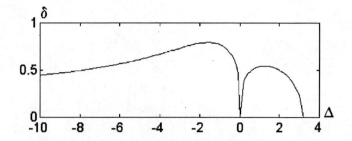

Fig. 1.4.3. The growth rate $\delta=$Im $D$ versus the parameter $\Delta$ in the case of a deep modulated electron beam.

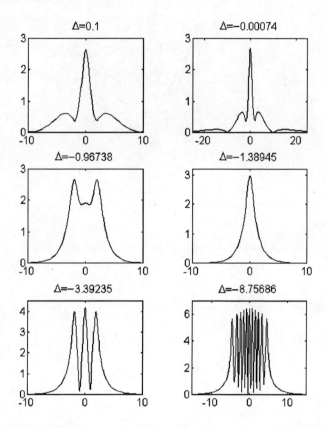

Fig. 1.4.4. Solitary pulses of the electric-field amplitude $w=|y|$ in a plasma with a deep modulated electron beam.

The system (1.4.5) reduces to the form

$$2y' - i(\Delta + |y|^2)y = i\rho, \quad \rho'' = 2\,\mathrm{Re}\,y \qquad (1.4.9)$$

and has the integral

$$-\frac{\rho'^2}{2} + \left(\Delta + \frac{|y|^2}{2}\right)|y|^2 + 2\rho\mathrm{Re}\,y = 0. \qquad (1.4.10)$$

Accordingly (1.4.9), in the linear stage, $|y|^2 \ll |\Delta|$, the exponential growth of a small

41

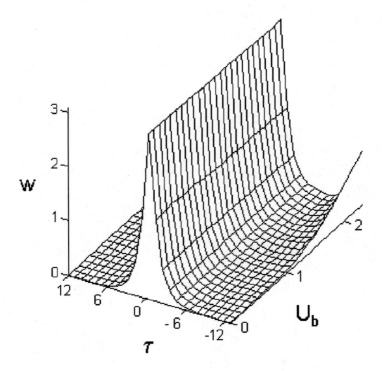

Fig. 1.4.5. Time evolution of the electric-field amplitude $w = |y(\tau, U_b)|$ with $U_b = 10^2/\alpha$ for the beam modulation frequency $\Delta = -1.38945$ in a plasma with a highly modulated electron beam.

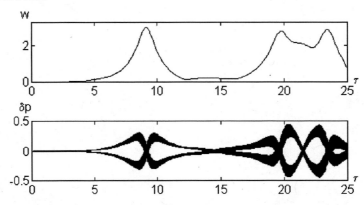

Fig. 1.4.6. Electric-field amplitude $w$ and relative momentum spread $\delta p$ for $U_b = 1.85$ and $\Delta = -1.38945$.

perturbation, $y \sim \exp(-iD\tau)$, occurs with the complex growth rate,

$$D = \begin{cases} i\sqrt{\sqrt{\Delta^4/64 - \Delta/2} - \Delta^2/8}, & \Delta < 0 \\ \left(\dfrac{\Delta}{8}\right)^{1/4}\left[\sqrt{1 + \sqrt{\Delta^3/32}} + i\sqrt{1 - \sqrt{\Delta^3/32}}\right], & \Delta > 0 \end{cases} \qquad (1.4.11)$$

(see Fig.1.4.3).

Figs. 1.4.4 show the electric-field amplitude obtained by integrating equations (1.4.9) numerically for $\nu = 10^{-2}$ and for different values of the parameter $\alpha$.

In comparison with the symmetric solutions (1.4.9) considered above, the asymmetry of the kinetic solutions of set (1.4.3) appears that is introduced by asynchronous oscillations of the beam electrons in the wave field. The results of numerical integration carried out for two groups of electrons are represented in Fig. 1.4.5 and Fig. 1.4.6. By analogy with the continuous beam, the solutions are always asymmetric because of the irreversible heating of beam electrons under the action of the field of the unstable mode. At the same time the beam energies (at which the kinetic solutions closed to the symmetric solutions are realized) are much less in comparison with a continuous beam, $U_b \approx U/10$.

## §1.5. Numerical model of plasma with a relativistic electron beam.

The plasma heating that cannot be described in the hydrodynamic approach because of the asynchronous nature of electron oscillations was taken into account in numerical experiments [9]. Particle-in-cell simulations of the plasma and beam showed that the nonlinear relaxation of a monoenergetic relativistic electron beam (REB) in a dense plasma occurs in three stages. In the first stage, the exponential growth of a small perturbation with the highest hydrodynamic rate results in the trapping of resonant electrons by the wave and in the saturation of an unstable mode at the first maximum of the field amplitude. The nonlinear phase oscillations of the trapped-electron bunches that arise in the second stage are accompanied by oscillations of the field amplitude and the onset of an oscillatory (modulational) instability, during which the short-wavelength perturbations $k\lambda_d \simeq 1$, where $\lambda_d$ is the Debye length) are damped via Landau damping and the wave field energy is converted into the energy of plasma electrons (ions) [10,25]. In the final stage of the instability, energy exchange between the beam and the plasma is almost completely absent.

We solve the Vlasov–Poisson set of equations [17] in order to reproduce the results of numerical experiments [9] and to compare them with the results obtained in the single-mode model [14], which takes into account the plasma nonlinearity in the hydrodynamic approach. Our calculations show that the numerical and theoretical time evolutions of the field energy coincide in the initial stage (when the field energy is growing exponentially) but differ substantially in the stage in which the primary spectrum decays. However, in the asymptotic limit $t > 10^3 \, \omega_p^{-1}$, the energy losses of the beam in the kinetic and hydrodynamic models are found to be nearly the same.

*1. Single-mode model.*

In the single-mode model of the instability of a plasma with a low-density electron beam, all of the beam and plasma electrons interact with a narrow Langmuir wave packet:

$$E = \mathrm{Re}\left[E(t,x)\exp(ik_0 x - \omega_0 t)\right]. \tag{1.5.1}$$

The interaction of an REB with a plasma is described by a nonlinear parabolic equation for the field amplitude $E(t,x)$ and the set of equations for the motion of the beam electrons

$$\frac{v_g'}{2}\frac{\partial^2 E}{\partial x^2} + i\left(\frac{\partial}{\partial t} + v_g\frac{\partial}{\partial x}\right)E + \frac{\omega_p}{2}\left(\epsilon + \frac{|E|^2}{E_{NL}^2}\right)E = -i\frac{4\pi e n_b v_0}{S}\sum_{s=1}^{S}\exp(-i\psi_s),$$

$$\ddot{\psi}_s = \frac{ek_0}{m\gamma_s^3}\mathrm{Re}[E(t,x_s)\exp(i\psi_s)], \quad \gamma_s = (1 - \dot{x}_s^2/c^2)^{-1/2}, \tag{1.5.2}$$

where $\psi_s = k_0 x_s(t) - \omega_0 t$, $x_s(t)$ is the coordinate of the $s$-th beam electron, $S$ is the number of beam electrons per wavelength $\lambda = 2\pi/k_0$,

$$\epsilon = 1 - \omega_L^2/\omega_0^2, \quad \omega_L^2 = \omega_p^2 + (k_0 v_T)^2, \quad v_g = k_0 v_T^2/\omega_0,$$

$$v_g' = dv_g/dk_0, \quad v_T = (T/m)^{1/2}, \quad \omega_0 = k_0 v_0,$$

43

$T$ is the plasma electron temperature.

In a homogeneous cold plasma, the nonlinear correction to the plasma frequency arises due to the change in the electron velocity in the wave electric field $E_{NL} = E_p$, [where $E_p^2 = (32\pi/3)n_p mc^2$] [14], whereas, in an inhomogenious heated plasma, the high-frequency pressure gradient is balanced by the kinetic pressure gradient and the static electric field when $E_{NL} = E_T$, (where $E_T^2 = 32\pi n_p T$) [25].

For a cold plasma, $T = 0$, equations (1.5.2) are ordinary differential equations with respect to time. In order to pass over to the limiting case of a linear beam dynamics, we expand the argument of the exponential function in small perturbations of the electron trajectories: $\psi_s = \psi_{s0} + \tilde{\psi}_s$, where $|\tilde{\psi}_s| \ll \psi_{s0}$. If $S \gg 1$ and the electrons are distributed uniformly over the spatial period of the wave ($\psi_{s0} = 2\pi s/S$), then equations (1.5.2) reduce to the hydrodynamic equations

$$\frac{dE}{dt} - i\frac{\omega_p}{2}\left(\epsilon + \frac{|E|^2}{E_p^2}\right)E = 4\pi e v_0 \rho, \quad \frac{d^2\rho}{dt^2} = \frac{iek_0}{2m\gamma_0^3}E, \quad (1.5.3)$$

where

$$\rho = \frac{n_b}{S}\sum_{s=1}^{S} i\tilde{\psi}_s \exp(-i\psi_{s0})$$

is the depth of modulation of the beam density.

According to (1.5.3), the growth rate of the small perturbations is

$$\delta = \sqrt{3}(\nu/16)^{1/3}\omega_p/\gamma_0, \quad \omega_0 = \omega_p(1 - \delta/\sqrt{3}). \quad (1.5.4)$$

In the nonlinear stage, the trapping of the beam electrons by the wave causes the perturbation amplitude to saturate when [1,2]

$$\rho/n_b \simeq ek_0 E_m/m\gamma_0^3\delta^2 \simeq 1, \quad E_m^2 \simeq \alpha 8\pi n_b m v_0^2 \gamma_0,$$
$$\alpha = \gamma_0\nu^{1/3} \ll 1, \quad \nu = n_b/n_p. \quad (1.5.5)$$

In the high-energy range, the first formula in (1.5.5) gives $\rho/n_b = \Omega_{\text{ph}}^2/\delta^2 \simeq \alpha^{-1/2}$ when the field amplitude reaches its maximum value $E_m \simeq (8\pi n_b m v_0^2 \gamma_0)^{1/2}$. Consequently, for $\sqrt{\alpha} \gg 1$, the beam motion satisfies linearized equations and the instability is suppressed by the plasma nonlinearity. In this energy range, passing over to dimensionless variables in the nonlinear equations (1.5.3) yields $E_m \sim \alpha^{-1}$.

The dotted curves in Figs. 1.5.1 and 1.5.2 illustrate numerical solutions to equations (1.5.2) obtained with 50 particles per wavelength for a cold plasma and a monoenergetic beam.

## 2. Numerical experiment.

The set of equations for a plasma with an REB in a self-consistent field includes Poisson's equation and the Vlasov equations for the distribution functions $f_\alpha$ of the plasma and beam electrons

$$\frac{\partial f_\alpha}{\partial t} + v_\alpha\frac{\partial f_\alpha}{\partial x} + e_\alpha E\frac{\partial f_\alpha}{\partial p} = 0, \quad \frac{\partial E}{\partial x} = 4\pi\sum_\alpha e_\alpha \int f_\alpha, dp, \quad (1.5.6)$$

with summation over the electron species. Under the periodic boundary conditions, the energy integral is

$$\left\langle \frac{E^2}{8\pi} + \sum_\alpha \int mc^2\gamma(f_\alpha - f_\alpha^{(o)})\, dp \right\rangle = 0, \quad (1.5.7)$$

44

where the angular brackets denote averaging over the perturbation wavelength $\lambda = 2\pi/k_0$, $f_{\alpha 0}$ is the initial distribution function of electrons of species $\alpha$, and $\gamma = \sqrt{1 + p^2/m^2c^2}$.

At the initial time $t = 0$, the electrons of a monoenergetic beam and the plasma electrons and ions are assumed to be distributed uniformly in space:

$$f_b^{(o)} = \sigma e \sum_{s=0}^{N_b} \delta(x - x_{bs})\delta(p - p_0), \quad f_{e,i}^{(o)} = \pm\sigma e \sum_{s=0}^{N_{e,i}} \delta(x - x_{ps})\delta(p), \qquad (1.5.8)$$

where $x_{\alpha s} = 2\pi s/N_\alpha k_0$, $N_\alpha$ is the number of particles of species $\alpha$ (the number of charged planes) per wavelength, $\sigma e$ is the surface electron density, and $p_0$ is the initial momentum of the beam electrons. The number of plasma ions, which are assumed to be immobile, is determined by the charge neutrality condition for the beam–plasma system, $N_i = N_e + N_b$.

The distribution functions

$$f_\alpha^{(o)} = \sigma e \sum_s \delta\left[x - x_{\alpha s}\left(t\right)\right]\delta\left[p - p_{\alpha s}\left(t\right)\right], \qquad (1.5.9)$$

(where $x_{\alpha s}$ and $p_{\alpha s}$ are the coordinate and momentum of the $s$th particle of species $\alpha$) satisfy the initial conditions (1.5.8) and allow us to represent the energy integral (1.5.7) as

$$W_f + W_p + \delta W_b = 0,$$

$$W_f = \frac{E^2}{8\pi n_b mc^2\gamma_0}, \qquad W_p = \frac{1}{N_b\gamma_0}\sum_{s=1}^{N_p}\gamma_{ps}, \qquad (1.5.10)$$

$$\delta W_b = -1 + \frac{1}{N_b\gamma_0}\sum_{s=1}^{N_b}\gamma_{bs},$$

where $n_b = \sigma N_b/l$ is the beam density, $mc^2\gamma_0$ is the initial beam energy, and $l$ is the length of the space interval.

Below, we will present the results of numerical integration of the set of equations with a self-consistent field for a plasma with $N_p = 64512$ electrons and a monoenergetic beam consisting of $N_b = N_p/64$ electrons. Our computations were carried out with the dimensionless parameters

$$\tau = \omega_p t, \quad L = \omega_p l/c, \quad \beta_{\alpha s} = v_{\alpha s}/c.$$

The length $l$ of the space interval was chosen to be equal to the wavelength $\lambda$. We found that small perturbations $W_f(0) \sim 10^{-6}$ with a wavelength above the threshold value were unstable and grew exponentially with time at the growth rate (1.5.4).

The computation results shown in Fig. 1.5.1 enable us to analyze the time evolution of the energy density $-\delta W_b = W_f + W_p$ of Langmuir oscillations for different beam energies. As $\gamma_0$ increases, the energy lost by the beam grows and becomes maximum at $\alpha \simeq 1$. In the energy range $\alpha > 1$, the fraction of the energy that is transferred from the beam to the plasma is reduced and the oscillation energy in the plasma increases slower because of a relativistic decrease in the growth rate $\delta \sim 1/\gamma_0$.

Fig. 1.5.1b shows extremes of the function $-\delta W_{bm}$ that correspond to the first peak in the function $-\delta W_b(\tau)$. A comparison with the time-dependent solution to equations (1.5.2) shows that the oscillation energy obtained from the kinetic model of a plasma is lower than that evaluated from the hydrodynamic model.

An exponentially growing solution to equations (1.5.6) can be obtained under the conditions of the plasma and phase resonances,

$$\omega = \omega_p = \sqrt{2\sigma ke^2 N_p/m}, \quad \omega = kv_0.$$

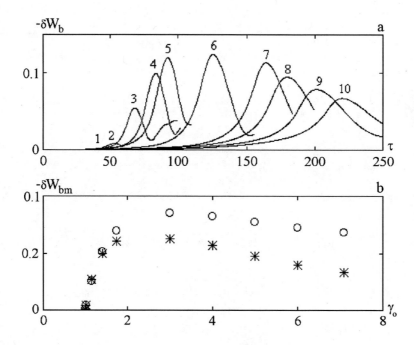

Fig. 1.5.1. (a) Energy density of Langmuir oscillations for a beam with the energy $\gamma_0$:
(1) 1.005, (2) 1.02, (3) 1.15, (4) 1.4, (5) 1.75, (6) 3, (7) 4, (8) 5, (9) 6, and (10) 7.
(b) Maximum oscillation amplitude versus the beam energy in the kinetic (asterisks)
and hydrodynamic (circles) plasma models.

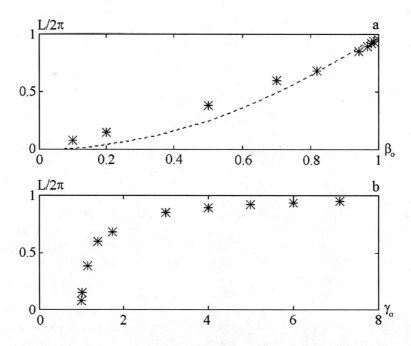

Fig. 1.5.2. The wavelength of the unstable mode as a function of (a) the beam velocity
and (b) the beam energy. The dashed curve corresponds to the function $\beta_0^2$.

These conditions imply that the plasma frequency and the perturbation wavelength both

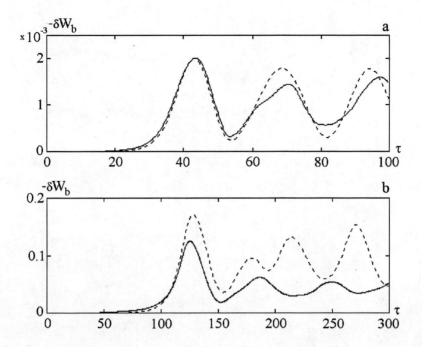

Fig. 1.5.3. Comparison between the results obtained in the kinetic (solid curves) and hydrodynamic (dashed curves) plasma models for a beam with the energy $\gamma_0 =$ (a) 1.005 and (b) 3.

depend on the beam velocity $v_0 = \beta_0 c$,

$$\omega_p = 2\pi c/\lambda, \quad \lambda = \lambda_R \beta_0^2, \quad \lambda_R = \pi mc^2/\sigma e^2 N_p. \qquad (1.5.11)$$

Unlike the hydrodynamic model (1.5.2), in which the density is constant and $\lambda \sim \beta_0$, in the kinetic model, the number of particles $N_p$ is fixed and the wavelength is proportional to the squared beam velocity, $\lambda \sim \beta_0^2$.

The computed profiles $L(\beta_0)$ and $L(\gamma_0)$ shown in Fig. 1.5.2 agree fairly well with formula (1.5.11).

For a nonrelativistic beam ($\beta_0 = 0.1$), the plasma heating during the instability is insignificant and the time evolutions of $-\delta W_b(\tau)$ in Fig. 1.5.3 obtained from the kinetic and hydrodynamic models are fairly close to each other. For an REB with $\gamma_0 = 3$, the amplitude of the nonlinear phase oscillations of the field is markedly lower (Fig. 1.5.3b).

At $\tau < 300$, the functions $W_f$ and $W_p$ are essentially the same and agree fairly well with the following formula in terms of macroscopic electrodynamics:

$$W_f + W_p = \frac{d}{d\omega}(\omega\epsilon)\,W_f \simeq 2W_f. \qquad (1.5.12)$$

According to (1.5.12), under the resonance condition $\epsilon = 0$, the energy lost by the beam is shared equally between the field and the plasma electrons. In a later stage, when the field energy $W_f$ is converted into the energy $W_p$ of the plasma electrons (Fig. 1.5.4), the evolutions of these functions differ greatly from one another. In this stage of instability, the energy exchange between the beam and the wave is almost completely absent. In the final stage ($\tau > 1000$), the function

$$-\delta W_b = W_f + W_p$$

47

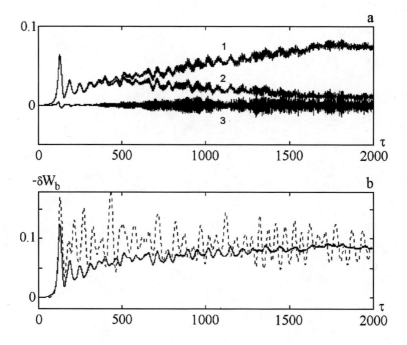

Fig.1.5.4. Transition of the beam–plasma system to a quasisteady regime for $\gamma_0 = 3$: (a) the time evolutions of (1) the plasma energy $W_p$, (2) the field energy $W_f$, and (3) the power and (b) the time evolutions of the energy $-\delta W_b$ lost by the beam. The solid curves are obtained in the kinetic model, and the dashed curves, in the hydrodynamic model.

approaches an asymptotic about which an analogous hydrodynamic solution to equations (1.5.2) oscillates nonlinearly (Fig. 1.5.4b). Note that the sum of the functions $W_f$ and $W_p$ shown in Fig. 1.5.4b experiences no small-scale oscillations, which means that they are periodic functions in $\tau$.

The average velocity of the particles of species $\alpha$ is defined as

$$\beta_\alpha = \frac{1}{N_\alpha} \sum_{\alpha s = 1}^{N_\alpha} \beta_{\alpha s}, \tag{1.5.13}$$

and their temperature coincides with the mean kinetic energy of a plasma flow moving with the velocity $\beta_\alpha$,

$$\frac{T_\alpha}{mc^2} = \frac{1}{N_\alpha} \sum_{s=1}^{N_\alpha} \frac{1 - \beta_{\alpha s} \beta_\alpha}{(1 - \beta_{\alpha s}^2)^{1/2}} - (1 - \beta_\alpha^2)^{1/2}. \tag{1.5.14}$$

For a nonrelativistic plasma such that $\beta_{ps} \ll 1$ and $\beta_p \ll 1$, formula (1.5.14) can be simplified to $T_p = mc^2 \left( \overline{\beta_{ps}^2} - \beta_p^2 \right) /2$, where $\overline{\beta_{ps}^2}$ is the mean squared particle velocity.

In Fig. 1.5.5 and Fig. 1.5.6, plots are given of the functions $\beta_{b,p}(\tau)$) and $T_{b,p}(\tau)$, which describe the time relaxation of the directed velocity and temperature of the beam and plasma. In Fig. 1.5.5, the dotted curve shows the nonlinear wave phase velocity calculated from the empirical formula

$$v_{\mathrm{ph}}/v_0 = 1 - 2^{-4/3} \left[ \nu^{1/3}/\gamma_0 + \sqrt{3} \left( W_f + W_p \right) \right]. \tag{1.5.15}$$

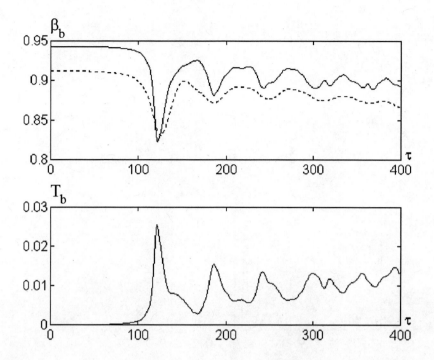

Fig. 1.5.5. Average beam velocity $\beta_b$ and beam temperature $T_b$ (in units of $mc^2$) for $\gamma_0=3$ as a function of time. The dashed curve corresponds to the phase velocity of a nonlinear wave.

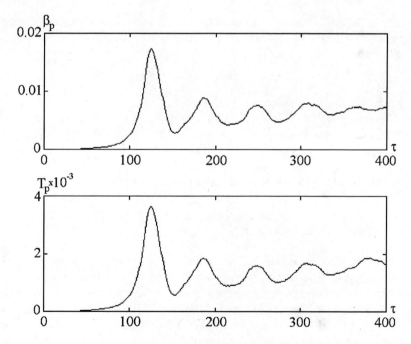

Fig. 1.5.6. Average plasma velocity $\beta_p$ and plasma temperature $T_p$ (in units of $mc^2$) for $\gamma_0=3$ as a function of time.

The plots shown in Fig. 1.5.7 illustrate the time evolution of the beam and plasma electron distribution functions, $f_b(\beta)$ and $f_p(\beta)$, in velocity space $\beta$ for a beam with the

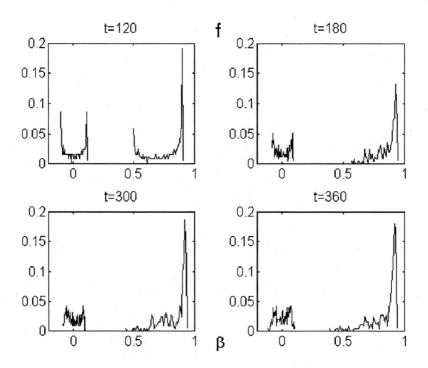

Fig. 1.5.7. Distribution functions of the (1) beam and (2) plasma electrons for $\gamma_0=3$ at different times $\tau =$ (a) 120, (b) 180, (c) 300, and (d) 360.

initial energy $\gamma_0 = 3$. The electrons of each species $\alpha$ are divided into 50 groups with nearly equal electron velocities $\beta_{\alpha j}$. If the number of electrons of species $\alpha_j$ is $n_{\alpha j}$, then, by definition, we have

$$f_{\alpha j} = \frac{n_{\alpha j}}{N_\alpha}, \qquad (1.5.16)$$

where $N_\alpha$ is the number of particles of species $\alpha$.

Before the time $\tau \approx 120$, at which the oscillation amplitude saturates, a tail of decelerated electrons forms on the beam distribution function. Thereafter, the oscillation energy is partially converted back into beam energy and the beam distribution function subsequently does not undergo any qualitative change (Fig. 1.5.7). The expansion of the plasma electron distribution function in velocity space illustrates electron heating by the field of the unstable mode.

### 3. Modulational instability .

Numerical simulations show that, during the oscillatory instability of the primary spectrum, the energy of the plasma ions grows [9] due to both the onset of a parametric instability and the decay of a pump wave into Langmuir and ion acoustic modes [10]. In the immobile-ion approximation, the field energy is converted into the energy of the plasma electrons because of the onset of the modulational instability of the wave (the self-compression of Langmuir wave packets in a heated plasma) [25].

Numerical integration shows that, in the stage of the high-frequency modulation of the wave amplitude, the energy exchange between the beam and the wave field is almost completely absent (see Fig. 1.5.4). Consequently, following [25], we can find the growth rate of the modulational instability from the nonlinear parabolic equation without allowance for the trapped particles.

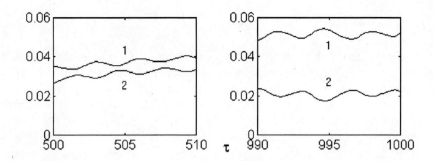

Fig. 1.5.8. Functions (1) $W_p(\tau)$ and (2) $W_f(\tau)$ over different time intervals (fragments of Fig. 1.5.4a).

Setting the right-hand side of the first of equations (1.5.2) to zero, we represent the solution as

$$E = (E_0 + E_1 + iE_2)\exp(-i\Omega_0 t), \qquad (1.5.17)$$

where

$$\Omega_0 = -(\omega_p/2)(\epsilon + E_0^2/E_T^2),$$

$$E_1, E_2 \sim \exp(-i\Omega t + ikx), \quad E_1 \ll E_0, \quad E_2 \ll E_0.$$

Substituting (1.5.17) into (1.5.2) yields the dispersion relation [25]

$$(\Omega - kv_g)^2 = \frac{\omega_p^2 k^2 \lambda_d^2}{4}\left(-Q_0 + k^2\lambda_d^2\right), \quad Q_0 = \frac{W}{n_p T} = \frac{E_0^2}{16\pi n_p T}, \qquad (1.5.18)$$

where $W$ is the wave energy density, $v_g' = \omega_p \lambda_d^2$, $\lambda_d = v_T/\omega_p$ is the Debye length, and $v_g \simeq v_T^2/v_0$.

The dispersion relation (1.5.18) implies that the modes with wavenumbers $k^2\lambda_d^2 < Q_0$ are unstable. The maximum growth rate $\delta_{\mathrm{mod}} = \mathrm{Im}\,\Omega$ of the modulational instability is

$$\delta_{\mathrm{mod}} = \frac{\omega_p Q_0}{4}, \quad k^2\lambda_d^2 = \frac{Q_0}{2}. \qquad (1.5.19)$$

Numerical experiments in which the condition $Q_0 \simeq W_f/W_p \simeq 1$ holds over a sufficiently long time interval $\tau \leq 1000$ showed that the functions $W_f(\tau)$ and $W_p(\tau)$ become modulated at a high frequency, the modulation period being about $T \approx \delta_{\mathrm{mod}}^{-1}$ (Fig. 1.5.8). An increase in the amplitude of electron oscillations with $k\lambda_d > 0.1$ results in the energy transfer from the wave with $k_0 = \omega_p/v_0$ to short-wavelength oscillations. The secondary waves are damped via Landau damping, and their energy is converted into the energy of the plasma electrons [25].

## 4. Main results.

Numerical integration of the Vlasov set of equations with a self-consistent field demonstrates the possibility of converting the energy of an relativistic beam into the energy of a regular Langmuir wave in a plasma. For a beam with $n_b/n_p = 1/64$ and $\gamma_0 = 3$ ($\alpha \approx 0.75$ and $\delta/\omega_p \approx 0.06$), the energy density of Langmuir oscillations amounts to approximately 12% of the beam energy density.

An analysis of the functions $W_p(\tau)$, $W_f(\tau)$ and $-\delta W_b(\tau)$ shows that the instability occurs in three stages (Fig. 1.5.4).

A small perturbation that increases exponentially with the hydrodynamic growth rate $\delta$ stops growing at the time $\tau \approx 120$, when the instability is suppressed by the trapping of beam electrons by the wave. By this time, the beam energy becomes minimum and a tail of decelerated electrons forms on the distribution function of the initially monoenergetic beam (Fig. 1.5.7). As the field passes through a maximum, the mean beam velocity and beam temperature undergo jumps, which correspond to the conversion of a fraction of the energy of directed electron motion into thermal plasma energy (Fig. 1.5.5).

In the second stage of the nonlinear relaxation of a beam in a plasma, the beam electrons are accelerated by the wave field and the field energy is partially converted back into beam energy. As a result, at $\tau > 120$, the mean beam velocity, first, increases and, then, experiences weakly damped oscillations. The oscillations of the beam and plasma temperatures are in antiphase with the oscillations of the directed beam velocity (Figs. 1.5.5 and 1.5.6). Analyzing the shape of the beam distribution function shown in Fig. 1.5.7 at different times, we can conclude that only a small fraction of decelerated electrons in the tail of the distribution function interact with the wave.

In the final stage of evolution of the beam–plasma system ($\tau > 1000$), the energy exchange between the beam and the wave is almost completely absent. The field energy is converted into the energy of plasma electrons and the functions $W_f(\tau)$ and $W_p(\tau)$ differ greatly from each other (Fig. 1.5.4). However, their sum $-\delta W_b(\tau) = W_f(\tau) + W_p(\tau)$, describing the energy lost by the beam, approaches an asymptotic about which the analogous nonlinear solution found in a single-mode model oscillates (Fig. 1.5.4b).

The fact that the sum of the functions $W_f(\tau)$ and $W_p(\tau)$ experiences no high-frequency oscillations indicates that the modulation of the primary wave amplitude is regular. Figure 1.5.8 shows the fragments of Fig. 1.5.4 at different times. The modulation period is on the order of the inverse growth rate of the modulational instability (1.5.18).

## §1.6. Nonlinear dispersion in an unstable plasma-beam system.

The beam–plasma instability is known to occur at the intersection of the beam and plasma branches of oscillations, when small Langmuir oscillations grow exponentially with time (see §1.5). The unstable mode saturates when the beam breaks up into bunches. During this process, the wave traps resonant electrons and the electromagnetic field becomes modulated with a period of about the reciprocal of the linear growth rate. Since, in the region behind the first maximum of the wave amplitude, there is (on average) essentially no energy exchange between the beam and the wave over the period of the nonlinear oscillations, the beam energy is converted into the energy of the plasma wave mainly in the initial (exponential) stage of the instability.

The most general way of investigating the nonlinear interaction between an electron beam and a plasma is to numerically integrate the kinetic equations for the beam and plasma electrons along with the Maxwell equations for the electromagnetic field. This approach makes it possible to determine all of the physical parameters of the beam–plasma system throughout the nonlinear evolution of the beam and plasma. However, in practice,

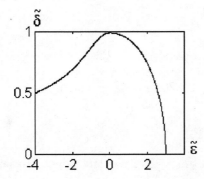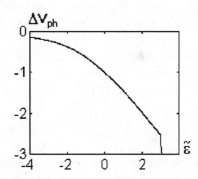

Fig. 1.6.1. The growth rate $\tilde{\delta}$ and the phase velocity $\Delta v_{ph}$ versus the permittivity of plasma $\tilde{\epsilon}$.

the solution of a spatially periodic onedimensional problem is restricted to the investigation of the time dependence of the energy of Langmuir oscillations and the construction of the phase diagram of a beam [27]. For this reason, most numerical studies in this area are in fact aimed at refining the maximum amplitude of oscillations in a plasma with a beam and yield less informative results in comparison with those obtained from linear theory.

Our purpose here is to interpret numerical solutions in more detail based on the extended notions of the physical quantities that are contained in the linear dispersion relation [18]. Clearly, this approach requires that the wave frequency, wavenumber, and wave phase velocity be generalized so that, in the limit of small-amplitude perturbations, they coincide with those used in linear theory. We analyze the beam–plasma instability using a composite model in which the growth rate is calculated from the corresponding linear hydrodynamic formula on the basis of the results obtained from a numerical kinetic model. This approach enables us to establish the applicability range of the hydrodynamic approximation for beams with different energies.

### 1. Linear theory.

For small Langmuir oscillations of the form $E(t, x) \sim \exp(ikx - i\omega t)$, the equations for a plasma with a monoenergetic relativistic electron beam (1.5.6) yield the following dispersion relation, which determines the dependence of the frequency $\omega$ on the wavenumber $k$ [28]:

$$1 - \frac{\omega_p^2}{\omega^2}\left(1 + \frac{3}{2}\frac{k^2 v_T^2}{\omega^2}\right) - \frac{\omega_b^2}{(\omega - kv_0)^2}\frac{1}{\gamma_0^3} = 0, \qquad (1.6.1)$$

where $\omega_p^2 = 4\pi n_p e^2/m$, $\omega_b^2 = (n_b/n_p)\omega_p^2$, $n_b$ and $n_p$ are the beam and plasma densities, $v_0$ is the initial beam velocity, $\gamma = \left(1 - v_0^2/c^2\right)^{-1/2}$, $v_T$ is the thermal velocity of the plasma electrons, and $(kv_T/\omega)^2 \ll 1$.

According to (1.6.1), small Langmuir oscillations grow exponentially with time. Figure. 1.6.1 show the growth rate of the perturbations $\tilde{\delta} = 2\,\mathrm{Im}\,\omega/\sqrt{3}\mu^{1/3}\omega_p$ and the phase velocity shift

$$\Delta v_{ph} = \frac{2}{\mu^{1/3}}\left(\frac{\mathrm{Re}\,\omega}{kv_0} - 1\right), \qquad \mu = \frac{n_b}{2n_p\gamma_0^3},$$

as a function of $\tilde{\epsilon} = \epsilon/(2\mu)^{1/3}$, $\epsilon = 1 - \omega_p^2/k^2 v_0^2$.

## 2. Wave frequency, wavenumber, and wave phase velocity in the nonlinear stage of instability.

We assume that all of the physical parameters to be calculated are periodic functions of the coordinate, the period being equal to the wavelength $\lambda = 2\pi v_0/\omega$ of the resonant mode. For time-dependent quantities, we introduce the averaging operation

$$f(t) \equiv\; < f(t,x) > = \frac{1}{\lambda} \int_0^\lambda f(t,x)\,dx. \tag{1.6.2}$$

The electric field $E(t,x)$ and its space and time derivatives are assumed to satisfy the Maxwell equations in the electrostatic approximation [29]:

$$\frac{\partial E}{\partial x} = -4\pi e(n_b + n_p - n_i), \qquad \frac{\partial E}{\partial t} = -4\pi e(j_b + j_p), \tag{1.6.3}$$

where $n_{b,p}$ and $j_{b,p}$ are the beam and plasma current densities and $n_i$ is the plasma ion density. Taking into account the fact that the ratios composed of the electric field and its space and time derivatives have the dimensionalities of length, time, and velocity, we define the wave frequency, wavenumber, and wave phase velocity in the nonlinear stage of the instability as

$$k(t) = \sqrt{\frac{\langle E_x^2\rangle}{\langle E^2\rangle}}, \qquad \omega(t) = \sqrt{\frac{\langle E_t^2\rangle}{\langle E^2\rangle}}, \qquad v_{ph}(t) = \frac{\omega(t)}{k(t)}. \tag{1.6.4}$$

We eliminate the derivatives $E_x$ and $E_t$ in (1.6.4) to obtain

$$\omega(t) = \sqrt{\frac{\langle (j_b + j_p)^2\rangle}{\langle E^2\rangle}}, \qquad k(t) = \sqrt{\frac{\langle (n_b + n_p - n_i)^2\rangle}{\langle E^2\rangle}},$$

$$v_{ph}(t) = \sqrt{\frac{\langle (j_b + j_p)^2\rangle}{\langle (n_b + n_p - n_i)^2\rangle}}. \tag{1.6.5}$$

These formulas imply that $k(t)$, $\omega(t)$, and $v_{ph}(t)$ depend on the averaged moments of the beam and plasma distribution functions, which should be calculated by integrating the self-consistent set of equations numerically. The Langmuir frequencies of the beam and plasma electrons and the beam velocity can be defined in a similar manner:

$$\omega_{b,p}(t) = \frac{\omega_p}{\sqrt{n_{b,x}}} \left\langle \sqrt{n_{b,p}(t,p)} \right\rangle,$$

$$v(t) = \frac{\langle j_b(t,x)\rangle}{\langle n_b(t,x)\rangle}, \qquad v_T^2(t) = \frac{2\langle T(t,x)\rangle}{m}. \tag{1.6.6}$$

We continue to draw an analogy with linear theory and switch to the dimensionless variables

$$\mathcal{K} = \frac{ck}{\omega_0}, \quad \Omega = \frac{\omega}{\omega_0}, \quad \Omega_p = \frac{\omega_p}{\omega_0}, \quad \Omega_b = \frac{\omega_b}{\omega_0},$$

$$V = \frac{v}{c}, \quad V_{ph} = \frac{v_{ph}}{c}, \quad V_T = \frac{v_T}{c}, \quad \Gamma = \frac{1}{\sqrt{1 - V^2}},$$

$$\tau = \omega_0 t, \quad \omega_0 = \omega_p(0).$$

54

Then, we define the nonlinear dielectric function of a plasma with a beam as

$$\varepsilon(\Omega, \mathcal{K}) = \left[\Omega^2 - \Omega_p^2 \left(1 + \frac{3}{2}\frac{V_T^2}{V_{\text{ph}}^2}\right)\right](\Omega - \mathcal{K}V)^2 - \frac{\Omega_b^2 \Omega^2}{\Gamma^3}. \tag{1.6.7}$$

### 3. Numerical experiment.

The dynamics of the instability of a beam–plasma system was investigated numerically by solving the Vlasov equations for the beam and plasma electrons by a particle-in-cell method. Formula (1.6.7) makes it possible to establish the limits of applicability of the linear dispersion relation $\varepsilon[\Omega(0), \mathcal{K}(0)] = 0$ in the nonlinear stage of the instability. The positive values of the dielectric function, $\varepsilon[\Omega(0), \mathcal{K}(0)] > 0$, indicate that the beam–plasma system has evolved into a nonlinear stage.

The electrons interact with the wave under the conditions of the Cherenkov and plasma resonances. Nonlinear variations of the beam modulation frequency $\mathcal{K}V$ and plasma frequency $\Omega_p$ cause the beam and plasma electrons to lose their synchronism with the wave; as a result, the wave amplitude grows at a lower rate. The role played by each of these effects in the nonlinear saturation of the instability is described by the functions

$$R_b = \Omega - \mathcal{K}V, \quad R_p = \Omega - \Omega_p \tag{1.6.8}$$

in formula ( 1.6.7). The time evolutions of these functions are illustrated by curves 1 and 2 in Fig. 1.6.2. Although the parameters $\mathcal{K}V$ and $\Omega_p$ correspond to the highly nonlinear stage, the dielectric function $\varepsilon(\Omega, \mathcal{K})$ (curve 3) changes only slightly as compared to the linear theory up to the time at which the energy density of Langmuir oscillations, $W = \langle E^2 \rangle / 4\pi n_b mc^2 \gamma_0$ (curve 4), reaches its maximum.

For a nonrelativistic beam ($\gamma_0 < 1.75$), the field amplitude saturates because of the deviation $R_b$ of the frequency of Langmuir oscillations from the Cherenkov resonance frequency. The higher the beam energy, the larger the deviation $R_p$ from the plasma resonance frequency. For $\gamma_0 = 3$, these deviations become comparable in magnitude when the function $W(\gamma_0)$ reaches its maximum (see Fig. 1.5.1a). For a higher energy beam, the plasma resonance is destroyed earlier than the Cherenkov resonance, and the growth of the oscillation amplitude slows because the wave properties of the plasma change. However, our simulations show that, in the energy range under consideration, the nonlinear saturation of the field amplitude is always associated with the separation of the beam into individual bunches.

The beam is unstable when its phase is synchronized with the phase of Langmuir oscillations; the phase synchronization condition holds for oscillations with phase velocities $V_{\text{ph}} < V_0$. This conclusion, which was drawn from linear theory, is confirmed by our numerical experiment, which, in particular, permits us to follow the evolution of the beam velocity $V$ and wave phase velocity $V$ph in the nonlinear stage of the instability. The numerical results obtained for beams of different energies are displayed in Fig. 1.6.3. An analysis of the results illustrated shows that, in the range $\gamma_0 \leq 3$, the beam velocity decreases more rapidly than the wave phase velocity and, in the range $\gamma_0 > 3$, the situation is the opposite. In the energy range $\gamma_0 \approx 3$, the conditions for the beam–plasma interaction are optimum: the phase resonance takes place throughout the nonlinear stage of the instability [26].

According to Fig. 1.6.2, the dispersion relation $\varepsilon(\Omega, \mathcal{K}) \approx 0$ is valid up to the time at which the beam breaks up into bunches and the "warm" correction to the electron plasma frequency in formula (1.6.7) remains small: $V_T^2/V_{\text{ph}}^2 \simeq 10^{-3}$. We thus conclude that, in the first stage of the instability, the nonlinear relaxation of the beam–plasma system is governed

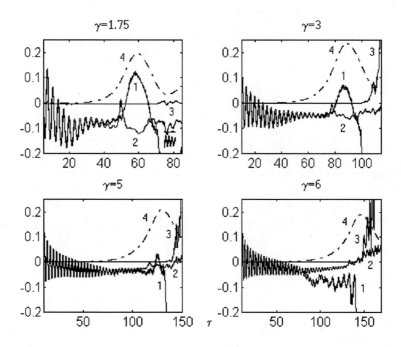

Fig. 1.6.2. Nonlinear evolution of (1) the deviation $R_b$ from the Cherenkov resonance frequency, (2) the deviation $R_p$ from the plasma resonance frequency, and (3) the dielectric function $\varepsilon(\Omega, \mathcal{K})$ of a plasma with a beam for different initial values of the relativistic factor $\gamma$. Curves 4 illustrate the evolution of the energy density $W$ of Langmuir oscillations.

by the time evolution of the linear growth rate. For a low-density beam propagating in a cold $(V_T = 0)$ plasma, the dispersion relation has the solution $\Omega = \mathcal{K}V + \Delta\Omega$, where the small correction to the frequency, $|\Delta\Omega|/\mathcal{K}V \ll 1$, satisfies the equation

$$\left(\frac{\Delta\Omega}{\mathcal{K}V}\right)^2 \left[\frac{\varepsilon(\mathcal{K}V)}{2} + \frac{\Delta\Omega}{\mathcal{K}V}\right] = \frac{\Omega_b^2}{2\Omega_p^2\Gamma^3} \equiv \mu. \tag{1.6.9}$$

For fixed $\mu$, the complex roots of (1.6.9) depend only on the plasma dielectric function at the beam modulation frequency, $\varepsilon(\mathcal{K}V) = 1 - \Omega_p^2/\mathcal{K}^2V^2$.

In Fig. 1.6.4, curves 1 illustrate the behavior of the nonlinear plasma dielectric function

$$R = R_p - R_b = \mathcal{K}V - \Omega_p \approx \varepsilon(\mathcal{K}V)/2. \tag{1.6.10}$$

and curves 2 reflect time evolutions of the hydrodynamic growth rate. The dotted curves 3 refer to the growth rate obtained by numerical integration.

In our hybrid model of the beam–plasma instability, the growth rate is calculated with the help of formula (1.6.9) from linear theory on the basis of the results obtained from the nonlinear numerical model. This approach makes it possible to interpret the nonlinear saturation of the oscillation amplitude as the evolution of the hydrodynamic growth rate from its maximum value $\delta_m = (\sqrt{3}/2)\mu^{1/3}\Omega_p$ to the instability threshold

$$\delta_+ = (2\mu)^{1/3} \left[1 - \frac{\varepsilon(\mathcal{K}V)}{\varepsilon_c}\right]^{1/2} \Omega_p, \quad \varepsilon_c = 3(2\mu)^{1/3}. \tag{1.6.11}$$

56

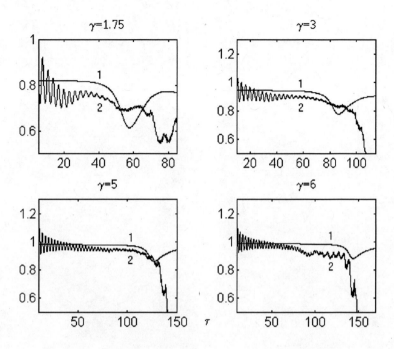

Fig. 1.6.3. Nonlinear evolution of (1) the beam velocity $V$ and (2) the wave phase velocity $V_{\mathrm{ph}}$ for beams with different energies.

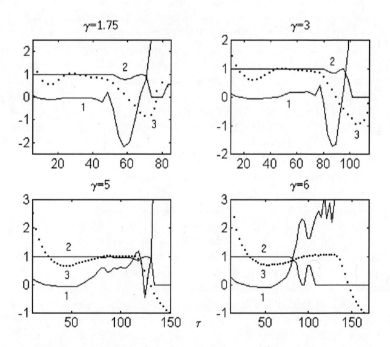

Fig. 1.6.4. Nonlinear evolution of (1) the plasma dielectric function $\tilde{R}=R/\mu^{1/3}$ and (2) the hydrodynamic growth rate $\delta(R)/\delta_{\mathrm{m}}$. Curves 3 refer to the numerically calculated growth rate $\delta(\tau)/\delta_{\mathrm{m}}=(2\tau)^{-1}\ln(W/W_0)$.

57

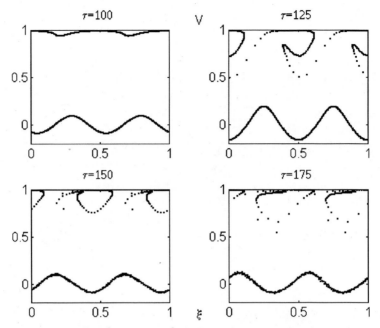

Fig. 1.6.5. Phase plane $[(\xi,V),\ \xi=x/\lambda]$ of a beam for $\gamma_0=6$ and a plasma electrons.

For a nonrelativistic beam, the plasma density can be assumed to be linear, $\Omega_p \approx 1$, and the nonlinear saturation of the oscillation amplitude is governed by the time evolution of the beam modulation frequency $\mathcal{K}V$. Numerical integration shows that, as Langmuir oscillations grow, the mean beam velocity $V$ decreases and the wavenumber $\mathcal{K}$ increases monotonically. The first effect dominates in the initial stage of the instability, when a significant fraction of the beam energy is expended on the excitation of Langmuir oscillations. Consequently, due to the increase in the amplitude of the fastest growing mode, the plasma dielectric function becomes negative, $\varepsilon(\mathcal{K}V) < 0$, in which case the oscillations grow at a rate $\delta_- \approx \sqrt{2\mu/|\varepsilon(\mathcal{K}V)|}\Omega_p$. The onset of the satellite modes that grow at low rates is accompanied by the redistribution of the oscillation energy among the modes in a narrow Langmuir wave packet [27] and the increase in the effective wavenumber $\mathcal{K}$. Accordingly, the beam modulation frequency $\mathcal{K}V$, first, falls off to its minimum value and, then, starts to increase. As the instability saturates, the dielectric function $\varepsilon(\mathcal{K}V)$ becomes positive, in which case the growth rate equals zero.

In the high-energy range $\gamma_0 \approx 3 - 5$, in which the nonlinear decrease in the Langmuir frequency of the plasma electrons is already observed [14], Langmuir oscillations also saturate in an essentially hydrodynamic fashion. However, for $\gamma_0 = 6$, the correlation between hydrodynamic and kinetic (numerical) solutions is found to worsen. According to our numerical experiment, the dielectric function $\varepsilon(\mathcal{K}V)$ increases monotonically and rapidly becomes positive, in which case unstable Langmuir oscillations should saturate nonlinearly by the time $\tau \approx 100$, at which the hydrodynamic growth rate vanishes. However, the oscillations continue to grow up to the time $\tau \approx 150$. An analysis of the phase diagram of a beam shows that multiflow motion already appears at $\tau \approx 125$; consequently, the instability saturates after the beam has broken up into bunches.

Hence, we have developed a numerical model in which the key parameters in linear theory–the wave frequency, wavenumber, and wave phase velocity–are extended to the nonlinear stage of the instability and are used, together with the field energy density, as the global parameters of an unstable beam–plasma system.

We have shown that the beam energy is converted into the energy of Langmuir oscillations and the mean beam velocity decreases. The averaged (over the spatial period of the Langmuir wave) frequency of the fastest growing mode oscillates around the initial Langmuir frequency $\Omega \approx 1$ of the plasma electrons. Consequently, when the Langmuir wave spectrum shifts toward shorter wavelengths ($\mathcal{K} > 1$), the wave phase velocity in the nonlinear stage of the instability decreases. The asynchronous nature of these processes is the reason for the saturation of unstable Langmuir oscillations in a plasma with a nonrelativistic beam.

In the case of relativistic beams, the oscillation energy density is higher, the plasma density is modulated, and the Langmuir frequency averaged over the spatial period is lower. The last effect destroys the plasma resonance, thereby limiting the ability of the plasma to serve as a resonant waveguide structure for high-energy electron beams.

During nonlinear relaxation, the beam and plasma parameters under consideration change. However, the physical state of a beam–plasma system depends on the particular combination of these parameters that enters the plasma dielectric function: $\epsilon(0) = 1 - \omega_p^2/k^2 v_0^2$. The instability saturates nonlinearly when the dielectric function, which is initially equal to $\varepsilon(0) = -\mu^{1/3}$, starts to increase and becomes positive, $\epsilon(t) > 0$. In the case $n_b/n_p = 1/64$, which we have analyzed above, the time scale on which the instability saturates depends on the beam energy. For $\gamma_0 \approx 3$ (when the ratio of the energy density of Langmuir oscillations to the beam energy density is maximum), this time is the longest.

**Appendix 1.1**

We use the following set of equations to describe the interaction of a relativistic electron beam with plasma [12]:

$$\frac{\partial f}{\partial t} + v\frac{\partial f}{\partial x} + eE\frac{\partial f}{\partial p} = 0,$$

$$\frac{\partial^2 E}{\partial t^2} + \omega_L^2 E = -4\pi e\frac{\partial}{\partial t}\int_{-\infty}^{\infty}(f - f_0)\,v\,dp,$$

(A1.1.1)

where $f(t, x, p)$ is the function describing the electron momentum distribution in the beam. We seek the solution of Eq. (A1.1.1) in the form

$$f(t, x, p) = f_0(t, p) + \operatorname{Re} f_k(t, x, p)\exp(-ikx),$$
$$E(t, x) = \operatorname{Re} E_k(t)\exp(-ikx),$$

(A1.1.2)

where $f_0$ is the electron distribution function and $f_k$ are the amplitudes of small oscillating additional terms.

Substituting Eq. (A1.1.2) in Eq. (A1.1.1) and averaging with respect to $x$, we obtain

$$\frac{\partial f_k}{\partial t} - ikv f_k = -eE_k\frac{\partial f_0}{\partial p},$$

(A1.1.3)

$$\frac{\partial f_0}{\partial t} = -\frac{e}{2}\operatorname{Re} E_k^*\frac{\partial f_k}{\partial p},$$

(A1.1.4)

$$\frac{\partial^2 E_k}{\partial t^2} + \omega_p^2 E_k = -4\pi e\int_{-\infty}^{\infty} v\frac{\partial f_k}{\partial t}\,dp.$$

(A1.1.5)

Let us transform Eq. (A1.1.4) by eliminating the time derivative on the right hand-side with the add of Eq. (A1.1.2)

$$\frac{\partial^2 E_k}{\partial t^2} + \omega_p^2\left(1 + \frac{1}{n_p}\int_{-\infty}^{\infty}\frac{f_0}{\gamma^3}\,dp\right)E_k = -i4\pi ek\int_{-\infty}^{\infty} v^2 f_k\,dp,$$

(A1.1.6)

59

where $n_p$ is the plasma density, and $dp/dv = m\gamma^3$. Comparing the right-hand sides of (A1.1.4) and (A1.1.6), we find the following integral

$$mc^3 \int_{-\infty}^{\infty} \left( \frac{p}{mc} - \arctan \frac{p}{mc} \right) f_0 dp - \frac{1}{8\pi k} \mathrm{Re} \left( i E_k^* \frac{dE_k}{dt} \right) = S_0. \qquad (A1.1.7)$$

We suppose that at the initial time the thermal spread in the beam is sufficiently small and the distribution function is closed to delta-function:

$$f_0(t, p) = n_b \delta[p - p(t)] = \delta(p'),$$

where $p(t)$ is the monotonic part of electron momentum. The solution of Eq. (A1.1.3) can be represented in the form

$$f_k = -e \int_0^t E_k(\tau) \frac{\partial f_0}{\partial p'} \exp i \left[ \Phi(t) - \Phi(\tau) + kv'(t - \tau) \right] d\tau, \qquad (A1.1.8)$$

where $\Phi(t) = k \int_0^t v(t') \, dt'$. Substituting for $f_k$ from Eq. (A1.1.8) into Eq. (A1.1.6) and evaluating the integrals with respect to the momenta by means of $\delta(p')$, we find

$$\frac{\partial^2 E_k}{\partial t^2} + \omega_p^2 \left( 1 + \frac{\nu}{\gamma^3} \right) E_k$$
$$= \omega_b^2 \int_0^t E_k(\tau) \frac{kv}{\gamma^3} \left[ -2i + kv(t - \tau) \right] \exp i \left[ \Phi(t) - \Phi(\tau) \right] d\tau, \qquad (A1.1.9)$$

where $v \equiv v(\tau)$, $\omega_b^2 = 4\pi e^2 n_b / m$, and $\nu = n_b / n_p$.

We now introduce the slowly varying field amplitude $E(t)$ by substituting

$$E_k(t) = E(t) \exp[i\Phi(t)], \quad \dot{E} \ll \dot{\Phi} E$$

and if we neglect the second derivatives of the amplitude we can write Eq. (A1.1.7) and Eq. (A1.1.9) in the form

$$n_b mc^3 \left( \frac{p}{mc} - \arctan \frac{p}{mc} \right) - v \frac{|E|^2}{8\pi} = S_0,$$
$$ik \frac{d^2}{dt^2} \left( 2v \frac{dE}{dt} + E \frac{dv}{dt} \right) + \frac{d^2}{dt^2} \left[ (\omega_p^2 - k^2 v^2) E \right] - \omega_b^2 \frac{k^2 v^2}{\gamma^3} E = 0. \qquad (A1.1.10)$$

The terms on the order of $\nu^{1/3}/\gamma_0 \ll 1$ is omitted. At ultrarelativistic beam energies, $p/mc \approx \gamma \gg 1$ and $\arctan(p/mc) \approx \pi/2$, the first relation (A1.1.10) transforms to the energy conservation law (1.2.4).

**Appendix 1.2**

Working in the linear approximation,

$$\eta' = 0, \quad N' = \delta N, \quad a' = \delta a$$

we find from the equations of system (1.2.10) the following results for an exponentially growing solution, $\sim \exp(\delta\tau)$:

$$\cos \eta = h(\alpha - 2h^2), \quad \delta^2 = h^2(\alpha - h^2 - 2h^{-1} \cos \eta),$$
$$\delta^2 = (2h)^{-2}(1 - \cos^2 \eta), \qquad (A1.2.1)$$

where $\delta$ is the growth rate and $h = a/N$. Excluding $\cos\eta$ from the second and third equations (A1.2.1), we find

$$\delta^2 = h^2(3h^2 - \alpha), \quad h^2(4h^2 - \alpha)^2 - 1 = 0. \tag{A1.2.2}$$

Taking into account

$$(4h^3 - \alpha h - 1)(4h^3 - \alpha h + 1) = 0,$$

and solving cubic equation, we find $h = h_+ + h_-$, where functions $h_+$ and $h_-$ are given by (1.2.11).

**Appendix 1.3**

The roots of the fourth-degree equation

$$U(w) = (1/4)(w^2 - h)^2 - 4\rho^2(1 + w), \quad h = 1 + 2\rho'^2 \tag{A1.3.1}$$

are

$$w_{1,2} = \sqrt{z_1} \pm 2\mathrm{Re}\sqrt{z_2}, \ b = -\sqrt{z_1}, \ d = 2\mathrm{Im}\sqrt{z_2}, \tag{A1.3.2}$$

where $z_{1,2}$ are given by

$$
\begin{aligned}
z_1 &= h/3 + R + S, \\
z_2 &= h/3 - (R + S)/2 + i\sqrt{3}(R - S)/2, \\
R, S &= (h^3/27 - 2\rho^2 h/3 + 2\rho^4 \pm \sqrt{Q})^{1/3}, \\
Q &= 4\rho^4(\rho^4 + 16\rho^2/27 - 2\rho^2 h/3 \\
&\quad + h^3/27 - h^2/27) > 0.
\end{aligned}
\tag{A1.3.3}
$$

**Appendix 1.4**

To determine the asymptotic behavior of $T_E$ near the critical point $\rho_c \to 2/\sqrt{27}$, $\rho'_c \to 1/\sqrt{6}$, we seek a solution of (1.3.22) in the form

$$\rho \approx 2/\sqrt{27} + \xi/\sqrt{6} + L\xi^2/2 + M\xi^{5/2}, \tag{A1.4.1}$$

where the coefficients $L$ and $M$ are to be determined. Substituting (A1.4.1) into the right-hand side of (1.3.22), and using (A1.3.3) and (1.3.18), we find $L = 1/\sqrt{3}$. For $M$ we find the transcendental equation

$$
\begin{aligned}
M &= \frac{\sqrt{3}}{5}\left\{R_0 + S_0 + \sqrt{3(R_0^2 + S_0^2 + R_0 S_0)}\left[1 - 2E(k)/K(k)\right]\right\}, \\
k_0^2 &= \frac{1}{2}\left[1 - \frac{\sqrt{3}(R_0 + S_0)}{2\sqrt{R_0^2 + S_0^2 + R_0 S_0}}\right], \\
R_0, S_0 &= 5^{1/3}2^{7/6}3^{-3/2}\left[M \pm \sqrt{M^2 + 2^{7/2}5^{-2}3^{-3}}\right]^{1/2},
\end{aligned}
\tag{A1.4.2}
$$

from which we find $M \approx -0.12$. Substitution of (A1.4.1) into (1.3.19) leads to

$$T_E \approx \frac{4K(k_0)}{[3(R_0^2 + S_0^2 + R_0 S_0)]^{1/4}}\mu\xi^{-1/4} \approx 9.72\mu\xi^{-1/4}, \tag{A1.4.3}$$

which yields estimate (1.3.20).

A convenient way to evaluate the integral on the right side of (1.3.22) is to eliminate $\cos \eta$ and to transform to the variable $w$ with the help of (1.3.11), (1.3.17), and the relation $d\xi = dw/w'$, where $w'$ is given by (1.3.17). As a result we find

$$\frac{1}{T_E} \int_{\xi}^{T_E} A \cos \eta \, d\xi'$$
$$= \int_{w_1}^{w_2} \left( \frac{\rho'^2}{2} + \frac{1 - w^2}{4} \right) \frac{dw}{\sqrt{-U(w)}} \Big/ \int_{w_1}^{w_2} \frac{dw}{\sqrt{-U(w)}}. \tag{A1.5.1}$$

The trigonometric substitution

$$w = \frac{w_1 + w_2}{2} - \frac{w_1 - w_2}{2} \frac{\lambda - \cos \varphi}{1 - \lambda \cos \varphi} \tag{A1.5.2}$$

[see (1.3.18)] puts the integral in a standard form. This integral is expression (1.3.22).

## CHAPTER 2

## SELF-FOCUSING OF AN INTENSE RELATIVISTIC ELECTRON BEAM DUE TO THE BEAM–PLASMA INSTABILITY

The underlying base of the physical mechanism of the beam plasma instability is energy exchange between resonance beam particles and the longitudinal component of the plasma perturbation electric field that is accompanied by the beam drag and transformation of its kinetic energy into one of Langmuir oscillations. For this reason, the original papers on the beam plasma instability [1,2] considered infinite beam and plasma, allowing to obtain the solution in the approximation of 1D Poisson equation.

The instability of finite beams in an infinite plasma was considered in many papers [3-13]. For the finite radius beam the inhomogeneity of particle distribution in the transverse cross-section causes a "hook-up" of potential and electromagnetic oscillation branches and an energy transformation of the beam-excited Langmuir wave into electromagnetic energy on the density gradient. That is why the instability increment is determined not only by the energy, yielded per unit volume of beam and plasma, but it also depends on the Pointing vector flow across the beam boundary into plasma, which complicates drastically the calculations even in the approximation of small (linear) oscillations.

It should be noted that for the arbitrary geometry beam the current density remains undetermined in the Maxwell equations. This is why we discuss below a relatively common case of cylindrical beam with a smooth boundary that represents a "washed-out step". Conducted within the framework of this model, the numeric integration of plasma hydrodynamic equations with a kinetic pressure gradient and monovelocity electron beam equations indicated that the transverse oscillation spectrum singularity, existing for cold plasmas and a beam with the smoothed-out boundary [4,9], is eliminated while considering the plasma thermal dispersion, and the electric field of small electromagnetic perturbations goes on continuously across the beam boundary into the plasma.

The dispersion equation is obtainable for the beam with a sharp boundary, while "joining together" the homogeneous equation solutions, found inside the beam and outside of it, on the boundary with the plasma [4]. This method is employed in the paper [8] to analyze waveguide properties of the plasma with a relativistic electron beam, surrounded by dielectric. Under the conditions of the beam instability a slow Langmuir wave is excited in the plasma that is transformed on the plasma boundary into electromagnetic wave, propagating in the dielectric when the following inequality is satisfied: $\epsilon^e > c^2/v_0^2$ ($\epsilon^e$ is the wall permittivity and $v_0$ is the beam velocity). The radiation carries away the pulse, and the beam surface electrons are subject to the action of radiation drag. In this way, the actual possibility of radial beam focusing by coherent radiation is confirmed [14-17].

This Chapter discusses in the linear approximation the physical mechanism of beam focusing by electromagnetic radiation pressure, the problem of transverse beam spectrum with a smoothed-out boundary and the coherent radiation of a curvilinear relativistic electron beam in a plasma [18].

### §2.1. Electrostatic self-focusing of intense relativistic electron beams in dense plasma.

We will here present an elementary theory of electrostatic self-focusing proposed by Winterberg [15] with somewhat different assumptions and which can be easily generalized to the relativistic case. It is found that electrically neutralized beams in dense plasmas can exhibit strong self-focusing even if one neglects the focusing effect of the self-magnetic beam field. The self-magnetic beam field can be partially neutralized. In the nonrelativistic case the frequency of the plasma oscillations excited by a monoenergetic nonrelativistic electron beam with a cold collisionless plasma is given by [see Eqs. (2.5.3)–(2.5.5)]

$$\omega = \left(1 - 2^{-4/3}\nu^{1/3}\right)\omega_p, \quad \nu = n_b/n_p \qquad (2.1.1)$$

where $n_b$ and $n_p$ are the electron number densities in the beam and the plasma ; $\omega_p$ is the electron plasma frequency.

It is furthermore assumed that the beam has a well defined boundary due to its self-magnetic field. The longitudinal plasma oscillations excited by the two-stream instability are connected with longitudinal electrostatic plasma waves. Such longitudinal waves propagating obliquely to the beam axis will be reflected at the beam boundary formed by the self-magnetic field of the beam. It is well known that the reflection of a longitudinal wave propagating in an anisotropic medium is accompanied by the excitation of transverse waves. The same occurs if the medium is permeated by a magnetic field and the reflection occurs at some boundary. In the case where the boundary is caused by a magnetic field effects both the wave reflection and the coupling of the longitudinal with the transverse wave modes. The coupling between longitudinal and transverse modes caused by the reflection from a boundary of a wave propagating in an anisotropic medium is known to be very strong which means that a few such reflections suffice to have complete equipartition in between longitudinal and transverse modes. In summary one can state that as a consequence of the two-stream instability strong transverse beam oscillations are excited with a frequency which is the same as for the longitudinal oscillations given by Eq. (2.1.1).

Furthermore, a plasma can be described by a dielectric constant $\varepsilon$ which for transverse oscillations of frequency $\omega$ is given by

$$\varepsilon = 1 - \frac{\omega_p^2}{\omega^2}. \qquad (2.1.2)$$

For the sake of simplicity and better understanding we have omitted in Eq. (2.1.2) the effect of the self-magnetic field since it will only lead to a correction of the dispersion relation Eq. (2.1.2) which is insignificant for our result. From Eq. (2.1.2) it follows that if $\omega < \omega_p$, which is the case for oscillations with a frequency given by Eq. (2.1.1), $\varepsilon$ becomes negative and the Coulomb repulsion $F_C = e^2/\varepsilon r^2$ in between two electrons gives way to an attraction. The effect is somewhat similar to the attraction of electrons due to electron-phonon interaction in a solid which at low temperatures leads to the phenomena of superconductivity. Here in our case the electrons are attracted by the electron-plasmon interaction but the

64

phenomena is otherwise entirely classical. By inserting the value for $\omega$ given by Eq. (2.1.1) into Eq. (2.1.2) one has

$$\varepsilon = -\left(\nu/2\right)^{1/3}. \tag{2.1.3}$$

If $v_\perp^2$ defines the average of the square of the transverse beam electron velocity this equilibrium condition is given by

$$\varepsilon \frac{E^2}{8\pi} + \frac{1}{2} n_b m v_\perp^2 = 0, \tag{2.1.4}$$

where $E$ is the self-electric field by the beam and thus $\varepsilon E^2/8\pi$ the electric pressure in a medium with a dielectric constant $\varepsilon$. We furthermore have

$$\operatorname{div} \varepsilon E = 4\pi e n_b. \tag{2.1.5}$$

which yields at the beam surface with radius $r_0$

$$\varepsilon E = 2\pi e n_b r_0. \tag{2.1.6}$$

From Eqs. (2.1.4) and (2.1.6) follows

$$r_0 = \frac{2 v_\perp}{\omega_b} \sqrt{-\varepsilon} \tag{2.1.7}$$

where $\omega_b = \left(4\pi e^2 n_b/m\right)^{1/2}$ is the beam electron plasma frequency. Inserting into Eq. (2.1.7) the value of $\varepsilon$ given by Eq. (2.1.3) results in

$$r_0 = 2^{5/6} \frac{v_\perp}{\omega_p} \frac{1}{\nu^{1/3}}. \tag{2.1.8}$$

The generalization to the relativistic case can be easily established. This is being done by substituting in Eq. (2.1.4) for $n_b m v_\perp^2/2$ the corresponding relativistic value for the beam stagnation pressure. The value of this stagnation pressure is obtained by the following reasoning: the stagnation pressure is a result of the reflection of the beam electrons impinging onto the beam boundary resulting only in a change of the direction but not in a change of the magnitude of the electron velocity. It thus follows that in the expression for the stagnation pressure the transverse relativistic electron mass enters. For $\gamma \gg 1$ therefore, the stagnation pressure is given by $n_b m c^2 \gamma$ hence Eq. (2.1.4) has to be replaced by

$$\varepsilon \frac{E^2}{8\pi} + n_b m c^2 \gamma = 0, \quad \gamma \gg 1. \tag{2.1.9}$$

Eq. (2.1.1) for the frequency of the beam oscillations has to be replaced in the relativistic case by

$$\omega = \left[1 - 2^{-5/3} \left(\nu/\gamma\right)^{1/3}\right] \omega_p, \quad \gamma \gg 1 \tag{2.1.10}$$

and thus Eq. (2.1.3) to be replaced

$$\varepsilon = -\left(\nu/4\gamma\right)^{1/3}. \tag{2.1.11}$$

From Eqs. (2.1.6), (2.1.9), and (2.1.11) one thus has

$$r_0 = 2^{3/2} \frac{c}{\omega_b} \gamma^{1/2} \sqrt{-\varepsilon} = 2^{7/6} \frac{c}{\omega_p} \left(\frac{\gamma}{\nu}\right)^{1/3}. \tag{2.1.12}$$

65

Note that an expression (2.1.10) corresponds to the angle between the wave vector and the beam velocity vector $\theta = \arctan k_\perp/k_\parallel = \pi/4$ when the growth rate of instability reaches a maximum [26].

If we express $n_b$ by the total electron current in amps for a relativistic electron beam (drift velocity $v \simeq c$) and by the beam radius $r_0$, that is $n_b = 6.4 \times 10^7 \, \mathrm{I}/r_0^2$, and furthermore assumed that the beam interacts with a plasma of $n_p = 5 \times 10^{22} \mathrm{cm}^{-3}$ corresponding to a plasma of solid state density we have

$$r_0 = 0.13 \, \gamma/\mathrm{I} \ (\mathrm{cm}) \,,$$
$$n_b = 3.8 \times 10^9 \, \mathrm{I}^3/\gamma^2 \ (\mathrm{cm}^{-3}) \,. \tag{2.1.13}$$

If we assume for example $\mathrm{I} = 10^6 A$ and $\gamma = 10$ (5 MeV electrons) we have $r_0 \simeq 1.3 \times 10^{-5}$ cm and $n_b \simeq 3.8 \times 10^{22}$ cm$^{-3}$ and the beam focuses down to a very small radius.

In order to a self-focusing effect occurs, which is described in our theory phenomenologically by a negative value of $\varepsilon$, the energy density in the beam must increase radially outwards. Physically, transverse beam oscillations are excited by the two-stream instability and the requirement for the a radial increase in $\varepsilon$, the energy density simply means that the transverse wave energy increases radially outward. This condition seems to be fulfilled by considering the fact that the magnetic beam field effecting the coupling between longitudinal and transverse waves also increases radially outward.

The effect of electrostatic self-focusing may be of importance in a scheme for the release of controlled thermonuclear energy where an intense relativistic electron beam is shot onto a dense thermonuclear target to ignite a sequence of thermonuclear micro-explosions [26]. In this scheme the concentration of the beam energy into a small volume is important and this might be achieved with the described electrostatic beam focusing.

## §2.2. Radiation pressure at the surface of a beam of coherently radiating electrons.

The radial divergence of an electron beam in a plasma due to an instability can be reduced by placing the beam–plasma system in a retarding medium with a high dielectric constant $\epsilon^e > c^2/v_{\mathrm{ph}}^2$ [$v_{\mathrm{ph}}$ is the phase velocity of the wave excited by the beam in the plasma (Ref. [16])]. The radiative energy flux which crosses the lateral surface of the beam under these conditions carries off part of the power evolved in the plasma volume and reduces the instability growth rate. As a result, a regime can be set up in which the fluctuation perturbations of the field are stabilized by radiation and the instability is essentially suppressed.

In the present section we analyze the case in which the instability does occurs and energy accumulates in the beam–plasma volume. At the same time, some of the power evolved by the beam surface experience a force which is quadratic in the field amplitude. This is the radiative-damping force, which is proportional to the beam density, since the radiation is coherent. Since the field amplitude increases exponentially in time, this process is highly transient.

We derive an equation for the radiation force acting on the surface of a rectilinear beam of arbitrary cross section moving in a plasma-filled waveguide with dielectric walls. The conditions for the onset of this effect are studied, and the possible use of this effect for focusing relativistic beams in a plasma is discussed.

### 1. Radiative-damping.

We assume that a rectilinear beam of relativistic electrons of arbitrary cross section, bounded by a contour $\Gamma$, moves in an anisotropic dispersive medium along interface.

We assume that the dielectric constant in the volume occupied by the beam is $\epsilon^i(\omega)$, while that outside the beam is $\epsilon^e(\omega)$ [17].

In the hydrodynamic approximation the beam can be described by the continuity equation and the equation motion with the self-consistent field. If the density and velocity (momentum) of the beam are written as the sums of fixed and oscillating parts,

$$n = n_o + \tilde{n}(t, \mathbf{r}), \quad \mathbf{v} = \mathbf{v}_o + \tilde{\mathbf{v}}(t, \mathbf{r}), \tag{2.2.1}$$

the system of equations is

$$\frac{d\tilde{n}}{dt} + \operatorname{div}(n_o \tilde{\mathbf{v}}) = 0, \tag{2.2.2}$$

$$\frac{d\tilde{\mathbf{p}}}{dt} = e(\mathbf{E} + \frac{1}{c}\mathbf{v}_o \times \mathbf{H}), \tag{2.2.3}$$

$$n_o \left( \frac{\partial \mathbf{p}_o}{\partial t} + < (\tilde{\mathbf{v}} \nabla \tilde{\mathbf{p}} >) \right) = \left\langle \tilde{n} \left( e\mathbf{E} + \frac{e}{c}\mathbf{v}_o \times \mathbf{H} - \frac{d\tilde{\mathbf{p}}}{dt} \right) \right\rangle, \tag{2.2.4}$$

where

$$\frac{d}{dt} = \frac{\partial}{\partial t} + \mathbf{v}_o \nabla.$$

We assume the interface and the beam boundary to be "geometric", i.e., we neglect the structure of the transition layer. Expanding the functions $\tilde{n}$ and $\tilde{\mathbf{v}}$ in Fourier series, we consider the harmonic $\sim \exp(i\omega t - ik_{\parallel} r_{\parallel})$, where

$$k_{\parallel} = \mathbf{l}k, \quad r_{\parallel} = \mathbf{l}r,$$

and $\mathbf{l} = \mathbf{v}_0/v_0$ is the unit vector parallel to the longitudinal beam axis.

Substituting the function

$$n_k = \frac{i}{\Delta_k} \left[ -ik_{\parallel} v_{\parallel k} n_0 + \nabla_{\perp}(n_0 v_{\perp k}) \right], \quad \Delta_k = \omega - \mathbf{k}v_0 \tag{2.2.5}$$

from (2.2.2) into (2.2.4), and integrating both sides over the cross section of a layer, of width $2\delta$, on the two sides of the beam surface $\Gamma$, we let $\delta \to 0$ tend toward zero. Since the integrals are only affected by terms containing derivatives with respect to the normal to the surface, by introducing the unit vector the normal to the surface, $\mathbf{n}$, we can write the result of the integration as

$$\left( \frac{\partial \mathbf{pr}}{\partial t} \right)_k = \frac{1}{4} e n_0 \operatorname{Re} \oint (\mathbf{nv}_k^*) \left[ \mathbf{p}_k - \frac{ie}{\Delta_k} \left( \mathbf{E}_k^e + \frac{1}{c} \mathbf{v}_0 \times \mathbf{H}_k^e \right) \right] d\Gamma. \tag{2.2.6}$$

The quantity in parentheses on the right side of (2.2.4) takes on different values at the external and internal boundaries of the beam since the fields are discontinuous because of the discontinuity in the dielectric constant of the medium and because of the surface charge and surface current:

$$\mathbf{t}(\mathbf{H}_k^e - \mathbf{H}_k^i) = -\frac{4\pi i e n_0 v_0}{c\Delta_k}(\mathbf{nv}_k),$$

$$\mathbf{n}(\epsilon^e \mathbf{E}_k^e - \epsilon^i \mathbf{E}_k^i) = -\frac{4\pi e i n_0}{\Delta_k}(\mathbf{nv}_k), \tag{2.2.7}$$

where $\mathbf{t} = \mathbf{l} \times \mathbf{n}$ is the unit vector tangent to the surface. Accordingly, when the integral along the normal to the surface is evaluated, this factor should be replaced by half the sum

67

of its values at the external and internal boundaries. The terms which are proportional to the "internal" fields $\mathbf{E}_k^i$ and $\mathbf{H}_k^i$ vanish by virtue of (2.2.3).

The quantity $\mathbf{p}_\Gamma$ is the momentum acquired by the electrons at the beam boundary under the influence of the field. The first term on the right side is the surface component of the ponderomotive force [20], while the second can be interpreted as the radiative-damping force [3],

$$\mathbf{F}_k^e = -\frac{1}{4} \, en_0 \mathrm{Re} \, \frac{i}{\Delta_k} \oint (\mathbf{n}\mathbf{v}_k^*) \left( \mathbf{E}_k^e + \frac{1}{c} \, \mathbf{v}_0 \times \mathbf{H}_k^e \right) d\Gamma. \tag{2.2.8}$$

The longitudinal momentum component $\mathbf{lp}_\Gamma$ is related in a simple manner to the energy flux across the beam boundary. Specifically, by integrating both sides of the equation [19]

$$-\mathbf{jE} = \frac{1}{4\pi} \left( \mathbf{E}\frac{\partial \mathbf{D}}{\partial t} + \mathbf{H}\frac{\partial \mathbf{H}}{\partial t} \right) + \mathrm{div}\,\mathbf{S}, \quad \mathbf{S} = \frac{c}{4\pi}\mathbf{E} \times \mathbf{H} \tag{2.2.9}$$

over the cross section of the surface layer, we find by analogy with (2.2.6),

$$\lim_{\delta \to 0} \int \mathbf{jE}_k df_\Gamma = - \oint \mathbf{n}(\mathbf{S}_k^e - \mathbf{S}_k^e) \, d\Gamma. \tag{2.2.10}$$

For a low-density beam, $|\Delta_k| \ll \omega$, we have $(\mathbf{jE})_k \cong e\tilde{n}\mathbf{v}_0\mathbf{E}_k$; using the definition of $\mathbf{lF}_k^e$, we find

$$\left( \frac{\partial \mathbf{lp}_\Gamma}{\partial t} \right)_k = -\frac{1}{2v_0} \oint \mathbf{n}(\mathbf{S}_k^e - \mathbf{S}_k^i) \, d\Gamma. \tag{2.2.11}$$

The longitudinal component of the ponderomotive force vanishes, since the beam does not experience directional damping (loss of momentum density [8]) in the absence of radiation. When radiation does occurs, there is an additional damping of the surface electrons. In other words, part of the directed momentum of the beam is converted into the longitudinal component of the field momentum flowing across the lateral surface of the beam. This effect is due to force (2.2.8).

For a physical interpretation (2.2.6), we will use the solution of Eq. (2.2.3),

$$\mathbf{p}_k = -\frac{ie}{\Delta_k} \left( \mathbf{E}_k^i + \frac{1}{c}\mathbf{v}_0 \times \mathbf{H}_k^i \right).$$

Taking into account that the components of the electromagmetic field takes on different values at the external and internal the boundaries of the beam, we introduce fields averaged over the plasma–beam boundary

$$\overline{\mathbf{E}}_k = \frac{1}{2} \left( \mathbf{E}_k^i + \mathbf{E}_k^e \right), \quad \overline{\mathbf{H}}_k = \frac{1}{2} \left( \mathbf{H}_k^i + \mathbf{H}_k^e \right), \tag{2.2.12}$$

and we represent Eq. (2.2.6) in the form

$$\left( \frac{\partial \mathbf{p}_\Gamma}{\partial t} \right)_k = -\frac{1}{2} \, en_0 \mathrm{Re} \, \frac{i}{\Delta_k} \oint (\mathbf{n}\mathbf{v}_k^*) \left( \overline{\mathbf{E}}_k + \frac{1}{c}\mathbf{v}_0 \times \overline{\mathbf{H}}_k \right) d\Gamma \tag{2.2.13}$$

which determines the motion of the electron layer at the beam boundary.

## 2. Dispersion relation.

Let us examine the self-consistent system of equations consisting of (2.2.2) and (2.2.3) and Maxwell's equations [19]:

$$\operatorname{curl}\mathbf{E} = -\frac{1}{c}\frac{\partial\mathbf{H}}{\partial t}, \quad \operatorname{curl}\mathbf{H} = \frac{1}{c}\frac{\partial\mathbf{D}}{\partial t} + \frac{4\pi e}{c}(n_0\tilde{\mathbf{v}} + \tilde{n}\mathbf{v}_0). \tag{2.2.14}$$

In the E-wave excited by the beam ($\mathbf{l}\mathbf{H} = 0$) all the field components and the variable beam parameters can be expressed in terms of the longitudinal component of the electric field, $E_{\|} = \mathbf{l}\mathbf{E}$.

We separate the transverse component of the second equation in (2.2.14) by using the vector identity

$$(\operatorname{curl}\mathbf{H})_{\perp} = -\mathbf{l}\times\mathbf{l}\times\operatorname{curl}\mathbf{H} = (\mathbf{l}\nabla)(\mathbf{l}\times\operatorname{curl}\mathbf{H}),$$

and then eliminate magnetic field $\mathbf{H}$ by using the first equation in (2.2.14):

$$\mathbf{l}\times\mathbf{H}_k = -k_{\|}\mathbf{E}_{\perp k} + i\nabla_{\perp}E_{\|k}.$$

Using the transverse component of (2.2.3) we can write

$$\mathbf{v}_{\perp k} = \frac{e}{m\gamma_0\Delta_k q_1}\left(k_{\|} - \frac{\omega}{c}\beta_0\epsilon^i\right)\nabla_{\perp}E_{\|k},$$

$$\mathbf{E}_{\perp k} = \frac{i}{q_1}\left(k_{\|} - \frac{\omega_b^2}{c\Delta_k}\beta_0\right)\nabla_{\perp}E_{\|k}. \tag{2.2.15}$$

Thus, the magnetic field is

$$\mathbf{H}_k = \frac{\omega}{q_1 c}\left(\epsilon^i - \frac{\omega_b^2}{\omega\Delta_k}\right)[\mathbf{l}\nabla_{\perp}E_{\|k}],$$

$$q_1 = k_{\|}^2 - \frac{\omega^2}{c^2}\epsilon^i + \frac{\omega_b^2}{c^2}, \tag{2.2.16}$$

where $\omega_b^2 = 4\pi e^2 n_0/m\gamma_0$ is the plasma frequency of the beam.

The component of the force $\mathbf{F}_k^e$ normal to the beam surface, which determines the radiation pressure, is proportional to the field components $\mathbf{E}_k^e$ and $\mathbf{H}_k^e$, which are related to the "internal" fields by (2.2.20). Working from Eqs.(2.2.7), (2.2.15) and (2.2.16), after some straightforward calculations we find

$$\mathbf{n}\mathbf{F}_k^e = -\frac{k_{\|}^2}{16\pi\epsilon^e}\operatorname{Re}\oint\frac{\omega_b^2}{\Delta_k^2}\left(1 - \epsilon^i\beta_0\beta_{\mathrm{ph}}\right)\left(1 - \epsilon^e\beta_0\beta_{\mathrm{ph}}\right)^*$$

$$\times\frac{q_2^*}{|q_1|^2}|\mathbf{n}\nabla E_{\|k}|^2\,d\Gamma, \tag{2.2.17}$$

where

$$q_2 = \epsilon^i - \frac{\omega_b^2}{\Delta_k^2}(1 - \beta_0^2\epsilon^i), \quad \beta_{\mathrm{ph}} = \frac{v_{\mathrm{ph}}}{c},$$

and $v_{\mathrm{ph}}$ is the phase velocity of the wave.

The frequency and wave vector which appear on the right side of Eq. (2.2.17) are related by the dispersion relation, which can be found by using the boundary conditions at the beam surface.

69

Substituting $\mathbf{E}_{\perp k}$ and $\mathbf{H}_k$ from Eqs.(2.2.15) and (2.2.16) into the longitudinal component of Eq. (2.2.14), we find the following equation for $E_{\|k}$:

$$\Delta_\perp E_{\|k} + \alpha^2 E_{\|k} = 0,$$
$$\alpha^2 = -\frac{q_1}{q_2}\left(\epsilon^i - \frac{\omega_b^2}{\Delta_k^2 \gamma_0^2}\right). \tag{2.2.18}$$

The equation for the field outside the beam can be found from (2.2.18) by setting $\omega_b = 0$, i.e., by using the substitution $\alpha^2 \to -\sigma^2 = \omega^2 \epsilon^e/c^2 - k_\|^2$. Correspondingly, the equations for the fields, (2.2.15) and (2.2.16), change.

We assume that

$$E_{\|k}^i = E_k^i \psi_\alpha, \quad E_{\|k}^e = E_k^e \psi_\sigma,$$

and we assume that $\psi_\alpha^i$ and $\psi_\sigma^e$ are the eigenfunctions of the two-dimensional Laplacian ($\psi_\sigma^e$ falls off monotonically with distance from the beam boundary). Using the condition $E_{\|k}^i = E_{\|k}^e$ at the plasma boundary, and noting that the magnetic field has a discontinuity, (2.2.7), we find

$$\left\{\frac{\psi_\alpha^i}{(\mathbf{n}\nabla)\psi_\alpha^i} + \frac{\sigma^2}{\alpha^2}\frac{1}{\epsilon^e}\left(\epsilon^i - \frac{\omega_b^2}{\Delta_k^2 \gamma_0^2}\right)\frac{\psi_\sigma^e}{(\mathbf{n}\nabla)\psi_\sigma^e}\right\}_\Gamma = 0. \tag{2.2.19}$$

We can simplify Eq. (2.2.18) by making use of the smallness of the parameter $|\Delta_k|/\omega \ll 1$ and by retaining of the terms which are proportional to the beam density only those which contain the factor $\Delta_k^{-2}$:

$$\epsilon^i - \frac{\omega_b^2}{\Delta_k^2}\frac{1}{k^2}\left(\alpha^2 + \frac{k_\|^2}{\gamma_0^2}\right) = 0, \quad k^2 = \alpha^2 + k_\|^2. \tag{2.2.20}$$

The parameter $\alpha$ is governed by (2.2.19), in which we must set $\omega = \omega(\alpha)$.

The most unstable harmonic is that for which

$$|\epsilon^i| = \left|1 - \frac{\omega_p^2}{\omega^2}\right| \ll 1, \quad \omega \cong \omega_p = k_\| v_0. \tag{2.2.21}$$

For these modes we find from (2.2.17)

$$\mathbf{n}\mathbf{F}_k^e = \frac{\Gamma}{16\pi}\beta_0^2|\epsilon^i|^2|\mathbf{n}\nabla E_{\|k}|^2 \mathrm{Re}\left(\frac{1}{\epsilon^e} - \beta_0\beta_{ph}\right)^* \frac{k^2}{|\alpha^2 + k_\|^2 \gamma_0^{-2}|^2}. \tag{2.2.22}$$

In working from (2.2.17) to (2.2.22) we omit the integral, since the dispersion relation used here is derived for the case in which the density of the beam and plasma electrons is constant over the cross section ($\Gamma$ is the circumference of the cross section).

### 3. Cylindrical beam.

The physical conditions for the appearance of a radiative energy flux across the beam surface, on the one hand and the dynamics of the surface layer of electrons in the course of the instability, on the other, are analyzed on the basis of the model of a cylindrical beam. The longitudinal component of the electric field is

$$E_{\|k}^i = E_k^i J_s(\alpha r)\exp(-is\varphi), \quad E_{\|k}^e = E_k^e K_s(\sigma r)\exp(-is\varphi),$$

70

where $J_s(\alpha r)$ and $K_s(\sigma r)$ are the Bessel functions. Correspondingly, dispersion relation (2.2.19) becomes [9,10]

$$\frac{J_s(\alpha R)}{J_s'(\alpha R)} = -\frac{1}{\epsilon^e}\left(\epsilon^i - \frac{\omega_b^2}{\Delta_k^2\gamma_0^2}\right)\frac{\sigma}{\alpha}\frac{K_s(\sigma R)}{K_s'(\sigma R)}, \tag{2.2.23}$$

where $R$ is the beam radius.

For a broad beam, $\sigma R \gg 1$, we can use the asymptotic relation $K_s' \simeq -K_s$. We also adopt the condition $k_{\parallel}/|\alpha|\gamma_0^2 \ll 1$, and omit terms containing longitudinal mass $m\gamma_0^3$ [21]. The Eqs. (2.2.20) and (2.2.23) become

$$\epsilon^i - \frac{\alpha^2}{k^2}\frac{\omega_b^2}{\Delta_k^2} = 0, \quad \frac{J_s(\alpha R)}{J_s'(\alpha R)} = \frac{\epsilon^i}{\epsilon^e}\frac{\sigma}{\alpha}. \tag{2.2.24}$$

The transverse wave number is determined by the second formula relation (2.2.24). Under the conditions of plasma resonance, $|\epsilon_i| \ll 1$, this relation can be reduced to the form

$$\alpha = k_{\perp} + \alpha^{(1)}, \quad \alpha^{(1)} = \frac{\epsilon^i}{\epsilon^e}\frac{\sigma}{\lambda_{ns}},$$

$$k_{\perp} = \frac{\lambda_{ns}}{R}, \quad J_s(\lambda_{ns}) = 0. \tag{2.2.25}$$

Omitting the small additional term $\alpha^{(1)}$ in first formula (2.2.24), we find for a low-density beam

$$\omega = \omega_p - \left(\frac{1}{\sqrt{3}} + i\right)\delta_k, \quad \delta_k = \frac{\sqrt{3}}{2}\left(\frac{k_{\perp}^2}{2k^2}\frac{\nu}{\gamma_0}\right)^{1/3}\omega_p, \tag{2.2.26}$$

where $\delta_k \ll \omega_p$ is the growth rate of instability, and $\nu = n_0/n_p$.

As noted above, the surface current causes a discontinuity in the field component $H_\varphi$, as described (2.2.7). Correspondingly, the radial component of the Poynting vector $S_r = (-c/4\pi)E_{\parallel}H_\varphi$ has different values inside and outside the boundary. Using Eqs. (2.2.7), (2.2.15), and (2.2.16), along with (2.2.24), we can show that

$$S_{rk}^i = \frac{v_0}{8\pi\epsilon^e}|\epsilon^i E_k J_s(\alpha R)|^2 \mathrm{Re}\left(\frac{i\sigma}{k_{\parallel}}\frac{\alpha}{\alpha^*}\right),$$

$$S_{rk}^e = -\frac{v_0}{8\pi\epsilon^e}|\epsilon^i E_k J_s(\alpha R)|^2 \mathrm{Re}\left(\frac{i\sigma k_{\parallel}}{|\alpha|^2}\right). \tag{2.2.27}$$

The quantity $S_{\mathrm{rk}}$ is strongly affected by the parameter

$$\sigma^2 = k_{\parallel}^2 - \frac{\omega_b^2}{c^2}\epsilon^e,$$

which determines the wave-propagation conditions in the region $r > R$. In the case of an instability the complex frequency is defined by Eq. (2.2.26).

In vacuum ($\epsilon^e = 1$) the slow wave

$$v_{\mathrm{ph}} = \frac{\mathrm{Re}\,\omega}{k_{\parallel}} = \left(1 - \frac{\delta_k}{\sqrt{3}}\right)v_0$$

is exponentially damped over a distance

$$\Delta r \simeq \gamma_0 k_{\parallel}^{-1} \simeq \gamma_0 c/\omega_p$$

71

from the plasma boundary, and the energy flux is proportional to the small parameter $\delta_k/\omega_p \ll \gamma_0^{-2}$.

The conditions for emission of radiation from the beam change if $\beta_{\text{ph}}^2 \epsilon^e > 1$. We assume $\beta_0^2 \epsilon^e - 1 \gg \delta_k/\omega_p$ and

$$\sigma = \pm i k_\| \sqrt{\beta_0^2 \epsilon^e - 1} \left( 1 - \frac{i \beta_0^2 \epsilon^e}{\beta_0^2 \epsilon^e - 1} \frac{\delta_k}{\omega_p} \right). \tag{2.2.28}$$

A diverging cylindrical beam corresponds to the plus sign "+" in (2.2.28). Setting $\sigma = i k_r = i k_\| \sqrt{\beta_0^2 \epsilon^e - 1}$ in (2.2.27), we find $S_{\text{rk}}^e > 0$ and $S_{\text{rk}}^i < 0$.

Far from the beam, $|\sigma| r \gg 1$, the small imaginary increment in the wave vector in (2.2.28) becomes important. Note that [25]

$$K_s(\sigma r) \sim \exp\left( -\sqrt{\sigma^2 r^2 + s^2} \right),$$

for harmonics $\sigma^2 r^2 \gg s^2$, which correspond to the maximum radiation, we find the following functional dependence of the field amplitude on $t$ and $r$:

$$|E_k^e|^2 \sim \exp\left[ 2\delta_k \left( t - \beta_0^2 \epsilon^e \frac{k_\|}{k_r} \frac{r}{v_0} \right) \right]. \tag{2.2.29}$$

It follows from (2.2.29) that as time elapses the radius of the zone in which the field amplitude is finite increases in accordance with

$$r^e = u_r t, \quad u_r = \frac{v_0}{\beta_0^2 \epsilon^e} \frac{k_r}{k_\|}, \tag{2.2.30}$$

where $u_r$ is the radial propagation velocity of the energy flux.

This equation is derived on the basis of simple geometric considerations. Using $\mathbf{u}^i \simeq \mathbf{v}_0$ for the velocity of the energy flux from the inner side of the boundary, and taking the angle between the wave vector in the dielectric, $\mathbf{k}^e = (k_r, s/r, k_\|)$, and the beam surface to be $\vartheta \simeq \arctan(k_r/k_\|)$, we find that $\mathbf{u}^e = |\mathbf{u}^i| \cos \vartheta \, \mathbf{k}^e/|\mathbf{k}^e|$. Correspondingly, we have the radial projection of $\mathbf{u}^e$,

$$u_r = |\mathbf{u}^e| \sin \vartheta = \frac{v_0 \tan \vartheta}{1 + \tan^2 \vartheta}, \quad \tan \vartheta = \frac{k_r}{k_\|},$$

in agreement with (2.2.30).

The longitudinal component of the radiative-damping force is governed by Eqs. (2.2.11) and (2.2.27), while the radial component is governed by (2.2.22). The azimuthal projection can be found from general equation (2.2.8). For a cylindrical beam we have

$$F_{\|k}^e = -\frac{R}{8\epsilon^e} |\epsilon^i E_k J_s(\alpha R)|^2 \frac{k_r}{k_\|} \operatorname{Re}\left( \frac{k^2}{|\alpha|^2} \right),$$

$$F_{\text{rk}}^e = \frac{k_r}{k_\|} F_{\|k}^e, \quad F_{\varphi k}^e = \frac{s}{k_\| R} F_{\|k}^e. \tag{2.2.31}$$

The components of force $\mathbf{F}_k^e$ can be expressed in terms of the elements of the Maxwell stress tensor derived for a dispersive medium [19]

$$\sigma_{\alpha\beta}(\omega) = \frac{1}{8\pi} \operatorname{Re}\left( \frac{1}{2} \delta_{\alpha\beta} [\epsilon(\omega) \mathbf{E}^2 + \mathbf{H}^2] - \epsilon(\omega) E_\alpha E_\beta^* - H_\alpha^* H_\beta^* \right).$$

The corresponding expression is [3]

$$F_{\alpha k}^e = -\pi \operatorname{Re} n_\beta \left( \sigma_{\alpha\beta}^e - \frac{k_\alpha}{k_\parallel} \sigma_{\parallel\beta}^i \right)_k, \qquad (2.2.32)$$

where $\sigma_{\alpha\beta}^e$ corresponds to $\epsilon^e$, and $\sigma_{\parallel\beta}^i$ is governed by the dielectric constant of the plasma, $\epsilon^i$. Also, $n_\alpha$ is the unit vector along the normal to the surface. The second term in the (2.2.32) is due to the discontinuity of the magnetic field (2.2.7), resulting from the surface current.

For a cylindrical beam, the surface components of ponderomotive force, governed by the first term on the right side of Eq. (2.2.6), is

$$f_{rk}^\Gamma = \frac{R}{8} |\epsilon^i| \, |E_k J_s(\alpha R)|^2 \frac{|k|^2}{k_\parallel^2}, \quad f_{\varphi k}^\Gamma = f_{\parallel k}^\Gamma = 0. \qquad (2.2.33)$$

The sum of the forces in (2.2.31) and (2.2.33) determines the motion of the electron layer at the beam boundary. The corresponding equation of motion is

$$\sigma m \gamma_0 \frac{d^2 R}{dt^2} = f_{rk}^\Gamma(R) + F_{rk}^e(R), \qquad (2.2.34)$$

where $R$ is the beam radius, and $\sigma$ is the surface charge density. Substituting the radial forces from Eqs. (2.2.31) and (2.2.33), and taking account only of the lowest harmonic, $n = s = 0$, we find a nonlinear equation for the function $R(t)$:

$$\sigma m \gamma_0 \frac{d^2 R}{dt^2} = \frac{R}{8} |\epsilon^i| \, |E_k(t) J_1(\lambda_{00})|^2 \frac{k^2}{k_\parallel^2} \left( 1 - \frac{|\epsilon^i|}{\epsilon^e} \frac{k_r^2}{k_\perp^2} \right), \qquad (2.4.13)$$

where $k_\perp = \lambda_{00}/R$, $J_0(\lambda_{00}) = 0$, and $\lambda_{00} \simeq 2.4$.

For a broad beam, $k_\parallel R \gg 1$ ($k \simeq k_\parallel$), the width of boundary layer, $\Delta R \simeq k_\parallel^{-1} \simeq c/\omega_p$, and

$$\sigma \simeq n_0 2\pi R \Delta R \simeq n_0 2\pi R c/\omega_p,$$

the equation (2.2.35) becomes

$$\frac{d^2 R}{dt^2} = \frac{\Omega_0^2 R_{eq}}{2} e^{2\delta_k t} \left( 1 - \frac{R^2}{R_{eq}^2} \right), \qquad (2.2.36)$$

where

$$\Omega_0^2 = \omega_p^2 \sqrt{|\epsilon^i|^3 (\beta_0^2 - 1/\epsilon^e)} \, \frac{E_0^2}{4\pi n_0 m c^2 \gamma_0}, \quad R_{eq} = \frac{\lambda_n}{k_r} \sqrt{\frac{\epsilon^e}{|\epsilon^i|}},$$

and $E_0$ is the initial amplitude of the electric field, and $\delta_k$ is the growth rate of instability.

The quantity $R_{eq}$ is the radius at which the ponderomotive force is balanced by the radiation force; this is the point of a stable equilibrium. Setting in (2.2.36) $R = R_{eq} + \tilde{R}$ and assuming $\tilde{R} \ll R_{eq}$, we find

$$\tilde{R}(\tau) = \frac{\pi \Omega_0}{2\delta_k} \{ \tilde{R}_0 [N_0(\tau) J_1(1) - J_0(\tau) N_1(1)]$$
$$+ \dot{\tilde{R}}_0 [N_0(\tau) J_0(1) - J_0(\tau) N_0(1)] \}, \qquad (2.2.37)$$

73

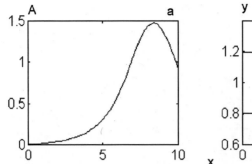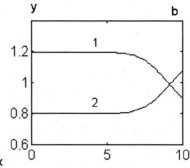

Fig. 2.2.1. (a) Field amplitude $A$ as a function of time; (b) Curve 1 and curve 2 reflect time evolutions of the beam radius [$y=R/R_{eq}$, $x=\delta_k t$, $A^2(0)=0.1$].

where $\tau = \exp(\delta_k t)$, $Z(\tau) \equiv Z(\Omega_0\tau/\delta_k)$ and $Z(1) \equiv Z(\Omega_0/\delta_k)$. The argument of the Bessel functions increases as time elapses, so that $|\tilde{R}|$ decreases.

For a broad beam, $R(0) \gg R_{eq}$, the radiation force is stronger than ponderomotive force, and Eq. (2.2.37) is linear. In this case the solution is that in (2.2.36) if the substitutions $\tilde{R}(0) \to R(0)$ and $\dot{\tilde{R}}(0) \to \dot{R}(0)$ are made. As the instability develops, $R(t)$ decreases, and the radiation is focused.

Since the field energy density toward the end of the hydrodynamic stage of the instability ($n_k \simeq n_0$) is

$$\frac{E_{\max}^2}{4\pi} \simeq \left(\frac{\nu}{\gamma_0}\right)^{1/3} n_0 mc^2 \gamma_0,$$

[see Eqs. (2.2.5), (2.2.15), (2.2.24) and (2.2.26)], we can use the definition of $\Omega_0$ to obtain in order of magnitude:

$$\frac{\Omega_m^2}{\delta_k^2} \simeq \left(\frac{\nu}{\gamma_0}\right)^{1/6}(\beta_0^2 - 1/\epsilon^e)^{1/2}. \tag{2.2.38}$$

In this approximation $|\epsilon_i| \simeq (\nu/\gamma_0)^{1/3}$ the equilibrium radius $R_{eq}$ is

$$R_{eq} \simeq \frac{\lambda_n c}{\omega_p}\left(\frac{\gamma_0}{\nu}\right)^{1/6}\frac{1}{\sqrt{\beta_0^2 - 1/\epsilon^e}}. \tag{2.2.39}$$

The radiative additional term in (2.2.25), $\alpha^{(1)}$, remains small (all over the harmonics $\lambda_n$) under the condition $(\nu/\gamma_0)^{1/6} \ll 1$.

At the limit of the range of validity of the linear approximation, $n_k \simeq n_0$, at which equation (2.2.36) is still valid qualitatively, we use the substitution

$$\frac{\Omega_0}{\delta_k} \exp(\delta_k t) \to A(t), \tag{2.2.40}$$

where function $A(t)$ takes into account a time-dependence of the field amplitude in the nonlinear stage of instability (see Ref. [22]).

The results of numerical integration of Eq. (2.2.36) which illustrate the behavior of the beam radius under the action of the radiative force are displayed in Fig. 2.2.1. An analysis of the linear stage of instability, $A(t) \sim \exp(\delta_k t)$, coincides with an analytical solution (2.2.37).

In conclusion we wish point out an interesting physical analogy with the self-focusing of relativistic electron beam in an unbounded plasma which is discussed in §2.1. At the limit of the range of validity of the above approximation, $\beta_0^2\epsilon^e - 1 \approx \delta_k/\omega_p \simeq (\nu/\gamma_0)^{1/3}$, at which equation (2.2.28) is still valid qualitatively, it follows from (2.2.39) that equilibrium radius, $R_{eq} \approx (c/\omega_p)(\nu/\gamma_0)^{-1/3}$, coincides with Eq. (2.1.8).

74

### §2.3. Radial self-focusing of an electron beam due to a high frequency pressure gradient in a plasma channel.

Below we consider one of the possible methods of obtaining of a self-focusing electron beam in a plasma. In the case we are considering the beam compression is caused by a high-frequency pressure due to plasma waves excited by the beam. Since the focusing force is "gradient" force, to realize radial focusing it is necessary that the amplitude of the focusing field increase in the radial direction. This condition must be satisfied if use is made of surface plasma waves, for example, by passing a beam through a channel in a plasma [14].

*1. System of quasilinear equations.*

The interaction of a low-density electron beam $n_b < n_p$ with a plasma can be described by a system of equations consisting Maxwell's equations for the fields, the linearized hydrodynamics equations of motion for the electrons in the plasma , and equations of motion for the beam electrons. In the linear approximation the amplitude of the electric field of the $k$-th harmonic of the slow axially symmetric E wave as a function of the coordinate $r$ inside the plasma channel is described by the following equation:

$$\frac{1}{r}\frac{d}{dr}\left(r\frac{dE_{zk}}{dr}\right) - k^2 E_{zk} = \frac{4\pi}{\omega}\left[ik^2 j_{zk} + k\frac{1}{r}\frac{d}{dr}(rj_{rk})\right] = g_k(r), \qquad (2.3.1)$$

where $j_{zk}$ and $j_{rk}$ are the Fourier components of the beam current density, which can be found from the linearized equations of motion of the beam. Since the beam density is assumed to be small, in solving Eq. (2.3.1) by successive approximations (in parameter $n_b/n_p \ll 1$) we find the field components $E_{zk}$, $E_{rk}$, and $B_{\varphi k} = \omega E_{rk}/ck$:

$$E_{zk} = AI_0(kr) + A\int_0^r [I_0(kr)K_0(k\xi) - I_0(k\xi)K_0(kr)]g_k(\xi)\xi d\xi,$$
$$E_{rk} = -iAI_1(kr) - i\int_0^r [I_1(kr)K_0(k\xi) - I_0(k\xi)K_1(kr)]g_k(\xi)\xi d\xi. \qquad (2.3.2a)$$

Correspondingly, the fields outside the beam are given by

$$E_{zk} = BK_0(kr); \quad E_{rk} = -iBK_1(kr) \qquad (2.3.2b)$$

[$A$ and $B$ in (2.3.2) are arbitrary constants].

The dispersion equation, which describes the dependence of the frequency $\omega$ on the wave number $k$, can be determined from the continuity conditions on the tangential components of the fields $E_z$ and $B_\varphi$ at the boundary between the beam and the plasma. Assuming that the beam radius $R_0$ is small compared with the wavelength ($kR_0 \ll 1$), we have

$$1 - \frac{\omega_p^2}{\omega^2} + \frac{4\pi ek^2}{\omega}\ln\left(\frac{C}{\pi kR_0}\right)\int_0^{R_0}\sigma_k(r)\,rdr = 0, \qquad (2.3.3)$$

where $\omega_p$ is the Langmuir frequency of the plasma, $j_{zk} = \sigma_k E_{zk}$, and $C = 0.577$ is the Euler constant. It should be noted that the dispersion equation (2.3.3) depends only on the longitudinal component of the beam current. This situation holds only for a thin beam $kR_0 \ll 1$ because under these conditions the transverse current appears in the integrand in (2.3.3) under the derivative sign and thus does not make a contribution to the integral when $\sigma_k(R_0) = 0$ that it to say, if the beam density vanishes at the boundary.

75

In describing the properties of the beam we make use of the kinetic equation for the beam distribution function $F(t, \mathbf{r}, \mathbf{v})$:

$$\frac{\partial F}{\partial t} + v_r \frac{\partial F}{\partial r} + \frac{v_\varphi}{r} \frac{\partial F}{\partial \varphi} + v_z \frac{\partial F}{\partial z} + \frac{e}{m} E_z \frac{\partial F}{\partial v_z}$$

$$+ \left( \frac{e}{m} E_r + \frac{v_\varphi^2}{r} \right) \frac{\partial F}{\partial v_r} - \frac{v_r v_\varphi}{r} \frac{\partial F}{\partial v_\varphi} = 0 \tag{2.3.4}$$

(the variables $v_r$, $v_\varphi$, and $\varphi$ are independent so that differentiation with the space angle $\varphi$ is subject to explicit dependence of $F$ on $\varphi$). This equation can be simplified by making use of the axial symmetry of the problem. Assuming that the distribution function can be written in the form $F(t, r, v) = f(t, r, z, v_r, v_z)\, \delta(v_\varphi)$, we average the above equation over the variable $v_\varphi$. Under these conditions the third and seventh terms vanish and the last term can be combined with the second. As a result we obtain the following equation for the distribution function $f$:

$$\frac{\partial f}{\partial t} + \frac{1}{r} \frac{\partial}{\partial r} (r v_r f) + v_z \frac{\partial f}{\partial z} + \frac{e}{m} E_r \frac{\partial f}{\partial v_r} + \frac{e}{m} E_z \frac{\partial f}{\partial v_z} = 0. \tag{2.3.5}$$

Writing the distribution function in the form

$$f = f_0 + f_1 = f_0 + \int f_k e^{i\Phi_k} dk, \quad \Phi_k = kz - \omega t$$

(where $f_0$ is the average, slowly varying function while $f_1$ is the ensemble of oscillations with random phases), we obtain the following system of equations for the functions $f_0$ and $f_k$:

$$\frac{\partial f_0}{\partial t} + \frac{1}{r} \frac{\partial}{\partial r} (r v_r f_0) + \frac{e}{m} \left\langle \mathbf{E} \frac{\partial f_1}{\partial \mathbf{v}} \right\rangle = 0,$$

$$-i(\omega - k v_z) f_k + \frac{1}{r} \frac{\partial}{\partial r} (r v_r f_k) - i \frac{e}{m} \frac{kr}{2} E_k \frac{\partial f_0}{\partial v_r} + \frac{e}{m} E_z \frac{\partial f_0}{\partial v_z} = 0 \tag{2.3.6}$$

[the formula for the fields that appears in (2.3.6) is assumed to be given and determined by Eq. (2.3.2)]. Solving the second equation in (2.3.6) we can write the function $f_k$ in the following form:

$$f_k = \frac{e}{m} E_k \sum_{s=0}^{\infty} (-i)^{s+1} \frac{v_r^s}{(\omega - k v_z)^{s+1}} \frac{\partial^s}{\partial r^s} \left[ \frac{\partial f_0}{\partial v_z} - i \frac{kr}{2} \frac{\partial f_0}{\partial v_r} \right]. \tag{2.3.7}$$

### 2. Hydrodynamic instability.

At the beginning of the process, when the beam is still monoenergetic, it is sufficient to determine the changes in time and coordinate of the moments of the function $f_0$ rather than the function itself, that is, the density $n_b$, the mean velocity $u_r$ and temperature $T_r$ [3]. To compute these quantities we make use of the following system of equations, which are obtained from the first equation in (2.3.6) [taking account of Eq. (2.3.7)]:

$$\frac{\partial n_b}{\partial t} + \frac{1}{r} \frac{\partial}{\partial r} (r n_b u_r) = 0,$$

$$\frac{\partial}{\partial t} (m n_b u_r) + \frac{2}{r} \frac{\partial}{\partial r} (r n_b T_\perp) = -\frac{e^2}{2m} \int \frac{k^2 \delta_k^2}{(\Delta_k^2 + \delta_k^2)^2} E_k^2 e^{2\delta_k t} dk\, r^2 \frac{\partial n_b}{\partial r}, \tag{2.3.8}$$

$$\frac{\partial T_\perp}{\partial t} = \frac{e^2}{4m} \int \frac{k^2 \delta_k}{\Delta_k^2 + \delta_k^2} E_k^2 e^{2\delta_k t} dk\, r^2.$$

The growth rate $\delta_k$ that appears in Eq. (2.3.8) are determined by Eq. (2.3.3), in which we substitute the current $j_{zk} = \sigma_k E_{zk} = e \int v_z f_k dv$:

$$1 - \frac{\omega_p^2}{\omega_k'^2} + \frac{4\pi e k^2}{m\left(\omega_k' - k v_0\right)^2} \ln\left(\frac{C}{\pi k R_0}\right) \int_0^{R_0} n_b(r)\, r dr = 0 \qquad (2.3.9)$$

($\omega_k' = \omega_k + i\delta_k$ is the complex frequency of the unstable $k$-th harmonic, and $v_0$ is the directed velocity of the beam).

The dispersion equation in (2.3.9) describes the interaction of a monoenergetic bounded beam moving through a plasma. Solving this equation we can find the growth rate of the most unstable mode $k_0 = \omega_0 / v_0$:

$$\delta_0 = \frac{\sqrt{3}}{2}\left[\frac{2\pi e^2}{m v_0^2} \ln\left(\frac{C v_0}{\pi \omega_p R_0}\right) \int_0^{R_0} n_b(r)\, r dr\right]^{1/3} \omega_p,$$

$$\Delta_0 = k v_0 - \omega_p = \frac{\delta_0}{\sqrt{3}}. \qquad (2.3.10)$$

In Eq. (2.3.9) and (2.3.10) only the mean beam linear density

$$\bar{n} = \int_0^{R_0} n_b(r)\, r dr,$$

and this quantity, in accordance with the first equation in (2.3.8), is independent of time. Thus, we can integrate the last equation in (2.3.8) and find the quantity $T_\perp$ in explicit form. Under these conditions the system in (2.3.8) reduces to the following equation for the function $n_b(t, r)$ :

$$\frac{\partial^2 n_b}{\partial t^2} - \frac{1}{2}\left(\frac{3e}{4m}\right)^2 \int \frac{k^2 E_k^2}{\delta_k^2} e^{2\delta_k t} dk\, \frac{1}{r}\frac{\partial}{\partial r}\left(r\frac{\partial r^2 n_b}{\partial r}\right) = 0. \qquad (2.3.11)$$

In deriving this equation we have taken $\Delta_k \approx \delta_k / \sqrt{3} \approx \delta_0 / \sqrt{3}$; this step is valid since the oscillation spectrum excited by a monoenergetic beam will be rather narrow $\Delta k \approx \delta_0 / \sqrt{3}$.

In (2.3.11) we carry out the substitution of variables

$$\tau = \delta_0 t, \quad x = \ln\left(r/R_0\right), \quad y = r^2 n_b$$

and then carry out the Fourier transform in terms of the variable $x$. We then obtain the following second-order equation for the function $y(\tau, q)$ :

$$\frac{d^2 y}{d\tau^2} + q^2 \alpha^2(\tau)\, y = 0,$$

$$\alpha^2(\tau) = \frac{1}{2}\left(\frac{3e\omega_p}{4m v_0 \delta_0^2}\right)^2 e^{2\tau} \int E_k^2\, dk. \qquad (2.3.12)$$

The solution of this equation can be expressed in terms of Bessel functions and is of the form [25]

$$y(\tau, q)\, J_0[q\alpha(0)] = y(0, q)\, J_0[q\alpha(\tau)] \qquad (2.3.13)$$

In order to determine the function we multiply both sides of (2.3.13) by $\exp\left(iqx\right)$ and integrate with respect to $q$. Since both sides of the relation are transformed in the same way we need only compute the integral in the right side:

$$Q = \frac{1}{2\pi}\int_{-\infty}^{\infty} y(0, x')\, dx' \int_{-\infty}^{\infty} J_0(q\alpha)\, e^{iq\left(x - x'\right)} dq. \qquad (2.3.14)$$

Carry out the integration with respect to $q$ by means of the relation

$$\frac{1}{2\pi} \int_{-\infty}^{\infty} J_0\left(\alpha q\right) e^{iq\left(x-x'\right)} dq = \frac{1}{\pi} \frac{1}{\sqrt{\alpha^2 - \left(x-x'\right)^2}} = \delta\left(\alpha - |x - x'|\right) \qquad (2.3.15)$$

and converting to the variables $r$ and $r'$, we can write the quantity $Q$ in the following form:

$$Q = \int_0^r n_b\left(0, r'\right) \delta\left(r' - re^{-\alpha}\right) r'^2 dr' + \int_r^{R_0} n_b\left(0, r'\right) \delta\left(r' - re^{\alpha}\right) r'^2 dr' \qquad (2.3.16)$$

The first integral in (2.3.16) is small $[\sim \exp\left(-2\alpha\right)]$ compared with the second and the second can be reduced to the following form if the left side of (2.3.13) is introduced:

$$n_b\left(t, r\right) = \begin{cases} n_b\left[0, rR_0/R\left(t\right)\right] R_0^2/R^2\left(t\right), & r \le R\left(t\right), \\ 0, & r > R\left(t\right). \end{cases} \qquad (2.3.17)$$

According to (2.3.17) the radius of the beam

$$R\left(t\right) = R_0 \exp\left[\alpha\left(0\right) - \alpha\left(t\right)\right]$$

diminishes in time while the density of particles in the region $r < R\left(t\right)$ increase uniformly over the entire volume of the beam.

The self-focusing of the beam which is analyzed above admits of a simple physical explanation: as the instability develops, surface plasma waves are generated and the amplitudes of the fields of these waves increase in the radial direction from the beam axis toward the periphery. Under these conditions the particles in the beam arc in a high-frequency potential well whose depth and wall curvature increase with time. As a result the particles collect at the bottom of the well. i.e., close to the point $r = 0$ where the force that acts on the beam vanishes.

In order to evaluate the efficiency of this focusing method we must estimate the quantity $\alpha\left(t\right)$ that appears in Eq. (2.3.17). Since the limits of applicability of the theory are bounded the inequality $E \ll mv_0\delta_0^2/e\omega_p$, strictly speaking this quantity is small $\alpha\left(t\right) \ll 1$. On the other hand, the formula given above only give a qualitative description of the process and cannot be used to make rigorous quantitative calculations. Extrapolating the results that have been obtained to the case of stronger fields $E \simeq mv_0\delta_0^2/e\omega_p$ we find $\alpha\left(t\right) \simeq 1$ so that the significant self-focusing of the beam occurs up to the end of the hydrodynamic stage in the development of the instability, that is to say, in a time $T \simeq \delta_0^{-1}$.

As the instability develops the longitudinal temperature of the beam

$$T_\| = \frac{e^2}{2m} \int \frac{dk}{\Delta_k^2 + \delta_k^2} E_k^2 e^{2\delta_k t}$$

increases and this violates the monoenergetic condition $\delta_0^2 \gg k_0^2 T_\|/m$. Further growth in the oscillation amplitudes is accompanied by strong smearing of the beam in longitudinal velocity and an expansion of the oscillation spectrum in wave vector. The interaction of the beam with the plasma in this stage of the development of the instability can be analyzed by solving (2.3.6) and (2.3.7) in the quasilinear approximation [3].

### 3. Quasilinear diffusion.

The quasilinear growth rate $\delta_k \ll \Delta_k$ can be determined by substituting the beam current in Eq. (2.3.3)

$$\delta_k = \frac{\pi}{2} \frac{\omega_p^3}{n_p} \ln\left(\frac{C}{\pi k R_0}\right) \int_0^{R_0} r\, dr \int k\delta\left(\omega_p - k v_z\right) \frac{\partial f_0}{\partial v_z}. \tag{2.3.18}$$

The equation in (2.3.6) together with the equation for the amplitude of the electric field

$$\partial E_k^2 / \partial t = 2\delta_k E_k^2, \tag{2.3.19}$$

represent a close system of equations for the problem. Noting that the function $f_0$ appears in the expression for growth rate (2.3.18) averaged over cross-section, we introduce the function

$$F(t, v_z) = k^2 \ln\left(\frac{C}{\pi k R_0}\right) \int_0^\infty dv_r \int_0^{R_0} f_0\, r\, dr. \tag{2.3.20}$$

Now, integrating the first equation in (2.3.6) with respect to $r$ and substituting (2.3.18) and (2.3.19) we obtain the following system of quasilinear equations:

$$\begin{aligned}
\frac{\partial F}{\partial t} &= \pi \frac{e^2}{m^2} \frac{\partial}{\partial v_z} \int E_k^2 \delta\left(\omega_p - k v_z\right) \frac{\partial F}{\partial v_z}\, dk, \\
\frac{\partial E_k^2}{\partial t} &= \pi \frac{\omega_p^3}{k n_p} \int E_k^2 \delta\left(\omega_p - k v_z\right) \frac{\partial F}{\partial v_z}\, dv_z,
\end{aligned} \tag{2.3.21}$$

which coincides with the system obtained for an infinite beam [3]. It is shown in this reference that in a time of the order of the reciprocal growth rate a plateau develops on the beam distribution function and a steady-state spectrum of superthermal oscillations is exited:

$$E_k^2 = \frac{m^2 \omega_p}{e^2 n_p} v_z^3 \int_{v_1}^{v_2} \left[F_\infty\left(v_z\right) - F_0\left(v_z\right)\right] dv_z, \tag{2.3.22}$$

where $F_0$ is the initial distribution function while $F_\infty$ is the height of the plateau, which is known quantity and is determined from the conservation of the total number of particles:

$$F_\infty = \frac{\bar{N}_b}{v_2 - v_1}, \quad \bar{N}_b \approx \bar{k}^2 \ln\left(\frac{C}{\pi k R_0}\right) N_b, \tag{2.3.23}$$

($v_2$ and $v_1$ are respectively the upper and lower limits of the plateau), $\bar{k} \approx \omega_p / \bar{v}$, and $\bar{v} = (v_1 + v_2)/2$.

It follows from (2.3.21) that the spectrum of oscillations excited by the beam in the plasma is determined completely by the longitudinal motion of the particles and is independent of the transverse motion. In view of this feature, by solving (2.3.21) we can determine $E_k^2$ and then treat the transverse motion in the specified field.

The equation that describes the dependence of the beam distribution function on transverse velocity $v_r$ and coordinate $r$ can be obtained by substituting $f_0 = f_\perp(r, v_r) F_\infty$ in (2.3.6) and averaging over the longitudinal velocity

$$\begin{aligned}
\frac{\partial f_\perp}{\partial t} &+ \frac{1}{r} \frac{\partial}{\partial r}\left(r v_r f_\perp\right) - D r^2 \frac{\partial^2 f_\perp}{\partial v_r^2} = 0, \\
D &= \frac{\pi e^2}{4 m^2 (v_2 - v_1)} \int_{v_1}^{v_2} dv_z \int k^2 E_k^2 \delta\left(\omega_p - k v_z\right) dk.
\end{aligned} \tag{2.3.24}$$

79

Substituting $E_k^2$ from Eq. (2.3.22) to Eq. (2.3.24) and carrying out the integration over the variables $k$ and $v_z$ we can find the coefficient $D$:

$$D = \frac{\pi}{8} \frac{\bar{N}_b}{n_p} \omega_p^3 \qquad (2.3.25)$$

(we have made use of the estimates of [3]: $v_2 - v_0 \simeq (\bar{n}_b/n_p)^{1/3} v_0$, $v_2 \gg v_1$).

The interaction of the beam with the transverse plasma field in this stage of the development of the instability can be analyzed qualitatively by solving Eq. (2.3.24) which describes an electron motion in the quasilinear approximation. To proceed, we assume that

$$\frac{v_r}{r} \ll D^{1/3} \approx \left(\frac{n_b}{n_p}\right)^{1/3} \omega_p \qquad (2.3.26)$$

and we reduce (2.3.24) to the diffusion equation. Furthermore, taking into account that an evolution of the distribution function in a velocity space (in accordance with the diffusion law) is accompanied by its evolution in the coordinate space, we seek the solution in the form

$$f_\perp (t, v_r) = \frac{n_b}{\sqrt{4\pi D r^2 t}} \exp\left[-\frac{(v_r - u)^2}{4 D r^2 t}\right] \qquad (2.3.27)$$

where the beam density $n_b(t, r)$ and the directed transverse velocity $u(t, r)$ are defined by the system of moment equations:

$$\begin{aligned}
&\frac{\partial n_b}{\partial t} + \frac{1}{r}\frac{\partial}{\partial r}(r u n_b) = 0, \\
&n_b\left(\frac{\partial u}{\partial t} + u\frac{\partial u}{\partial r}\right) = -\frac{2Dt}{r}\frac{\partial}{\partial r}(r^3 n_b).
\end{aligned} \qquad (2.3.28)$$

Substituting $u = r\dot{R}/R$ and $n_b = n(t)$ if $r \le R$, we get the ordinary equation

$$R'' + \tau R = 0, \qquad (2.3.29)$$

where $R(t)$ is the radius of the beam, and the prime means the derivative with respect to $\tau = (3\pi n_{b0}/4n_p)^{1/3} \omega_p t$ (see in more detail in §3.2).

The solution of this equation can be expressed in terms of Bessel functions and is of the form [25]

$$R(\tau) = \sqrt{\tau}\left[C_1 J_{-1/3}\left(2\sqrt{\tau^3}/3\right) + C_2 J_{1/3}\left(2\sqrt{\tau^3}/3\right)\right],$$
$$C_1 = 3^{-1/3}\Gamma(2/3)R(0), \quad C_2 = 3^{-2/3}\Gamma(1/3)R'(0) \qquad (2.3.30)$$

It is obvious that this solution has physical meaning only when inequality $R > 0$ is satisfied, since the Bessel functions $J_{\pm 1/3}$ change its sign if $\tau$ becomes large enough. We may therefore use the Coulomb repulsion of beam electrons to obtain a correction of equation Eq. (2.3.29). Adding the Coulomb force, $F_C = 2\pi n_b e^2 r$, in the right hand side of second equation (2.3.28), we find

$$y'' + \tau y - \mu/y = 0, \qquad (2.3.31)$$

where $y = R/R(0)$ and $\mu = (n_b/6\pi n_p)^{1/3}$. The results of numerical integration of Eq. (2.2.31) which illustrate the behavior of the beam radius under the action of the gradient force are displayed in Fig. 2.3.1 [$y(0) = 1$, $y'(0) = 0.3$, and $\mu = 0.3$]. Since the limits of

80

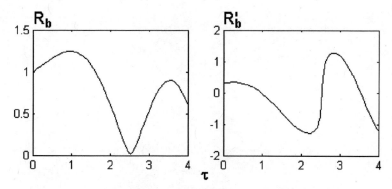

Fig. 2.3.1. Beam radius $R_b$ and radial velocity $R'_b$ vs time $\tau$.

applicability of the theory are bounded the inequality (2.3.26), strictly speaking the ratio $R'/R$ must be small.

*4. Relativistic beam.*

In the above we have considered self-focusing for the case of nonrelativistic beam. If the beam velocity is closed to the velocity of light $\gamma_0^{-2} = \left(1 - v_0^2/c^2\right)^{1/2} \ll 1$ it is necessary to consider the relativistic increase in the mass of the beam particles $m_{\parallel} = m\gamma_0^3$, which leads to a reduction in the growth rate: $\delta^* = \delta/\gamma_0$ and an increase in the focusing time: $T \approx 1/\delta^*$. At the same time the focusing efficiency is increased since the energy transferred by the beam to the field is increased.

As the beam density increases the Coulomb repulsion forces also increase: $F_C = 2\pi n_b e^2 r$. However, if the density of the beam is small compared with the density of the plasma, the force $F_C$ is found to be small compared with the high-frequency focusing force $F_{HF} \approx F_C \left(n_p/n_b\right)^{1/3}$ [the right hand side of the second equation in (2.3.8) and polarization effects can be neglected. Furthermore, the Coulomb forces are reduced by means of self-magnetic field in a relativistic beam $F_C^* \approx F_C/\gamma_0^2$.

The stationary solution of Eq. (2.3.24) $(\partial f_\perp/\partial t = 0)$ is presented in **Appendix 2.1** [founded by V.G. Dorofeenko, (2002)].

## §2.4. Beam with a smooth boundary.

It should be noted above that the fields are discontinuous [see (2.2.7)] because of the discontinuity in the dielectric constant of the beam and because of the surface charge and surface current. This is why we discuss below a relatively common case of cylindrical beam with a smooth boundary that represents a "washed-out step". Conducted within the framework of this model, the numeric integration of plasma hydrodynamic equations with a finite kinetic pressure gradient and monovelocity electron beam equations indicated that the transverse oscillation spectrum singularity, existing for cold plasmas and a beam with the smoothed-out boundary [4,9], is eliminated while considering the plasma thermal dispersion, and the electric field of small electromagnetic perturbations goes on continuously across the beam boundary into the plasma.

Let us use the density distribution

$$n_b = \begin{cases} n_0 & r \leq R \\ 0 & r \geq R \end{cases} \qquad (2.4.1)$$

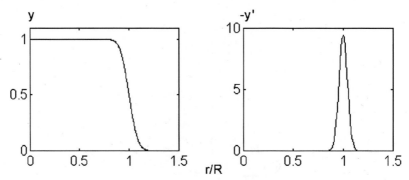

Fig. 2.4.1. Beam density $y=n_b/n_0$ and its derivative $y'$ for $\sigma/R=0.06$.

[$R$ is the radius of the beam]. The derivative of function $n(r)$ has a singular point on the surface $r = R$,

$$\frac{dn}{dr} = -n_0\,\delta(r - R), \tag{2.4.2}$$

which can be eliminating with the help of the following relation:

$$\delta(r - R) = \lim_{\sigma \to 0} \frac{2r}{\sigma^2} \exp\left(-\frac{r^2 + R^2}{\sigma^2}\right) I_0\left(\frac{2rR}{\sigma^2}\right). \tag{2.4.3}$$

Omitting the sign "lim" and comparing (2.4.2) and (2.4.3), we find a density distribution in the relatively common case of cylindrical beam with a smooth boundary of width $\sigma$,

$$n(r) = \frac{2n_0}{\sigma^2} \int_r^\infty \exp\left(-\frac{r^2 + R^2}{\sigma^2}\right) I_0\left(\frac{2rR}{\sigma^2}\right) r\,dr, \tag{2.4.4}$$

which can be used for $\sigma/R < 1$ (Fig. 2.4.1). In the limiting case $\sigma/R \to 0$, this function transforms to a "step" (2.4.1).

Let us consider the Poisson's equation

$$\nabla \mathbf{D}_{\perp k} - ik_\parallel D_{\parallel k} = 4\pi e \tilde{n}_k, \tag{2.4.5}$$

where the small perturbation of beam density $\tilde{n}_k$ is determined by (2.2.5) and (2.4.4).

By analogy with §2.1, integrating both sides of (2.4.5) over the cross section of a layer, of width $2\delta$, on the two sides of the beam surface, we let $\delta \to 0$ tend toward zero. Since the integrals are only affected by terms containing derivatives with respect to the normal to the surface, we can write the result of the integration as

$$E_{\mathrm{r}k}(R + \delta)\big[1 - \eta\Phi(\delta/\sigma)\big] = E_{\mathrm{r}k}(R - \delta)\big[1 - \eta\Phi(-\delta/\sigma)\big], \tag{2.4.6}$$

where

$$\eta = \frac{\omega_b^2}{\epsilon\Delta_k^2}, \quad \omega_b^2 = \frac{4\pi e^2 n_0}{m\gamma_0}, \quad \Phi(x) = \frac{1}{\sqrt{\pi}} \int_x^\infty \exp(-\xi^2)\,d\xi,$$

$n_0$ is the density at the beam axis, $\Delta_k = \omega - k_\parallel v_0$, and $\epsilon = 1 - \omega_p^2/\omega^2$ is the plasma permittivity. In passing from (2.4.5) to (2.4.6) we have used the asymptotic formula [25]

$$I_0\left(\frac{2rR}{\sigma^2}\right) \simeq \frac{\sigma}{\sqrt{4\pi rR}} \exp\left(\frac{2rR}{\sigma^2}\right), \quad \frac{\sigma}{R} \ll 1.$$

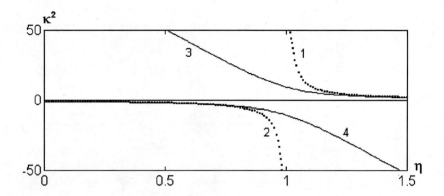

Fig. 2.4.2. Transverse spectrum $\kappa_\perp^2$ in the cases of cold (curves 1 and 2) and heated (curves 3 and 4) plasma electrons for $\kappa_\parallel = 10^{-2}$ .

It follows from (2.4.6) that the behavior of the field amplitude $E_{rk}(R)$ changes radically as a function of the parameter $\delta/\sigma$. In the limiting case

$$\Phi\left(\frac{\delta}{\sigma}\right) \ll 1, \quad \Phi\left(-\frac{\delta}{\sigma}\right) \simeq 1, \quad \frac{\delta}{\sigma} \gg 1$$

discontinuity (2.2.7) occurs. In the opposite case $\delta/\sigma \to 0$ the electric field goes on continuously across the beam boundary into the plasma.

As an example of an actual model of non-equilibrium beam with a smooth boundary we consider an electrostatic field distribution in a plasma with a finite temperature $T_p$. We use the substitution $\mathbf{D}_k \to \hat{\epsilon}\mathbf{E}_k$ in equation (2.4.5) where function

$$\hat{\epsilon} = \epsilon_L + \frac{v_T^2}{\omega_p^2}\Delta_r, \quad \epsilon_L = 1 - \frac{\omega_p^2}{\omega^2}\left(1 + \frac{k_\parallel^2 v_{Tp}^2}{\omega^2}\right) \tag{2.4.7}$$

takes into account the thermal velocity of plasma electrons, $v_T = \sqrt{T/m}$, and $\Delta_r$ is the radial part of the Laplace operator.

Introducing the potential $\mathbf{E} = -\nabla\varphi$ in Eqs. (2.4.5) and (2.4.7), we arrive at the following equation:

$$\mu^2\Delta_x^2\varphi + [(1 - \mu\kappa_\parallel^2 - \eta y)\varphi']' - \kappa_\parallel^2(1 - \eta y/\gamma_0^2)\varphi = 0, \tag{2.4.8}$$

where

$$\mu = \left(\frac{v_{Tp}}{\omega_p R}\right)^2 \frac{1}{\epsilon_L}, \quad \eta = \frac{\omega_b^2}{\Delta_k^2\epsilon_L}, \quad \kappa_\parallel = k_\parallel R,$$

$$y(x) = \frac{2}{s^2}\int_x^\infty \exp\left(-\frac{\xi^2 + 1}{s^2}\right) I_0\left(\frac{2\xi}{s^2}\right)\xi\,d\xi,$$

$$\Delta_x = \frac{1}{x}\frac{d}{dx}\left(x\frac{d}{dx}\right), \quad s = \frac{\sigma}{R},$$

and the prime means the derivative with respect to $x = r/R$.

It is convenient to consider the case of a infinite beam ($y = 1$) which propagates in a cold plasma ($\mu = 0$). In this case Eq. (2.4.8) may be solved by setting $\varphi(x) \sim J_0(\kappa_\perp x)$; we obtain

$$\kappa_\perp^2 = \frac{1 - \eta\gamma_0^{-2}}{\eta - 1}\kappa_\parallel^2 \tag{2.4.9}$$

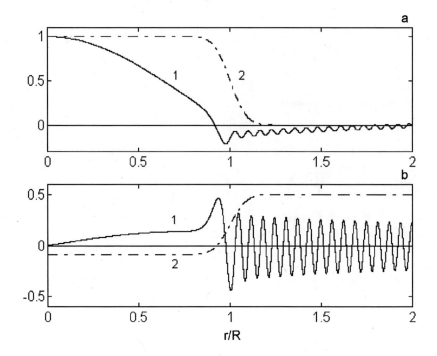

Fig. 2.4.3a. Longitudinal electric field (1) $E_z$ and (2) beam density distribution $y = n_b/n_0$

Fig. 2.4.3b. (1) Radial electric field $E_r$ and (2) function $Y$.

where an evident condition $\kappa_\perp^2 > 0$ fulfills if the inequality $\eta > 1$ holds.

In the case of the smoothly decaying density of the beam at its periphery, the equation (2.4.8) has a singular point, $y(x_c) = 1/\eta$, arising in going from a wave-packet to the single Fourier component [9]. Assuming $y$ to be a slowly changing function of $x$, one can modify (2.4.9) by means of the WKB method: $\eta \to \eta(x) = \eta y(x)$. Then, it follows from this formula that a singularity appears as a result of discontinuity in the spectrum of transverse wave numbers $\kappa_\perp(x)$.

We can use the asymptotic solution (2.2.23) in the special case when a beam density can be represented by (2.4.1). Since the function $y(x)$ reduces from $y(0) = 1$ to $y(x) \ll 1$ in the narrow range, $\sigma \ll R$, then $x_c$ lies interior to this layer. Because of the peculiarity of equation (2.4.8) is displaced to the beam boundary $(x_c \to 1)$, it can be taken into account by means of relations (2.2.7).

According to Eq. (2.4.8), the problem becomes simpler in the case of heated plasma $\mu > 0$ when a singular point disappears

$$\kappa_\pm^2 = \frac{1 - \mu^2\kappa_\parallel^2 - \eta \pm \sqrt{(1 - \mu^2\kappa_\parallel^2 - \eta)^2 + 4\mu^2\kappa_\parallel^2}}{2\mu^2}. \tag{2.4.10}$$

The spectrums (2.4.9) and (2.4.10) are represented in Fig. 2.4.2. The region of transparency $(\kappa_\perp^2 > 0)$ in a cold plasma $(\mu = 0)$ is determined by a curve 1. Thus, a crossing a frontier $\eta = 1$ correspond to the discontinuous going from an oscillating solution of Eq. (2.4.9) to a monotonic one $(\kappa_\perp^2 < 0)$ (curve 2). A heated plasma $(\mu > 0)$ is transparent for a radiation because of an oscillating mode uninterruptedly continues in the region $\eta < 1$ (curve 3).

The above results are confirmed by the numerical solutions of equation (2.4.8) which are represented in Fig. 2.4.3 and Fig. 2.4.4 in the case $\mu\kappa_\parallel = 10^{-2}$. Curves coincide with a

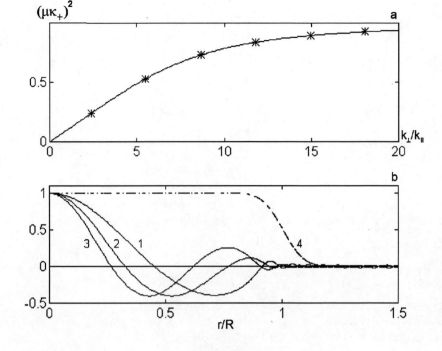

Fig. 2.4.4a. Transverse spectrum of heated plasma $\mu^2\kappa_+^2$, $\mu\kappa_\parallel=10^{-2}$. Here stars correspond to a discrete spectrum of cold plasma $J_0(\kappa_\perp x)=0$.

Fig. 2.4.4b. Radial distribution of potential for different $\kappa_\perp$: (1) 5.54, (2) 8.66, (3) 11.8. Dashed curve 4 correspond to a beam density distribution.

Bessel function $J_0(\kappa_\perp x)$ at the beam axis and uninterruptedly continue in the region $r > R$. It follows from Fig. 2.4.4b that the fields of higher transverse modes $\kappa_\perp \simeq (\eta-1)^{-1/2} \gg 1$ are localized interior to a beam volume.

## §2.5. Beam with a finite temperature.

We consider now the conditions of two-stream instability under which the thermal energy of the beam, $T_b > 0$, must be taken into account. Appending the kinetic pressure gradient, $-T_b\nabla n_b$, in a monovelocity electron beam equations, we obtain

$$i\Delta_k n_0 m\gamma_0 \mathbf{v}_{\perp k} = en_0\mathbf{E}_\perp - T_b\nabla_\perp n_k, \quad i\Delta_k n_0 m\gamma_0^3 v_{\parallel k} = en_0 E_{\parallel k} + ik_\parallel T_b n_k \qquad (2.5.1)$$

(we have retained the notation of §2.1). Introducing the potential $\mathbf{E} = -\nabla\varphi$, we find from Eqs. (2.5.1), (2.2.5) and (2.5.2)

$$(v_{Tb}^2\Delta_\perp - k_\parallel^2 v_T^2\gamma_0^{-2} + \Delta_k^2)(\Delta_\perp - k_\parallel^2)\varphi_k = \frac{\omega_b^2}{\epsilon n_0}[\nabla_\perp(n_0\nabla_\perp\varphi_k) - n_0\gamma_0^{-2}\varphi_k], \qquad (2.5.2)$$

where $\Delta_\perp$ is the radial part of the Laplace operator,

$$\Delta_o = \omega - k_\parallel v_0, \quad \omega_b^2 = 4\pi n_0 e^2/m\gamma_0, \quad v_{Tb}^2 = T_b/m\gamma_0.$$

Substituting $\nabla_\perp\varphi_k = \mathbf{k}_\perp\varphi_k$ in (2.5.2) (homogeneous beam), we get the following dispersion relation

$$\epsilon - \frac{\omega_b^2 F}{\Delta_k^2 - v_{Tb}^2 k^2 F} = 0, \qquad (2.5.3)$$

85

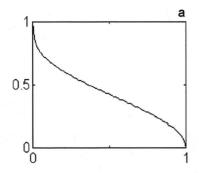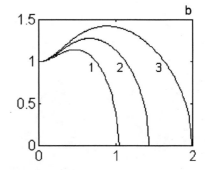

Fig. 2.5.1a. Growth rate $\delta_T/\delta_0$ as a function of the thermal parameter $g$;

Fig. 2.5.1b. Growth rate $\delta_T(k_\perp)/\delta_T(0)$ as a function of the transverse wave number $k_\perp/k_\parallel$ for different value $g_\parallel$: (1) 0.2; (2) 0.05; (3) 0.01, and $\gamma_0=3$.

where the form factor,

$$F = \frac{k_\perp^2}{k^2} + \frac{k_\parallel^2}{k^2\gamma_0^2}, \quad k^2 = k_\parallel^2 + k_\perp^2,$$

takes into account a polarization of unstable plasma wave.

For a low-density beam, omitting the small terms of the next order in the small parameter $\delta/\omega_p$ in (2.5.3), we can reduce this equation to the form

$$\Delta_k^3 - k^2 v_{Tb}^2 F \Delta_k - \frac{1}{2}\omega_b^2\omega_p F = 0. \qquad (2.5.4)$$

Substituting in (2.5.4)

$$\omega = \omega_p + \Delta_k, \quad \omega_p = k_\parallel v_0, \quad |\Delta_k| \ll \omega_p,$$

we find the growth rate $\delta_T = \mathrm{Im}\,\Delta_k$,

$$\delta_T = \frac{\delta_0}{2^{1/3}}\left[\left(1+\sqrt{1-g}\right)^{1/3} - \left(1-\sqrt{1-g}\right)^{1/3}\right], \qquad (2.5.5)$$

where

$$\delta_0 = \frac{\sqrt{3}}{2}\left(\frac{\nu F}{2\gamma_0}\right)^{1/3}\omega_p, \quad g = \frac{16}{27}\left(\frac{\gamma_0}{\nu}\right)^2\left(\frac{kv_{Tb}}{k_\parallel v_0}\right)^6 F, \quad \nu = \frac{n_b}{n_p}.$$

The equation (2.5.4) has complex roots for $0 \le g < 1$. The increase in the beam temperature accompanies by a growth rate decrease so that the instability saturates if a thermal velocity becomes $g = 1$,

$$\frac{v_{Tb}}{v_0} = 2^{1/3}\frac{\delta_0}{F}\frac{k^2}{k_\parallel^2}. \qquad (2.5.6)$$

(see Fig. 2.5.1a).

The growth rate of a cold beam, $\delta_0$, has a minimum value for a longitudinal wave, $k_\perp = 0$ ($F_k = \gamma_0^{-2}$), and increasing monotonically with respect to $k_\perp$. Function $\delta_T(k_\perp)$ reaches its maximum in the range of $k_\perp \gg k_\parallel$, $F \simeq 1$.

The growth rate $\delta_T(k_\perp)/\delta_T(0)$,

$$\delta_T(0) = \frac{\delta_\parallel}{2^{1/3}}\left[\left(1+\sqrt{1-g_\parallel}\right)^{1/3} - \left(1-\sqrt{1-g_\parallel}\right)^{1/3}\right],$$
$$\delta_\parallel = \frac{\sqrt{3}}{2^{4/3}}\frac{\nu^{1/3}}{\gamma_0}, \quad g_\parallel = \frac{16}{27}\frac{1}{\nu^2}\left(\frac{v_{Tb}}{v_0}\right)^6. \qquad (2.5.7)$$

as a function of $v_{Tb} > 0$ has a maximum (Fig. 2.5.1b) and, hence, the transverse plasma mode with a maximum growth rate exists [21].

By analogy with a heated plasma (2.4.8), the thermal motion in a beam eliminates the singularity of transverse spectrum which occurs in the case of a cold beam.

### §2.6. Instability of a monoenergetic beam of electrons with curve orbits in a plasma.

The excitation of collective plasma waves by a relativistic electron beam propagating in the absence of a guide field is considered in Ref. [22]. A dispersion relation is derived which takes into account the curved trajectories of the beam electrons. The appearance of curvature is shown below to be equivalent to the presence of a "negative" temperature in the beam–plasma system.

Let us examine the excitation of electrostatic waves in a plasma by a cold (hydrodynamic) beam of relativistic electrons which are propagating across an external magnetic field $B_0$ (a cylindrical slab). We assume that the static fields of the beam are neutralized by the charge and current of the plasma particles, and we write the equilibrium condition for the forces acting on the beam electrons in the form

$$\omega_B v = \frac{\gamma v^2}{r}, \quad \gamma = 1 + \frac{\omega_B^2 r^2}{c^2} \tag{2.6.1}$$

where $r$ is the radial coordinate of the cylindrical coordinate system, whose $z$ axis is parallel to $B_0$; also $\omega_B = eB_0/mc$, $v = (1 - \gamma^{-2})^{-1/2}$, and $\gamma$ is the relativistic factor.

Introducing the potential $\mathbf{E} = -\nabla\varphi$, we find the Poisson equation

$$\nabla(\hat{\epsilon}\nabla\varphi) = -4\pi e\tilde{n}_b, \tag{2.6.2}$$

in which the operator $\hat{\epsilon}$ models the effect of the plasma, and the oscillatory component of the beam density, $\tilde{n}_b$, can be expressed in terms of the potential using the equations of motion of the beam.

For perturbation of the type $\exp(is\theta - i\omega t)$, the necessary equations are [23]

$$v_{rs} = \frac{e}{m\gamma\delta_s^2}\left(i\Delta_s E_{rs} - \frac{v}{r}E_{\varphi s}\right), \quad v_{\varphi s} = \frac{e}{m\gamma\delta_s^2}\left(i\Delta_s E_{\varphi s} - \frac{v}{r}E_{rs}\right),$$

$$n_s = \frac{1}{\Delta_s r}\left[sn_b v_{\varphi s} - i\frac{d}{dr}(rn_b v_{rs})\right], \tag{2.6.3}$$

where $v_{rs}$ and $v_{\varphi s}$ are the amplitudes of the radial and azimuthal velocity perturbations, $\Delta_s = \omega - sv/r$, and $\delta_s^2 = \Delta_s^2 - v^2/r^2$.

Substituting (2.6.3) into (2.6.2) and using (2.6.1), we find

$$\frac{1}{r}\frac{d}{dr}\left[r\left(\epsilon - \frac{\omega_b^2}{\delta_s^2}\right)\frac{d\varphi_s}{dr}\right] - \frac{s^2}{r^2}\left(\epsilon - \frac{\omega_b^2}{\delta_s^2\gamma^2} + \frac{2\omega_b^2}{\delta_s^4}\frac{v^3\omega}{src^2}\right)\varphi_s = 0, \tag{2.6.4}$$

where the functions $v(r)$ and $\gamma(r)$ are defined by Eqs. (2.6.1), and

$$\epsilon = 1 - \frac{\omega_p^2}{\omega^2 - \omega_B^2}, \quad \omega_p^2 = \frac{4\pi n_p e^2}{m}, \quad \omega_b^2 = \frac{4\pi n_p e^2}{m\gamma}.$$

We restrict the analysis to a "thin" slab; in other words, we assume that the beam thickness $h$ is small compared with the average beam radius $R$: $h \ll R$. Then Eq. (2.6.4) can be simplified by using the substitution $r = R + x$ and by retaining the small terms proportional to $x/R$ only near the resonance:

$$\frac{d}{dx}\left[\left(\epsilon - \frac{\omega_b^2}{\delta_s^2}\right)\frac{d\varphi_s}{dx}\right] - k_\parallel^2\left[\epsilon - \frac{\omega_b^2}{\delta_s^2\gamma_0^2}\left(1 - 2\frac{\omega_B^2 v_0^2}{\delta_s^2 c^2}\right)\right]\varphi_s = 0, \qquad (2.6.5)$$

where

$$\delta_s^2 = \Delta_s^2 - \omega_B^2/\gamma_0^2, \quad \Delta_s = \omega - k_\parallel v_0 + k_\parallel v_0 x/R,$$

$$k_\parallel = s/R, \quad \gamma_0 = \gamma(R).$$

In going from (2.6.4) to (2.6.5) we made use of the relations $\omega \simeq s v_0/R$ and $v_0 \simeq \omega_B R/\gamma_0$.

Furthermore, assuming that the magnetic field "weakly" curves trajectories of the beam electrons, we regard terms proportional to $\omega_B^2/\Delta_s^2 \ll 1$ as small in comparison with unity. This assumption is evidently valid if the characteristic time for the instability is smaller than the Larmor period of the beam electrons. As a result we find the equation [1]

$$\frac{d}{dx}\left[\left(\epsilon - \frac{\omega_b^2}{\Delta^2(x)}\right)\frac{d}{dx}\varphi(x)\right] - k_\parallel^2\left(\epsilon - \frac{\omega_b^2}{\gamma_0^2\Delta^2(x)}\right)\varphi(x) = 0, \qquad (2.6.6)$$

$$\Delta_s \equiv \Delta(x), \qquad \varphi_s \equiv \varphi(x)$$

which in comparison with Eq. (2.4.8) takes into account the velocity spread in the beam due to a curvature of the electron trajectories.

For small scale perturbations,

$$\frac{k_\parallel v_0}{R\Delta} \ll k_\perp \simeq \frac{\varphi'}{\varphi},$$

Eq. (2.6.6) can be solved by the WKBJ method [24]:

$$\varphi(x) = \varphi_0 \sin\left(\int_{-h/2}^x K(x)\,dx\right), \qquad (2.6.7)$$

where

$$K^2(x) = -k_\parallel^2\left(\epsilon - \frac{\omega_b^2}{\gamma_0^2\Delta^2(x)}\right) \Big/ \left(\epsilon - \frac{\omega_b^2}{\Delta^2(x)}\right).$$

Using the boundary conditions at the waveguide wall for a beam bounded by conducting planes, $\varphi(-h/2) = \varphi(h/2) = 0$, we find from (2.6.7) a dispersion relation incorporating the curvature of the beam electron trajectories:

$$\int_{-h/2}^{h/2} K(x)\,dx = n\pi, \qquad (2.6.8)$$

---

[1] Noting that the perturbation is not derivable from a potential we find instead of (2.6.6) an analogous equation for the azimuthal component of the electric field, $E_\varphi$:

$$\frac{d}{dx}\left\{\left[\epsilon - \frac{\omega_b^2}{\Delta^2(x)}(1 - \beta_0^2\epsilon)\right]\frac{dE_\varphi}{dx}\right\} - \left(k_\parallel^2 - \frac{\omega^2\epsilon - \omega_b^2}{c^2}\right)\left(\epsilon - \frac{\omega_b^2}{\gamma_0^2\Delta^2(x)}\right)E_\varphi = 0.$$

In the absence of curvature $R \to \infty$, this equation reduces to the corresponding equation (2.2.19).

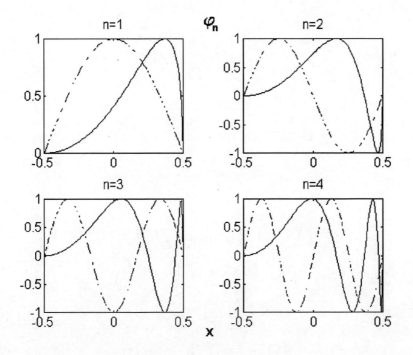

Fig. 2.6.1. Eigenfunctions of a beam with curved orbits $\varphi_n(x)/\varphi_0$ for $k_{\parallel}/k_{\perp}=1$ and $\mathcal{R}=2$, $\mathcal{R}=k_{\parallel}v_0 h/\Delta_0 R$. Dashed curves correspond to a rectilinear beam.

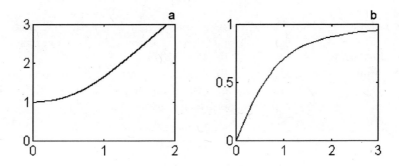

Fig. 2.6.2a. The growth rate of instability $\eta=2\mathrm{Im}\delta/\sqrt{3}\omega_p\nu$ as a function of the parameter $q^{1/6}$.

Fig. 2.6.2b. Electron velocity $v(\xi)/c$ as a function of the beam radius $\xi=\omega_B r/c$.

where $n$ are integer. The transverse eigenfunctions of the twodimensional Laplacian $\varphi_n(x)$ are represented in Fig. 2.6.1.

The integral in (2.6.8) reduces to an incomplete elliptic integral. If we look at the limit of a very relativistic beam, $\gamma_0^2 \gg 1$, however, and if we omit the corresponding term in (2.6.8), the result of the integration can be expressed in terms of elementary functions:

$$\sqrt{\frac{\omega_b^2}{\epsilon}-\left(\Delta_0-k_{\parallel}v_0\frac{h}{2R}\right)^2}-\sqrt{\frac{\omega_b^2}{\epsilon}-\left(\Delta_0+k_{\parallel}v_0\frac{h}{2R}\right)^2}=k_{\perp}v_0\frac{h}{R},$$

where $\Delta_0 = \omega - k_{\parallel}v_0$, $k_{\perp} = n\pi/h$. After some straighforward algebra, we can write

Eq. (2.6.9) as

$$\epsilon - \frac{k_\perp^2}{k^2} \frac{\omega_b^2}{\Delta_0^2 + (k_\perp v_0 h/2R)^2} = 0, \tag{2.6.9}$$

where $k^2 = k_\perp^2 + k_\parallel^2$.

We seek a solution of (2.6.9) in the form $\omega = k_\parallel v_0 + \Delta_0$ ($|\Delta_0| \ll k_\parallel v_0$), assuming that the condition for plasma resonance, $k_\parallel v_0 = \omega_p$ ($\omega_p \gg \omega_B$), holds simultaneously. Then the imaginary part of $\Delta_0$, $\delta = \operatorname{Im} \Delta_0$, which determines the instability growth rate, satisfies

$$\delta = \frac{\delta_0}{2^{1/3}} \left[ \left(1 + \sqrt{1+q}\right)^{1/3} + \left(-1 + \sqrt{1+q}\right)^{1/3} \right], \tag{2.6.10}$$

where

$$\delta_0 = \frac{\sqrt{3}}{2^{4/3}} \left( \frac{n_b}{n_p \gamma_0} \frac{k_\perp^2}{k^2} \right)^{1/3} \omega_p, \quad q = \left( \frac{1}{2^{5/3}} \frac{k_\perp}{k_\parallel} \frac{h}{R} \frac{\omega_p}{\delta_0} \right)^6.$$

For small values of the parameter $q$, we find from Eq. (2.6.10)

$$\delta \simeq \delta_0 \left[ 1 + \left( \frac{q}{4} \right)^{1/3} \right], \quad q \ll 1, \tag{2.6.11}$$

which actually corresponds to straight-line orbits. In the opposite limiting case we find

$$\delta = \delta_0 \left( \frac{q}{4} \right)^{1/6} \left( 2 + \frac{1}{9q} \right), \quad q \gg 1. \tag{2.6.12}$$

Figure 2.6.2a shows a curve of the growth rate as a function of the parameter $q^{1/6}$; the quantity plotted along the ordinate is $\eta = \delta/\delta_0$, so that the value $\eta = 1$ corresponds to the case of a straight beam.

It follows from (2.6.10) and (2.6.12) that the curved trajectories lead to an increase in the instability growth rate. In the extreme case of highly curved orbits [Eq. (2.6.12)], the growth rate becomes essentially independent of the beam density,

$$\delta \simeq k_\perp v_0 \frac{h}{2R}, \tag{2.6.13}$$

so that the motion of an individual electron ($\omega_b \to 0$) is "unstable".

An increase in the instability growth rate as a function of the curvature of the beam $\mathcal{R} = k_\parallel v_0 h / \Delta_0 R$ (Fig. 2.6.1) occurs because of the peaks of eigenfunctions of a beam with curved orbits, $\varphi_n(x)$ are replaced to an external boundary of the layer $R + h$ (Fig. 2.6.1), i.e., in the velocity region $v > v_0 = v(R)$ (Fig. 2.6.2b). As a result, at $h > 0$, the integral

$$-\int_{-h/2}^{h/2} <\mathbf{jE}> d\mathbf{f}, \tag{2.6.14}$$

describing the energy lost by the beam, increases.

In conclusion we wish point out an interesting physical analogy with beams in which the electrons move in straight orbits but there is a finite gradient in the transverse pressure, in which case the dispersion relation has the form in (2.5.3),

$$\epsilon - \frac{k_\perp^2}{k^2} \frac{\omega_b^2}{\Delta_0^2 - k_\perp^2 v_{Tb}^2} = 0, \tag{2.6.15}$$

where $v_{Tb}^2 = T_b/m\gamma_0$. A comparison of Eqs. (2.5.3) and (2.6.15) shows that the curved-orbit case is equivalent to a system with a "negative" temperature $T_{eff} = -m\gamma_0 v_0^2(h/2R)^2$. It is also clear that this instability can be suppressed by a spread in transverse velocities (angles) of order $v_T \approx v_0(h/2R)$ in the initially cold beam.

**Appendix 2.1**

Let us consider the equation (2.3.24) in the special case $\partial f_\perp/\partial t = 0$,

$$\frac{1}{r}\frac{\partial}{\partial r}\left(rv_r f_\perp\right) - Dr^2\frac{\partial^2 f_\perp}{\partial v_r^2} = 0, \tag{A2.1.1}$$

Introducing new variables

$$x = Dr^3/3, \quad y(x, v_r) = rf_\perp(r, v_r),$$

we rewrite (A2.1.1) as

$$v_r\frac{\partial y}{\partial x} - \frac{\partial^2 y}{\partial v_r^2} = 0 \tag{A2.1.2}$$

We seek for a partial solution of (A2.1.2) in the form

$$y = x^\alpha\varphi(\xi), \quad \xi = v_r^3/x \tag{A2.1.3}$$

Substituting (A2.1.3) into (A2.1.2), we find

$$\xi\left(9\varphi'' + \varphi'\right) + 6\varphi' - \alpha\varphi = 0, \tag{A2.1.4}$$

where the prime means the derivative with respect to $\xi$.

The integral

$$\int\limits_{-\infty}^{\infty} y(x, v_r)\,dv_r = \frac{x^{\alpha+1/3}}{3}\int\limits_{-\infty}^{\infty}\varphi(\xi)\,\xi^{-2/3}d\xi, \tag{A2.1.5}$$

is finite and possesses singularity at $x \to 0$ no greater, than $x^{-1/3}$ (otherwise the beam cross-section would contain an infinite number of electrons). Consequently, the inequality $\alpha > -2/3$ holds. On the other hand, the decreasing asymptote of the solution of (A2.1.4) at $\xi \to -\infty$ is $\varphi(\xi) \sim (-\xi)^\alpha$, therefore the integral in the right hand side of (A2.1.5) converges, if $\alpha < -1/3$. Combining these two inequalities, we obtain the region of feasible solutions of type (A2.1.3)

$$-2/3 < \alpha < -1/3 \tag{A2.1.6}$$

The general solution of (A2.1.4) is expressed in terms of the confluent hypergeometric functions [25]

$$\varphi(\xi) = C_1{}_1F_1\left(-\alpha, \frac{2}{3}, -\frac{\xi}{9}\right) + C_2\xi^{1/3}{}_1F_1\left(\frac{1}{3} - \alpha, \frac{4}{3}, -\frac{\xi}{9}\right) \tag{A2.1.7}$$

Function $\varphi(\xi)$ always decreases at $\xi \to \infty$, but at $\xi \to -\infty$ there are exponentially growing and decaying as $(-\xi)^\alpha$ asymptotes. Physically correct solution (vanishing at $\xi \to \pm\infty$) can be obtained fitting one of the arbitrary coefficients $C_1$, $C_2$. Numerical solution for $\alpha = -0.4$

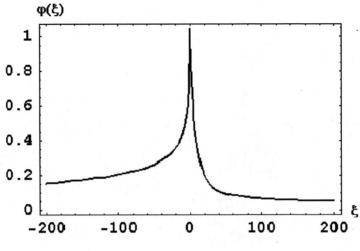

$\varphi(\xi)$

Fig. A2.1.1.

is presented in Fig. A2.1.1. Note that above solution has the irregularity $a + b\xi^{1/3}$ near the zero.

In the special cases of values $\alpha = -1/3$ and $\alpha = -2/3$ the solutions of Eq.(A2.1.4) can be obtained explicitly.

For $\alpha = -1/3$, the substitution

$$\zeta = \xi/18, \quad \varphi(\xi) = \zeta^{1/6} \exp(-\zeta) \psi(\zeta) \qquad (A2.1.8)$$

leads to the modified Bessel equation

$$\psi''(\zeta) + \zeta^{-1}\psi'(\zeta) - \left[1 + (6\zeta)^{-2}\right]\psi(\zeta) = 0 \qquad (A2.1.9)$$

which has the following solution

$$\psi(\zeta) = AK_{1/6}(\zeta),$$
$$y(x, v_r) = \frac{A}{\sqrt{x}} \operatorname{Re}\left[\sqrt{v_r} \exp\left(-\frac{v_r^3}{18x}\right) K_{1/6}\left(\frac{v_r^3}{18x}\right)\right]. \qquad (A2.1.10)$$

For $\alpha = -2/3$, the equation (A2.1.4) can be rewritten as the following system

$$9\varphi' + \varphi = \psi, \quad \psi' + \frac{2}{3\xi}\psi = 0 \qquad (A2.1.11)$$

which has an obvious solution

$$\psi = 9A\xi^{-2/3}, \quad \varphi = A\int_{-\infty}^{\xi} \exp\left[(\xi_1 - \xi)/9\right]\xi_1^{-2/3}d\xi_1. \qquad (A2.1.12)$$

The lower limit of integration in (A2.1.12) is chosen in such a way that the function $\varphi(\xi)$ remains finite at $\xi \to -\infty$].

It should be noted that both of solutions (A2.1.10) and (A2.1.12), $y \sim x^{-2/3}$ $[n_b \sim r^{-2}$ if $r > (v_0/\omega_p)(n_p/n_b)^{1/3}$ (see Ref. [14])] are not quite feasible because they result in an infinite number of electrons in the beam cross-section, $\int_0^\infty n_b r\, dr = \infty$.

# CHAPTER 3

## EXCITATION OF QUASITRANSVERSE PLASMA WAVES BY A BOUNDED ELECTRON BEAM

In the initial stage of the interaction of the monoenergetic relativistic electron beam with a plasma, there is a predominant growth of oblique plasma waves, according [1]. This growth leads to a spread in the transverse momenta of the beam electrons, but there is no substantial longitudinal slowing of the beam. The mathematical difficulties arising in the search for electromagnetic waves of this type do not allow to obtain an analytical nonlinear solution of this problem [2].

In general case, the electromagnetic radiation generated by the beam in a plasma is not localized in the volume occupied by the beam. But the radiation of the highest transverse harmonics, $k_\perp \gg k_\parallel$, which fields are closed to electrostatic one practically trapped in the beam volume (see Chap. 2). The quasitransverse plasma waves, $k_\perp \simeq R^{-1}$, can be excited in an unbounded plasma by thin beam which radius, $R$, is assumed small in comparison with the wavelength, $R \ll k_\parallel^{-1}$. Under the above approximation the radial coordinate $r$ is eliminated from the system of hydrodynamic equations which reduces to one-dimensional one [3], and we use the envelope equation [4] to describe the nonlinear transverse oscillations in a beam.

A method for decreasing the density in wave spectra is developed with the problem of controlled acceleration of particles in a plasma. The method involve applying, at the entrance of the beam–plasma system, a well-defined signal (or an initial modulation of the beam) which exceeds the fluctuation level in the plasma and which suppresses instability at all frequencies other than modulation frequency [5]. This type of interaction of a beam with a plasma, in which all the beam energy is pumped into a single mode determined by the external modulation, is also of interest for transporting high currents of relativistic electrons in a plasma, since the beam particles interact with a strong but regular field, and the nonlinear deviation from the wave-resonance particles and the saturation of the field amplitude are not accompanied by the excitation of adjacent harmonics in the spectrum.

The defocusing (focusing) effect of the field of the collective mode excited by a beam in a plasma are most apparent for small radius beams, since in this case the emission is polarized in the transverse direction and builds up in the beam volume even if there is a free boundary with the plasma. The calculations show that both in the absence of a longitudinal magnetic field (the transverse Cherenkov mode [6]) or in the strong magnetic field (anomalous Doppler effect [7]) the nonlinearity of the equations of motion of the beam does not suppress the instability, and the beam is defocused by the field, at least to a radius of order $k_\parallel^{-1}$.

In this chapter we discuss some nonlinear mechanisms which act to suppress the transverse Cherenkov and cyclotron instabilities in the "thin-beam" model [8-12]. Our investigation shows that the nonlinear evolution of the beam–plasma system depends on the growth rate of instability which is determined by the modulation frequency. This theory expands the results of one-dimensional nonlinear theory which is represented in Chap. 1.

### §3.1. Dispersion relation of the beam–plasma system in a longitudinal magnetic field.

The system of linear equations of motion and Maxwell's equations describing the propagation of a relativistic electron beam of radius $R$ in an unbounded plasma can be written for the region $r \leq R$ as follows:

$$\left(\epsilon_\perp \Delta_\perp + \frac{\omega^2}{c^2}\epsilon_1\epsilon_{||}\right)E_z = -i\frac{\omega^2}{c^2}\epsilon_3 B_z, \quad \left(\Delta_\perp + \frac{\omega^2}{c^2}\frac{\epsilon_1^2 - \epsilon_2^2}{\epsilon_1}\right)E_z = -i\frac{\epsilon_3}{\epsilon_1}\Delta_\perp E_z, \quad (3.1.1)$$

where $E_z$ and $B_z$ are the longitudinal components of the electric and magnetic fields of the wave of an axisymmetric perturbation with frequency $\omega$ and wave number $k_{||}$; $\Delta_\perp$ is the radial part of the Laplacian; and the rest of the notation is defined by

$$\epsilon_\perp = \epsilon_{11} - \frac{\omega_b^2}{\tilde{\Delta}^2}(1 - \epsilon_{11}\beta^2), \quad \epsilon_{||} = \epsilon_{33} - \frac{\omega_b^2}{\gamma^2\Delta^2}, \quad \epsilon_1 = \epsilon_{11} - \left(\frac{ck_{||}}{\omega}\right)^2 - \left(\frac{\omega_b\Delta}{\omega\tilde{\Delta}}\right)^2,$$

$$\epsilon_2 = \epsilon_{12} - \frac{\omega_b^2\omega_B\Delta}{\omega^2\tilde{\Delta}^2}, \quad \epsilon_3 = \epsilon_{12}\left(\frac{ck_{||}}{\omega} - \frac{\omega_b^2\Delta\beta}{\omega\tilde{\Delta}^2}\right) + \frac{\omega_b^2\omega_B}{\omega\tilde{\Delta}^2}\left(\frac{ck_{||}}{\omega} - \beta\epsilon_{11}\right),$$

$$\epsilon_{11} = 1 - \frac{\omega_p^2}{\omega^2 - \omega_{Bp}^2}, \quad \epsilon_{12} = \frac{\omega_p^2\omega_{Bp}}{\omega(\omega^2 - \omega_{Bp}^2)}, \quad \epsilon_{33} = 1 - \frac{\omega_p^2}{\omega^2},$$

$$\omega_p^2 = 4\pi e^2 n_p/m, \quad \omega_b^2 = 4\pi e^2 n_b/m\gamma, \quad \Delta = \omega - k_{||}v, \quad \tilde{\Delta}^2 = \Delta^2 - \omega_{Bp}^2,$$

$$\omega_{Bp} = eB_0/mc, \quad \omega_B = \omega_{Bp}/\gamma, \quad \gamma = (1 - \beta^2)^{-1/2}, \beta = v/c,$$

where $n_b$ and $n_p$ are the beam and plasma densities, and $\mathbf{B}_0$ is the external magnetic field, which is parallel to the beam velocity $\mathbf{v}_0$. The equations for the fields outside the beam $r > R$ follow from (3.1.1) in the case $\omega_b = 0$.

We write the boundary conditions at the beam surface $r = R$, as

$$\left[\frac{\epsilon_\perp(\epsilon_2^2 - \epsilon_1^2) - \epsilon_3^2}{\epsilon_1(\epsilon_1^2 - \epsilon_2^2)}E' + \frac{i\epsilon_3}{(\epsilon_1^2 - \epsilon_2^2)}B_z'\right]_i^e = 0, \quad \left[\frac{\epsilon_3 E_z' - i\epsilon_1 B_z'}{\epsilon_1^2 - \epsilon_2^2}\right]_i^e = 0, \quad B_z^i = B_z^e, \quad E_z^i = E_z^e.$$

$$(3.1.2)$$

We seek the solution of system (3.1.1) in the form

$$E_z = \begin{cases} A^i J_0(k_1^i r) + B^i J_0(k_2^i r), & r < R \\ A^e H_0^{(1)}(k_1^e r) + B^e H_0^{(1)}(k_2^e r), & r > R \end{cases} \quad (3.1.3)$$

where $A$ and $B$ are arbitrary constants, and the components of the transverse wave number are

$$2\epsilon_\perp\left(\frac{ck_{1,2}^i}{\omega}\right)^2 = \epsilon_{||}\epsilon_1 + \frac{\epsilon_\perp(\epsilon_1^2 - \epsilon_2^2) + \epsilon_3^2}{\epsilon_1}$$

$$\pm\left\{\left[\epsilon_{||}\epsilon_1 + \frac{\epsilon_\perp(\epsilon_1^2 - \epsilon_2^2) + \epsilon_3^2}{\epsilon_1}\right]^2 - 4\epsilon_{||}\epsilon_\perp(\epsilon_1^2 - \epsilon_2^2)\right\}^{1/2}, \quad (3.1.4)$$

94

where the subscripts 1 and 2 correspond to the "+" and "−". The wave numbers $k_{1,2}^e$ can be found from (3.1.2) by setting $\omega_b = 0$.

Substituting (3.1.3) into boundary conditions (3.1.2), and equating to zero the determinant of the system of homogeneous algebraic equations, we find the dispersion relation [9]

$$\frac{\omega}{ck_\parallel}\epsilon_\perp^i(\epsilon_1^2-\epsilon_2^2)^e J_0'(k_1^i R)J_0'(k_2^i R) + \frac{\omega}{ck_\parallel}\epsilon_\perp^e(\epsilon_1^2-\epsilon_2^2)^i J_0(k_1^i R)J_0(k_2^i R)\frac{H_0^{(1)'}(k_1^e R)H_0^{(1)'}(k_2^e R)}{H_0^{(1)}(k_1^e R)H_0^{(1)}(k_2^e R)}$$

$$+\frac{1}{(a_2-a_1)^e(a_2-a_1)^i}\left\{J_0(k_1^i R)J_0'(k_2^i R)\left[(a_1^i-a_2^e)\right.\right.$$

$$\times f(a_2,a_1)\frac{H_0^{(1)'}(k_1^e R)}{H_0^{(1)}(k_1^e R)} - (a_1^i-a_1^e)f(a_2,a_2)\frac{H_0^{(1)'}(k_2^e R)}{H_0^{(1)}(k_2^e R)}\right]$$

$$\left.\left.-J_0'(k_1^i R)J_0(k_2^i R)\left[(a_2^i-a_2^e)f(a_1,a_1)\frac{H_0^{(1)'}(k_1^e R)}{H_0^{(1)}(k_1^e R)} - (a_2^i-a_2^e)f(a_1,a_2)\frac{H_0^{(1)'}(k_2^e R)}{H_0^{(1)}(k_2^e R)}\right]\right\}\right. = 0,$$

$$(3.1.5)$$

where

$$a_{1,2} = \frac{1}{\epsilon_3}\left[\epsilon_\perp\left(\frac{ck_{1,2}^i}{\omega}\right)^2 - \epsilon_\parallel\epsilon_1\right],$$

$$f(a_1,a_2) = \left[a_1\epsilon_3 - \frac{\epsilon_\perp(\epsilon_1^2-\epsilon_2^2)+\epsilon_3^2}{\epsilon_1}\right]^i\left(\frac{\omega}{ck_\parallel}a_2\epsilon_1 - \epsilon_2\right)^e$$

$$+ (\epsilon_3 - a_1\epsilon_1)^i\left[a_2\epsilon_2 - \frac{ck_\parallel}{\omega}\epsilon_1 + \frac{\omega}{ck_\parallel}(\epsilon_2^2-\epsilon_1^2)\right]^e.$$

In the limit $B_0 \to 0$, in which $E$ and $H$ waves are independent, Eq. (3.1.5) becomes

$$\left[\epsilon_\perp^i\epsilon_1^e J_0'(k_1^i R) - \epsilon_\perp^e\epsilon_1^i J_0(k_1^i R)\frac{H_0^{(1)'}(k_1^e R)}{H_0^{(1)}(k_1^e R)}\right]\left[\epsilon_1^e J_0'(k_2^i R) - \epsilon_1^i J_0(k_2^i R)\frac{H_0^{(1)'}(k_2^e R)}{H_0^{(1)}(k_2^e R)}\right] = 0.$$

In a finite magnetic field, the E and H waves are coupled, and the dispersion relation is rather complicated; it does simplify significantly, however, in the limit $R \to 0$. For this case, retaining terms proportional to the derivatives of the Hankel functions, we find, in order of magnitude,

$$J_0(k_1^i R) \simeq k_1^e R\ln(k_1^e R), \quad |k_1^e R| \ll 1 \tag{3.1.6}$$

Note that the asymptotic solution at small radii exists only for the E wave, since the conditions $\epsilon_\perp \to 0$ and $k_1^e \to \infty$ as $R \to 0$ holds for only the "+" in (3.1.4). It follows from (3.1.6) that the dispersion relation in this case becomes

$$\epsilon_{11} + \frac{k_\parallel^2}{k_\perp^2}\frac{\omega_{Bp}^2}{\Omega_p^2}\left(1 + \frac{\Omega_p^2}{c^2 k_\parallel^2}\right) - \frac{\omega_b^2}{\tilde\Delta^2} = 0, \tag{3.1.7}$$

$$|\epsilon_{11}| \ll 1, \quad \frac{k_\parallel^2}{k_\perp^2} \ll 1, \quad k_\perp \simeq \frac{\lambda_n}{R}, \quad J_0(\lambda_n) = 0, \quad \Omega_p^2 = \omega_p^2 + \omega_{Bp}^2.$$

95

It follows from (3.1.6) and (3.1.7) that in this limiting case of thin beams,

$$(k_\parallel R)^2 \left| \frac{\epsilon_\parallel}{\epsilon_\perp} \right| << 1 \tag{3.1.8}$$

the dispersion relation is approximately the same as the corresponding relation for a beam in a plasma in a chamber with a conducting wall, in which case there is no radiation flux across the boundary of the system. Furthermore, the waves excited by the beam in the plasma are approximately electrostatic, as can be seen by comparing (3.1.7) with the dispersion relation of Ref. [2].

### §3.2. System of nonlinear equation in the "thin-beam" model.

The inequality (3.1.8) holds in the case of a thin beam which radius, $R$, is small in comparison with the wavelength, $R \ll k_\parallel^{-1}$. Under this condition the longitudinal components of the potential electric field, $\mathbf{E} = -\nabla \varphi$, curl $\mathbf{E} = 0$, satisfy the inequality

$$E_\parallel \simeq k_\parallel \varphi \ll E_r \simeq \varphi/R.$$

Because of the motion of the beam electrons is determined by the strong transverse electric field, $E_r$, there is no substantial longitudinal slowing of the beam, $v_z \equiv v \approx v_0$.

*1. Envelope equation.*

In the nonlinear approximation, the propagation of a relativistic electron thin beam in a plasma, is described by the following system of hydrodynamic equations [6-9]

$$m\gamma \frac{dv_r}{dt} + v_r \frac{\partial}{\partial r} = -e \frac{\partial \varphi}{\partial r} - \frac{1}{n_b} \frac{\partial}{\partial r}(n_b T), \tag{3.2.1}$$

$$\frac{dn_b}{dt} + \frac{1}{r} \frac{\partial}{\partial r}(r n_b v_r) = 0, \tag{3.2.2}$$

$$\frac{d}{dt}\left(\frac{T}{n_b}\right) + v_r \frac{\partial}{\partial r}\left(\frac{T}{n_b}\right) = 0, \tag{3.2.3}$$

where $v_r$, $n_b$, and $T$ are the radial velocity, density, and temperature of the beam; $d/dt = \partial/\partial t + v\, \partial/\partial z$; and we assume that $z$ axis of a cylindrical system of coordinates is directed along the beam axis. The relativistic factor, $\gamma = (1 - v^2/c^2)^{-1/2}$, does not depend on $v_r$ if the following inequality holds: $(\gamma v_r/c)^2 \ll 1$ .

Using the substitutions

$$v_r = \frac{\dot{R}(t,z)}{R(t,z)} r, \quad \varphi = N(t,z)\frac{r^2}{2} + C, \tag{3.2.4}$$

(where $C$ is arbitrary constant) we rewrite Eq. (3.2.1) in the form

$$m\gamma \left(\frac{\ddot{R}}{R} + eN\right) r = -\frac{1}{n_b} \frac{\partial}{\partial r}(n_b T). \tag{3.2.5}$$

From (3.2.2) we find the $R$ dependence of $n_b$

$$n_b = \begin{cases} n_0 R_0^2/R^2, & r \le R, \\ 0, & r > R, \end{cases} \tag{3.2.6}$$

where $R_0$ and $n_0$ are the initial radius and density of the beam.

We seek the solution of system (3.2.3) in the form

$$\frac{T}{n_b} = \mathrm{C}_1 + \mathrm{f}(t, z)r^2,$$

Carrying out the integration, $\mathrm{f}(t, z) = C_2/R^2(t, z)$, and choosing the arbitrary constants as $C_1 = 1$ and $C_2 = -1$, we represent the beam temperature as a function of radius

$$T = T_0 \frac{R_0^2}{R^2} \left(1 - \frac{r^2}{R^2}\right), \qquad (3.2.7)$$

where $T_0$ are the initial temperature (see **Appendix 3.1**).

Substituting (3.2.6) and (3.2.7) in (3.2.5) and eliminating the radial coordinate $r$, we find the envelope equation

$$\left(\frac{\partial}{\partial t} + v\frac{\partial}{\partial z}\right)^2 R + \frac{1}{m\gamma}\left(eN - \frac{2T_0 R_0^2}{R^4}\right)R = 0. \qquad (3.2.8)$$

The Poisson equation for transverse electrostatic waves, which describes the propagation of a thin beam in a plasma is written as

$$\left(\frac{\partial^2}{\partial t^2} + \omega_p^2\right)\frac{1}{r}\frac{\partial}{\partial r}\left(r\frac{\partial \varphi}{\partial r}\right) = -4\pi e \frac{\partial^2 n_b}{\partial t^2}. \qquad (3.2.9)$$

Substituting (3.2.4) and (3.2.6) in (3.2.9), we can reduce the equation to

$$\left(\frac{\partial^2}{\partial t^2} + \omega_p^2\right)N = -2\pi e n_0 R_0^2 \frac{\partial^2}{\partial t^2}\frac{1}{R^2}. \qquad (3.2.10)$$

Note that the function $R(t, z)$ introduced formally is the beam radius, and the density $n_b(t, z)$ is assumed independent of the coordinate $r$ within the beam and to vanish outside the beam [3]. Equations (3.2.8) and (3.2.10) constitute a closed system of nonlinear partial differential equations with derivatives with respect to $t$ and $z$ (in the absence of the external magnetic field).      *2. External magnetic field.*

The propagation of a thin beam in a plasma, in the direction parallel to the static external magnetic field $B_0$, is described by the following system of equations:

$$m\gamma\left(\frac{\partial v_r}{\partial t} + v_r\frac{\partial v_r}{\partial r} - \frac{v_\varphi^2}{r} + \omega_B v_\varphi\right) = -e\frac{\partial \varphi}{\partial r} - \frac{1}{n_b}\frac{\partial}{\partial r}(n_b T),$$

$$\frac{\partial v_\varphi}{\partial t} + v_r\frac{\partial v_\varphi}{\partial r} + \frac{v_r v_\varphi}{r} - \omega_B v_r = 0, \qquad (3.2.11)$$

$$\left(\frac{\partial^2}{\partial t^2} + \Omega_p^2\right)\frac{1}{r}\frac{\partial}{\partial r}\left(r\frac{\partial \varphi}{\partial r}\right) = -4\pi e\left(\frac{\partial^2}{\partial t^2} + \omega_{Bp}^2\right)(n_b - n_0),$$

where $\mathbf{v}_\perp = (v_r, v_\varphi)$ is the transverse velocity, and $\omega_B = eB_0/mc\gamma$ is the gyrofrequency of the beam electrons.

By analogy with (3.2.4), the equations (3.2.11) are conveniently transformed through the following change of variables:

$$v_r = \frac{\dot{R}(t, r)}{R(t, r)}r, \quad v_\varphi = \Omega r.$$

97

The second equation can be used to express $\Omega$ as a function of $R$:

$$\Omega = \frac{\omega_B}{2} + \frac{R_0^2}{R^2}\left(\Omega_0 - \frac{\omega_B}{2}\right).\tag{3.2.12}$$

Assuming that the initial angular velocity of the electrons is zero, $\Omega_0 = 0$, and using (3.2.6), (3.2.7), and (3.2.12), we can transform the system (3.2.11) to

$$\left(\frac{\partial}{\partial t} + v\frac{\partial}{\partial z}\right)^2 R + \left[\frac{\omega_B^2}{4}\left(1 - \frac{R_0^4}{R^4}\right) + \frac{1}{m\gamma}\left(eN - \frac{2T_0 R_0^2}{R^4}\right)\right]R = 0.\tag{3.2.13}$$

$$\left(\frac{\partial^2}{\partial t^2} + \Omega_p^2\right)N = 2\pi e n_0 R_0^2\left(\frac{\partial^2}{\partial t^2} + \omega_{Bp}^2\right)\left(1 - \frac{R_0^2}{R^2}\right),\tag{3.2.14}$$

where $\omega_p$ and $\omega_{Bp}$ are the plasma frequency and gyrofrequency of the plasma electrons, and $\Omega_p^2 = \omega_p^2 + \omega_{Bp}^2$.

### 3. Nonlinear plasma dispersion.

A system of non-potential equations corresponding to the electrostatic approximation (3.2.12) and (3.2.14) [see also (3.1.7)] and taking into account a plasma nonlinearity can be found from Maxwell's equations and the equations of motion of the beam electrons. Assuming

$$\left(\frac{k_\parallel}{k_\perp}\right)^2 \simeq (k_\parallel R)^2 \ll 1, \quad \frac{\partial}{\partial r} \gg \frac{\partial}{\partial z},\tag{3.2.15}$$

we can write the equations for the field as

$$\left[\left(\frac{\partial^2}{\partial t^2} + \Omega_p^2\right)\frac{\partial^2}{\partial t^2} - c^2\frac{\partial^2}{\partial z^2}\left(\frac{\partial^2}{\partial t^2} + \Omega_{Bp}^2\right)\right]E_r$$

$$+ c^2\frac{\partial^2}{\partial r\partial z}\left(\frac{\partial^2}{\partial t^2} + \Omega_{Bp}^2\right)E_z - \omega_p^2\omega_{Bp}\frac{\partial E_\varphi}{\partial t} = -4\pi e\frac{\partial}{\partial t}\left(\frac{\partial^2}{\partial t^2} + \Omega_{Bp}^2\right)n_b v_r,\tag{3.2.16}$$

$$\frac{\partial}{\partial r}\left[\frac{1}{r}\frac{\partial}{\partial r}(rE_\varphi)\right] = -\frac{\omega_{Bp}}{c^2}\frac{\partial E_r}{\partial t},\quad \left[\frac{1}{r}\frac{\partial}{\partial r}\left(r\frac{\partial}{\partial r}\right) + \frac{\omega_{Bp}^2}{c^2}\right]E_z = \frac{1}{r}\frac{\partial}{\partial r}\left(r\frac{\partial E_r}{\partial z}\right).$$

The first of these equations is exact, while the derivations of the second and third used the assumptions:

$$n_b \ll n_p, \quad \partial^2/\partial t^2 = -\Omega_p^2, \quad v_\varphi = 0.$$

Introducing the potential $E_r = -\partial\varphi/\partial r$, we can write (3.2.16) as

$$\left(\frac{\partial^2}{\partial t^2} + \Omega_p^2\right)\frac{1}{r}\frac{\partial}{\partial r}\left(r\frac{\partial\varphi}{\partial r}\right) - \frac{\omega_{Bp}^2\omega_p^2}{\Omega_p^2}\left(\frac{\partial^2}{\partial z^2} - \frac{\Omega_p^2}{c^2}\right)\varphi = -4\pi e\left(\frac{\partial^2}{\partial t^2} + \Omega_{Bp}^2\right)(n_b - n_0).$$

$$\tag{3.2.17}$$

Equation (3.2.17) can be derived, up to the term proportional to $1/c^2$ on the left side, from the Poisson equation.

In choosing the dependence of the potential on the coordinate $r$ we should note that in the absence of the last terms on the left side, which reflect the presence of field components $E_z$ and $E_\varphi$, and which are smaller by a factor $k_\parallel R \ll 1$ and the first terms, the problem has an exact solution: (3.2.4) and (3.2.6). This solution describes the field of a uniformly charged, infinite cylinder. The integration constant $C$ is arbitrary. When the succeeding terms in the expansion in $k_{||}R$ proportional to the tangential electric field are taken into account [this field

$$E_{z\mathbf{k}} = ik_{||}\varphi_{\mathbf{k}} = ik_{||}\varphi_{0\mathbf{k}}J_0(\lambda_n r/R)\exp(-ik_{||}z) \qquad (3.2.18)$$

vanishes at the beam boundary $r = R$], we must set $C = -R^2/2$ in (3.2.4):

$$\varphi = \frac{N}{2}(r^2 - R^2). \qquad (3.2.19)$$

From the orthogonality of the Bessel functions on the interval $0 \le r \le R$ we find

$$1 - \frac{r^2}{R^2} = 8\sum_{n=1}^{\infty}\frac{J_0(\lambda_n r/R)}{\lambda_n^2 J_1(\lambda_n)},$$

which tells us that choosing (3.2.19) as the potential corresponds to the excitation of a broad spectrum of transverse harmonics in the nonlinear stage. In the linear theory the various modes are independent, and system Eq. (3.2.17) is satisfied with $\varphi_{\mathbf{k}} \sim J_0(\lambda_n r/R)$.

Since the coordinate $r$ is not eliminated from the small terms on the left side (3.2.17) when the solution is chosen in the form (3.2.19), we take this terms into account qualitatively, multiplying the equations by $rdr$ and integrating from 0 to R. As a result we find

$$\frac{\partial^2 N}{\partial t^2} + \left[\Omega_p^2 - \frac{\omega_{Bp}^2\omega_p^2}{\Omega_p^2}\left(\frac{\partial^2}{\partial z^2} - \frac{\Omega_p^2}{c^2}\right)R^2\right]N = 2\pi e n_0 R_0^2\left(\frac{\partial^2}{\partial t^2} + \Omega_{Bp}^2\right)\left(1 - \frac{R_0^2}{R^2}\right). \qquad (3.2.20)$$

The term proportional to $\partial^2/\partial z^2$ in (3.2.20), which describes the nonlinear dispersion of the plasma, is important and depends only on the average value of the dispersion, so that $R^2$ can be taken through the "fast" derivative with respect to $z$. It is easy to see that in the linear approximation for the harmonic $\exp\left(i\omega t - ik_{||}z\right)$ we find from (3.2.13) and (3.2.20) the dispersion relation which coincides with (3.1.7) in the case $k_\perp^2 = 8/R^2$.

### §3.3. Dynamics of the formation of relativistic electron clumps in a plasma.

In this section we discuss a possible alternative way of producing equilibrium electron clumps in a dense plasma. The strong quasitransverse plasma wave in the frequency range $\omega < \omega_p$ which is need for this can be excited by a relativistic monoenergetic beam under the conditions for which the beam–plasma instability is present. For a beam of small radius $R \ll v/\omega_p$ the field of the unstable harmonic, depending on the wave phase, focuses or defocuses beam electrons, and after a time on the order of several growth times the beam breaks up into clumps. Numerical solutions reveal that the nonlinearity of the beam stabilizes the growth of the amplitude and phase of the wave and the instability goes over to steady state. Under these conditions the electron in the focusing phases of the field break up into dense clumps with longitudinal dimensions that depend on the frequency of the beam modulation [6,10].

### 1. System of "slow" equations.

The nonlinear development of the instability of a modulated beam is studied below on the basis of the Eqs. (3.2.8) and (3.2.10). We perform the change of variables

$$t, z \to t_1 = t, \ \Phi = \omega t - kz$$

99

and write these equations in the form

$$
\left[\frac{\partial}{\partial t} + (\omega - kv)\frac{\partial}{\partial \Phi}\right]^2 R + \frac{1}{m\gamma}\left(eN - \frac{2T_0 R_0^2}{R^4}\right)R = 0,
$$

$$
\left(\frac{\partial}{\partial t} + \omega\frac{\partial}{\partial \Phi}\right)^2 N + \omega_p^2 N = -2\pi e n_0 R_0^2 \left(\frac{\partial}{\partial t} + \omega\frac{\partial}{\partial \Phi}\right)^2 \frac{1}{R^2}. \tag{3.3.1}
$$

The wave number $k_{\|}$, which determines the wavelength of the unstable harmonic of the field, is an independent parameter of the problem. The parameter $\omega$ for the time-dependent problem in question only determines the origin of the frequency scale, and its choice is quite arbitrary. If we set $\omega = k_{\|}v$, then the derivative $\partial/\partial\Phi$ in the first equation drops out and it reduces formally to an ordinary differential equation in time which is more convenient for numerical simulations, in which the wave phase $\Phi$ enter as a parameter.

In the absence of the beam, $n_0 = 0$, the solution (3.3.1) has the form of a plane wave

$$
N = A_0 \exp[i\omega_p(t - z/v)], \quad \omega_p = k_{\|}v. \tag{3.3.2}
$$

For a low density beam

$$
\nu = \frac{n_0}{n_p\gamma} \ll 1
$$

the form of the solution is maintained, but the amplitude varies slowly with time:

$$
N(t, \Phi) = \frac{m\gamma\omega_p^2\nu^{2/3}}{e}\, \mathrm{Re}\,[Y(t)\exp(i\Phi)]. \tag{3.3.3}
$$

Hence we can drop the derivatives with respect to the "slow" time, $\partial/\partial t \ll \omega\partial/\partial\Phi$, in the field equation. Substituting (3.3.3) in (3.3.1) and averaging over the spatial period (wavelength) we find

$$
\ddot{X} + \left[\mathrm{Re}\,Y\exp(i\Phi) - \frac{\eta}{X^4}\right]X = 0, \quad 2i\dot{Y} - \kappa Y = \left\langle\frac{\exp(-i\Phi)}{X^2}\right\rangle, \tag{3.3.4}
$$

where we have written $Y = a\exp(-i\vartheta)$ and have introduced the dimensionless variables

$$
X = \frac{R}{R_0}, \quad \tau = \nu^{1/3}\omega_p t, \quad \kappa = \frac{1}{\nu^{1/3}}\left(1 - \frac{\omega_p^2}{k^2 v^2}\right),
$$

$$
\eta = \frac{2T_0}{R_0^2\omega_p^2\nu^{2/3}m\gamma}, \quad <\ldots> = \frac{1}{2\pi}\int_0^{2\pi}\ldots d\Phi.
$$

In the linear approximation the solution (3.3.4) for a cold beam $\eta = 0$ has the form [2].

$$
X = 1 + x_0\exp(-i\Delta\tau), \quad |x_0| \ll 1,
$$

Here $\Delta$ satisfies the dispersion relation

$$
2\Delta^3 - \kappa\Delta^2 + 1 = 0, \tag{3.3.5}
$$

[2] The thermal term in the dispersion relation can be dropped if the inequality $k_{\|}\sqrt{T/m\gamma}\ll|\Delta|$ holds. However, the term $\eta/X^4$ eliminates the singularity in the nonlinear stage of instability if the electron temperature is finite and the kinetic pressure gradient impedes the compression of the beam to zero radius ($X>0$). The minimum radius of the clump, $R_{\min}$, must not be larger than $R_{\min}\geq R_0\sqrt{n_0/n_p|\epsilon|}$, otherwise the condition that the plasma equation of motion be linear $n_b/n_p|\epsilon|\ll 1$ is violated.

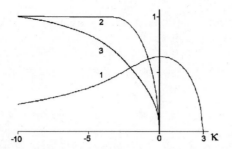

Fig. 3.3.1. Growth rate $\Gamma(\kappa)$ (1) and the phase width of the clumps $|\delta\Phi|/\pi$ for a solid beam (2) and bunched beam (3), as a function of the modulation frequency $\kappa$.

which for $\kappa < 3$ has the complex root

$$\Delta = \frac{\kappa}{6} + \frac{u+v}{2} + \frac{i\sqrt{3}}{2}(u-v) \tag{3.3.6}$$

$$u = (-q + \sqrt{D})^{1/3}, \quad v = (-q - \sqrt{D})^{1/3},$$
$$q = (\kappa/6)^3 - 1/4, \quad D = [1 - (\kappa/3)^3]/16,$$

which corresponds to exponential growth of the waves with growth rate $\Gamma(\kappa) = \operatorname{Im}\Delta(\kappa)$ (Fig. 3.3.1, curve 1). The asymptotic form of (3.3.6) is

$$\Delta = \begin{cases} \dfrac{1 + i\sqrt{3}}{2^{4/3}}, & |\kappa| \ll 1 \\[2mm] \dfrac{i}{|\kappa|}, & -\kappa \gg 1 \end{cases} \tag{3.3.7}$$

Numerical integration of Eq. (3.3.4) shows that the behavior of the nonlinear waves in a plasma with a beam depends sensitively on the parameter $\Delta(\kappa)$, which in turn is determined by the beam modulation frequency (see Fig. 3.3.2).

*2. Modulated beam.*

If there is an external modulation, the instability occurs only at the modulation frequency [5]. This case is slightly simpler to analyze, since the order of the differential equation for the field can be reduced.

If the modulation frequency is not the plasma frequency, $\kappa < 0$ and $\dot{Y} \ll |\kappa Y|$, the order of Eq. (3.3.4) lowers:

$$Y_1 = |\kappa| Y = \left\langle \frac{\cos \Phi}{X^2} \right\rangle. \tag{3.3.8}$$

We seek the solution of Eqs. (3.3.4) in the form

$$X = \sum_{n=0}^{\infty} X_n(\tau_1) \cos n\Phi,$$

$$\tau_1 = \frac{\omega_b}{|\epsilon|} t, \quad \omega_b^2 = \frac{4\pi e^2 n_0}{m\gamma}, \quad \epsilon = 1 - \frac{\omega_p^2}{\omega^2}. \tag{3.3.9}$$

101

Then we find an infinite system of differential equation with total derivatives with respect to the time for the functions $X_n$ and $Y_1$:

$$\ddot{X}_0 + \frac{1}{2} Y_1 X_1 = 0, \quad \ddot{X}_1 + Y_1 \left( X_0 + \frac{1}{2} X_2 \right) = 0,$$

$$\ddot{X}_n + \frac{1}{2} Y_1 (X_{n-1} + X_{n+1}) = 0, \quad n \geq 2. \tag{3.3.10}$$

In the linear approximation, the field amplitude increase with a growth rate $\omega_b |\epsilon|^{-1/2}$ if $\epsilon < 0$; it is sufficient to consider only the first harmonic, $R_1$.

In the nonlinear stage we must take into account more harmonics, since the system does not contain a small parameter, so that numerical methods must be used. Below we find the analytical solution in the approximation of the slight nonlinearity, so that we can analyze the beginning of the processes.

If the field amplitude is not too high, $X_1 \ll 1$, the system (3.3.10) reduced to

$$\ddot{X}_1 + Y_1 = 0, \quad Y_1 = \left\langle \frac{\cos \Phi}{(1 + X_1 \cos \Phi)^2} \right\rangle. \tag{3.3.11}$$

Taking the integral we find the second order equation:

$$\ddot{X}_1 - \frac{X_1}{(1 - X_1^2)^{3/2}} = 0. \tag{3.3.12}$$

We find an explanation for the nonlinear growth rate in (3.3.12) by noting that the electron density averaged over the period is

$$< n_b > = n_0 \left\langle \frac{1}{X^2} \right\rangle = \frac{n_0}{(1 - X_1^2)^{3/2}}.$$

Analyzis of the Eq. (3.3.12) shows that the nonlinearity does not suppress the instability at a low field amplitude, i.e., there are no turning points in the region $X_1 \ll 1$, where approximation (3.3.12) holds.

Numerical integration of Eq. (3.3.4) shows that in the range of frequencies $-\kappa \gg 1$ the growth rate of the field amplitude is not accompanied by any substantial phase shift of the wave relative to the beam. Hence, depending on the initial phase, electrons are focused or defocused by the field of of the instability and the beam breaks up into clumps with longitudinal dimensions $\delta z \leq \pi / k_{\parallel}$. The results of integration for $\kappa = -10$ are shown in Fig. 3.3.2a.

Because the oscillations of the electron disks in different cross sections of a thick clump on the order of half the wavelength are not synchronized, nonlinear oscillations of the amplitude and phase of the field, which constitute a global effect, are damped out. This regime of beam modulation, which keeps half the electrons, is the most favorable for beam transport in the dense plasma.

### 3. Maximum growth rate.

As a mismatch $\kappa$ increases (in the range $\kappa < 0$) and as we approach the maximum growth rate, $\Gamma_m = \sqrt{3}/2^{4/3}$, the correction to the frequency becomes complex

$$\Delta = \left( \frac{1}{\sqrt{3}} + i \right) \Gamma_m, \quad |\kappa| \ll 1 \tag{3.3.13}$$

102

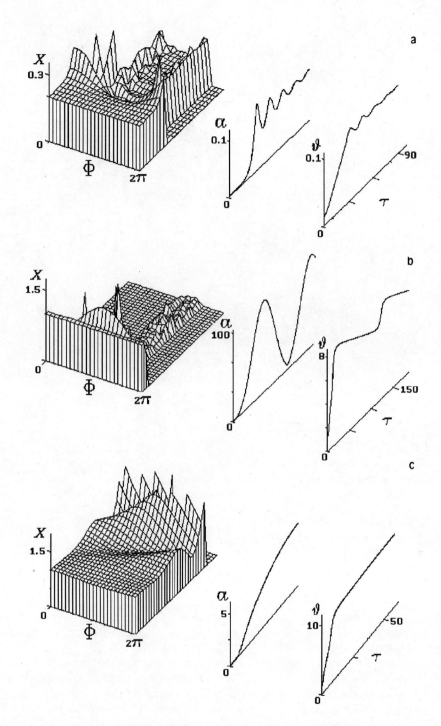

Fig. 3.3.2. Nonlinear development of the beam radius $X(\tau, \Phi)$, amplitude $a(\tau)$, and phase $\vartheta(\tau)$ of the field for (a) $\eta=0.01$, $\kappa=-10$; (b) $\eta=0.01$, $\kappa=-0.2$; and (c) $\eta=0.01$, $\kappa=0.2$.

and the increase in amplitude in the linear stage of the instability is accompanied by the large phase shift. Hence some of the beam particles are shifted from the focusing phases of the field to the defocusing phases and we are lost when the instability is stabilized by a nonlinearity. The size of the spread in phases in a clump as a function of the parameter $\kappa$

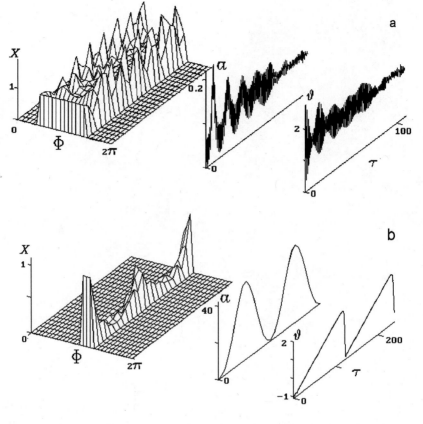

Fig. 3.3.3. Nonlinear development of the beam radius $X(\tau,\Phi)$, amplitude $a(\tau)$, and phase $\vartheta(\tau)$ of the field for performed clumps in a plasma with (a) $\eta=0.01$, $\kappa=-10$ and (b) $\eta=0.01$, $\kappa=-0.2$.

is shown in Fig. 3.3.1.

The results of numerical integration of the system

$$\ddot{X} + \left[a\cos(\Phi-\vartheta) - \frac{\eta}{X^4}\right]X = 0,$$

$$\dot{a} = -\frac{1}{2}\left\langle\frac{\sin(\Phi-\vartheta)}{X^2}\right\rangle, \quad \dot{\vartheta} = \frac{\kappa}{2} + \frac{1}{2a}\left\langle\frac{\cos(\Phi-\vartheta)}{X^2}\right\rangle \tag{3.3.14}$$

in the resonance case $\kappa = -0.2$ are represented in Fig. 3.3.2b.

Note that the decrease in the length of a clump is accompanied by synchronization of the oscillations in the electron disks in different cross sections and the appearance of strong modulation in the amplitude and phase of the field as a result of the development of coherent radial pulsations of narrow clumps.

In the range of modulation frequencies $\kappa > 0$ the nonlinear stabilization mechanism for the phase of the field no longer operates, and as the amplitude of the wave grows the beam breaks up (Fig. 3.3.2).

Some of the beam electrons remain in the focusing phases of the field, which implies a possible way of producing equilibrium clumps in a plasma.

The results of a numerical study of the nonlinear evolution of the clumps with parameters corresponding to the asymptotic solutions of Fig. 3.3.2ab injected into the plasma in the absence of the wave are shown in Fig. 3.3.3.

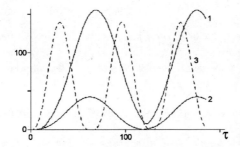

Fig. 3.3.4. Oscillations of the field amplitude $a(\tau)$ for (1) a solid beam, (2) bunched beam, and (3) charge sheets.

The stage in which the clumps are charge-neutralized and potential wells moving together with the beam develop is not associated with any great losses of particles (curve 3 in Fig. 3.3.1).

*4. Electron disks.*

Coherent radial pulsations of the thin disks can be qualitatively described using model of electron disks [3,10]. Representing the beam density in Eq. (3.2.9) in the form

$$n_b = \sigma_0 \frac{R_0^2}{R^2} \theta(r - R) \sum_{s=-\infty}^{\infty} (z - vt - sl) \tag{3.3.15}$$

(here $\sigma_0$ is the surface charge density and $l$ is the separation between disks) and averaging over space (for a single disk, over the spatial period of the wave), we find the following system of equations for the radius of a disk and for the field (at points $z_s = vt + sl$):

$$\ddot{X} + \left( \operatorname{Re} Y - \frac{\eta}{X^4} \right) X = 0, \quad 2i\dot{Y} - \kappa Y = \frac{\sigma_0}{n_p l} \frac{1}{X^2}, \tag{3.3.16}$$

where we have retained the notation of Eq. (3.3.4) and set $k = 2\pi/l$.

The results of numerical integration Eq. (3.3.16) are shown by the broken trace in Fig. 3.3.4 for $\sigma_0/n_p l = 1/8$. The qualitative agreement with oscillations of the wave field for three beam models confirms the assumption made above about their physical nature.

The applicability of Eq. (3.2.8) and (3.2.10) is determined by the inequality $R \ll c/\omega_p$, which yields the following restriction on the beam current:

$$I_b \ll \pi \frac{c^2}{\omega_p^2} e n_0 v = \frac{\nu}{4} I_a, \tag{3.3.17}$$

where $I_a = mc^3 \beta\gamma/e$ if the Alfven current.

When the amplitude of the oscillations in the plasma increases, the transverse beam electron velocity also increases according to

$$v_r/c \simeq \dot{R} \simeq \nu^{1/3} \omega_p R.$$

Hence the relativistic factor is constant only when the inequality $\gamma v_r \ll c$ holds. At the limit of the applicability of the theory $R \le c/\omega_p$ we have the inequality

$$\gamma \nu^{1/3} \ll 1, \tag{3.3.18}$$

105

which bounds the beam energy.

Thus, the field of the quasitransverse modes excited by the beam focuses or defocuses beam electrons, causing the beam to decay into clumps whose parameters remains unchanged after the instability undergoes nonlinear stabilization. It is shown above that the injection of such a bunched beam into a plasma does not result in significant particle losses.

### §3.4. Cherenkov instability in a weak magnetic field.

When there is no magnetic field the phase velocity of the unstable mode is closed to the beam velocity, and the transverse electric field focuses or defocuses beam electrons depending on the wave phase. This beam bunching effect in the nonlinear stage of the instability is maintained in a week magnetic field as well, when the electron gyrofrequency of the beam is less than or comparable to the growth rate [7,11].

In the strong magnetic field, when the beam electron gyrofrequency is greater than the growth rate and instability phase velocity can be very different from the beam velocity, radial oscillations of the beam electrons and plasma develop under cyclotron resonance conditions (the anomalous Doppler effect), leading to radial focusing of the beam as a whole [8,9,11]. The field of this mode is stabilized by the nonlinearity only in next order in the small parameter $k_\parallel R \ll 1$, due to the dependence of the plasma dielectric function on the beam radius.

*1. System of nonlinear equations.*

The nonlinear development of the instability of a thin modulated beam in the presence of longitudinal magnetic field can be analyzed by means of the self-consistent system of Eqs. (3.2.13) and (3.2.14). By analogy with §3.3, we perform the change of variables

$$t, z \rightarrow t_1 = t, \ \Phi = \omega t - kz, \tag{3.4.1}$$

after which it assumes the form

$$\left[\frac{\partial}{\partial t} + (\omega - kv)\frac{\partial}{\partial \Phi}\right]^2 R + \frac{1}{m\gamma}\left[eN + \frac{\omega_B^2}{4}\left(1 - \frac{R_0^4}{R^4}\right)\right] R = 0,$$

$$\left(\frac{\partial}{\partial t} + \omega\frac{\partial}{\partial \Phi}\right)^2 N + \Omega_p^2 N = 2\pi e n_0 \left[\left(\frac{\partial}{\partial t} + \omega\frac{\partial}{\partial \Phi}\right)^2 + \omega_{Bp}^2\right]\left(1 - \frac{R_0^2}{R^2}\right), \tag{3.4.2}$$

where $R(t, \Phi)$ is the beam radius, $R_0$ and $n_0$ are the initial radius and density of the beam, $\omega_B = eB_0/mc\gamma$ and $\omega_{Bp} = \gamma\omega_B$ are the beam and plasma gyrofrequencies, the plasma hybrid frequency is given by $\Omega_p^2 = \omega_p^2 + \omega_{Bp}^2$, and $\omega_p$ is the Langmuir frequency of the plasma.

The wave number $k$ determines the wavelength of the unstable mode, and the frequency $\omega$ is a formal parameter which remains to be determined. If the wave velocity is close to beam velocity, then it is convenient for purposes of numerical integration to set $\omega = kv$ and to eliminate the derivative, $\partial/\partial\Phi$, in the first of Eq. (3.4.2).

Substituting

$$N(t, \Phi) = \frac{m\gamma\Omega_p^2 \nu^{2/3}}{e} \text{Re}\left[Y(t)\exp i\Phi\right], \tag{3.4.3}$$

in the second of Eq. (3.4.2) and averaging over the spatial period, we find the following system of ordinary differential equations in time:

$$\ddot{X} + [\text{Re}\,Y\exp(i\Phi)] + \mu(X - X^{-3}) = 0, \quad 2i\dot{Y} - \kappa Y = \left\langle\frac{\exp(-i\Phi)}{X^2}\right\rangle, \tag{3.4.4}$$

106

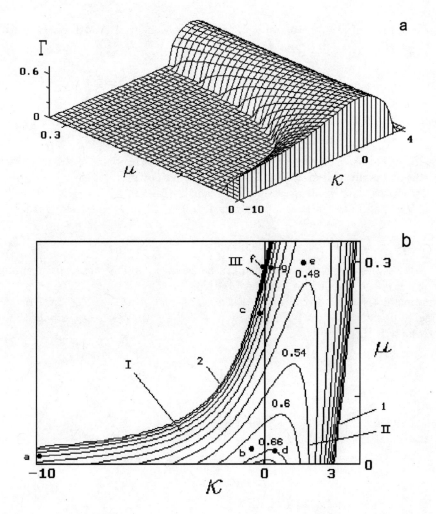

Fig. 3.4.1. Growth rate $\Gamma$ as a function of the modulation frequency $\kappa$ and magnetic field $\mu$: (a) surface plot, (b) contour plot.

which generalizes the equations (3.3.4) to the case in which a longitudinal magnetic field is present. In passing from (3.4.2) to (3.4.4) we have introduced the following dimensionless variables:

$$X = \frac{R}{R_0}, \quad \tau = \nu^{1/3}\Omega_p t, \quad \nu = \frac{\omega_b^2 \omega_p^2}{\Omega_p^4} \ll 1,$$

$$\kappa = \frac{k^2 v^2 - \Omega_p^2}{\Omega_p^2 \nu^{1/3}}, \quad \mu = \frac{\omega_B^2}{4\Omega_p^2 \nu^{2/3}}, \quad <\ldots> = \frac{1}{2\pi}\int_0^{2\pi} \ldots d\Phi. \tag{3.4.5}$$

2. *Instability growth rate.*

In the linear approximation

$$X = 1 + \tilde{x}, \quad |\tilde{x}| \ll 1 \tag{3.4.6}$$

for perturbation of the form $\exp(i\Delta\tau)$, Eq. (3.4.4) yields the dispersion relation

$$\Delta^3 + \frac{\kappa}{2}\Delta^2 - 4\mu\Delta - 2\kappa\mu - \frac{1}{2} = 0. \tag{3.4.7}$$

107

Complex roots of (3.4.7), corresponding to the exponential growth of the wave amplitude exist for $D > 0$, and are equal to

$$\Delta = -\frac{\kappa}{6} - \frac{u+v}{2} - i\frac{\sqrt{3}}{2}(u-v), \quad u = (-q+\sqrt{D})^{1/3}, \quad v = (-q-\sqrt{D})^{1/3},$$

$$D = \frac{\kappa}{3}\left[\mu - \left(\frac{\kappa}{12}\right)^2\right] - \frac{64}{27}\mu\left[\mu - \left(\frac{\kappa}{4}\right)^2\right]^2 + \frac{1}{16}, \quad q = \left(\frac{\kappa}{6}\right)^3 - \frac{2\kappa\mu}{3} - \frac{1}{4}, \tag{3.4.8}$$

The magnitude of the growth rate $\Gamma(\kappa,\mu) = \operatorname{Im}\Delta$ as a function of the mismatch $\kappa$ and magnetic field $\mu$ is illustrated in Fig. 3.4.1a, and a contour plot of this quantity is given in Fig. 3.4.1b. In parametric form the curves 1 and 2 (Fig. 3.4.1b), which follows from the condition $D = 0$ and determine the boundary $\Gamma = 0$ of the instability region, take the form

$$\mu_{1,2} = (\epsilon^{-4} - \epsilon^{-1})/4, \quad \kappa_{1,2} = \epsilon + 2\epsilon^{-2}, \tag{3.4.9}$$

where we have introduced the intermediate parameter $\epsilon$. The physical meaning of the parameter $\epsilon$ is spelled out below [see Eq. (3.4.13)]

Equation (3.4.9) has a discontinuity at the point $\epsilon = 0$; trace 1 in Fig. 3.4.1b corresponds to the region $0 < \epsilon < 1$, trace 2 to the region $-\infty < \epsilon < 0$. When $D$ is small, the solution (3.4.8) reduces to the form

$$\Delta = -\frac{\kappa_{1,2}}{6} + q^{1/3} - \frac{i}{q^{2/3}}\sqrt{\frac{D}{3}}, \tag{3.4.10}$$

$$D(\kappa,\mu) = D_{1,2} + \left(\frac{\partial D}{\partial \mu}\right)_{1,2}\tilde{\mu} + \left(\frac{\partial D}{\partial \kappa}\right)_{1,2}\tilde{\kappa},$$

$$D_{1,2} = D(\kappa_{1,2},\mu_{1,2}) \equiv 0, \quad \tilde{\mu} = \mu - \mu_{1,2}, \quad \tilde{\kappa} = \kappa - \kappa_{1,2},$$

$$|\tilde{\mu}| \ll 1, \quad |\tilde{\kappa}| \ll 1.$$

Taking into account Eq. (3.4.9), we find

$$q = \frac{(\epsilon^3 - 4)^3}{216\epsilon^6}, \quad \left(\frac{\partial D}{\partial \mu}\right)_{1,2} = -\frac{(\epsilon^3 - 4)^3}{108\epsilon^5}, \quad \left(\frac{\partial D}{\partial \kappa}\right)_{1,2} = \frac{(\epsilon^3 - 4)^3}{432\epsilon^7}.$$

As the result, the solution (3.4.10) assumes the form

$$\Delta = -\alpha - i\Gamma, \tag{3.4.11}$$

$$\alpha = \epsilon^{-2}, \quad \Gamma^2 = \epsilon\frac{\tilde{\kappa} - 4\tilde{\mu}\epsilon^2}{\epsilon^3 - 4} \simeq \epsilon^2\frac{(\alpha^2 - 4\mu)(\kappa - 2\alpha) - 1}{\epsilon^3 - 4}.$$

Since the relation

$$\epsilon = -2\alpha + \kappa_{1,2}, \tag{3.4.12}$$

holds, it follows from Eq. (3.4.5), (3.4.6) and (3.4.11) that the parameter

$$\epsilon = \frac{\omega_p^2}{\Omega_p^2\nu^{1/3}}\left(1 - \frac{\omega_p^2}{\tilde{\omega}^2 - \omega_{Bp}^2}\right) \tag{3.4.13}$$

is proportional to the plasma dielectric function (here $\tilde{\omega} = \omega - \nu^{1/3}\alpha\Omega_p$ is the wave frequency in the laboratory frame).

### 3. Weak nonlineariry.

Close to the instability threshold $\Gamma \ll \alpha$ is possible to have a regime where nonlinear stabilization occurs for waves with a small amplitude $|x| = |X - 1| \ll 1$, $|Y| \ll 1$ (Ref. [6]). In what follows we find the conditions under which this regime occurs.

Expanding Eqs. (3.4.4) in powers of $x$ and remaining terms $\sim x^3$, we can represent the solution in the form

$$x = x_0(\tau) + \operatorname{Re}[x_1(\tau)\exp i\psi) + x_2(\tau)\exp 2i\psi)],$$

$$Y = \tilde{Y}\exp(-i\alpha\tau), \quad \psi = \Phi/\nu^{1/3}\Omega_p - \alpha\tau. \tag{3.4.14}$$

Substituting (3.4.14) in (3.4.4) and averaging over the spatial period ($0 \le \Phi \le 2\pi$), we find

$$\left[\left(-i\alpha + \frac{d}{d\tau}\right)^2 + 4\mu\right]x_1 + \tilde{Y} + \mu\left[-6(2x_0x_1 + x_1^*x_2)\right.$$

$$\left. + \frac{15}{2}x_1^2x_1^*\right] + x_0\tilde{Y} + \frac{1}{2}x_2\tilde{Y}^* = 0,$$

$$\tilde{Y} = \frac{1}{\epsilon}x_1 + \frac{2i}{\epsilon^2}\dot{x}_1 - \frac{4}{\epsilon^3}\ddot{x}_1 + \frac{1}{\epsilon}\left(-3x_0x_1 - \frac{3}{2}x_1^*x_2 + \frac{3}{2}x_1^2x_1^*\right), \tag{3.4.15}$$

$$x_0 = \frac{6\mu\epsilon - 1}{8\mu\epsilon}x_1x_1^*, \quad x_2 = \frac{6\mu\epsilon - 1}{8(\mu - \alpha^2)\epsilon}x_1^2,$$

where the parameter $\epsilon$ is given by (3.4.13). In deriving the second equation we have used the expansion

$$(\epsilon - 2id/d\tau)^{-1} \simeq \epsilon^{-1} + 2i\epsilon^{-2}d/d\tau - 4\epsilon^{-3}d^2/d\tau^2.$$

After substituting $x_o$, $x_2$, and $\tilde{Y}$ in the equation for $x_1$, we find

$$\dot{x}_1 = \Gamma^2 x_1 - Ax_1^2x_1^*, \quad A = \frac{\epsilon^2(3\epsilon^6 - 5\epsilon^3 - 6)}{2(4 - \epsilon^3)(1 - \epsilon^3)(3 + \epsilon^3)}, \tag{3.4.16}$$

where $\Gamma^2$ is given by (3.4.11) and $\mu$ and $\alpha$ are expressed in terms $\epsilon$ by means of (3.4.9) and (3.4.11)

The stabilization condition $A(\epsilon) > 0$ is satisfied only for $\epsilon < 0$ in the following range of parameters:

$$\epsilon_1 < \epsilon < \epsilon_2, \quad \mu_1 < \mu < \mu_2, \quad \kappa_1 < \kappa < \kappa_2, \tag{3.4.17}$$

where $\epsilon_1 = -1,44$, $\epsilon_2 = 0,93$, $\mu_1 = 0,23$, $\mu_2 = 0,6$, $\kappa_1 = -0,48$, and $\kappa_2 = 1,37$. This section lies on trace 2 and is indicated in Fig. 3.4.1b by the heavy line.

Note that under hard excitation conditions $A(\epsilon) < 0$ the waves are unstable even when the linear stability condition $\Gamma^2 < 0$, is satisfied if the initial wave amplitude is above the threshold value.

In the special case $\kappa = 0$ treated in Ref. [9] we have $\epsilon = 2^{1/3}$. Then the wave frequency, gyrofrequency, and the growth rate are equal respectively to

$$\alpha = 2^{-2/3}, \quad 4\mu_2 = 3/2^{4/3}, \quad \Gamma = (\alpha^2 - 4\mu/3)^{1/2}.$$

Correspondingly, the energy density of the Langmuir oscillations is equal to

$$\frac{E_r^2}{4\pi} = n_o mv^2 \gamma \left(\frac{\nu}{2}\right)^{1/3} \left(\frac{3\delta}{4\alpha}k_{\|}R\right)^2.$$

## 4. Numerical simulation.

From the results of the numerical calculations the unstable region in the $(\kappa, \mu)$ (Fig. 3.4.1b) can be divided into three regions (I, II, II), in which the solutions behave in different ways. Regions I and II correspond to $\kappa < 0$ and $\kappa > 0$, excluding the "boundary layer" near the portion $\kappa_1 < \kappa < \kappa_2$, $\mu_1 < \mu < \mu_2$ of trace 2 (see Sec. 3, region III).

In region I the increase in the wave amplitude is stabilized by nonlinearity when $|\Delta R| \simeq R$ holds; this is accompanied by excitation of a plasma wave with a constant or periodically varying amplitude and a phase velocity close to the beam velocity (Fig. 3.4.2, points a, b c).

In region II there is no stabilization and the solution consists of a wave with an amplitude that grows without bound, whose phase velocity is less than the beam velocity. The amplitude of the beam modulation accordingly increases in time without bound (Fig. 3.4.3, points d, e).

Region III corresponds to the soft excitation regime, when the nonlinear shift in the growth rate is negative (see Sec. 3) and the instability is stabilized at small wave amplitudes (Fig. 3.4.4a, point f in Fig. 3.4.1b).

In the $(\kappa, \mu)$ plain of Fig. 3.4.1b the points corresponding to the solutions shown in the figures are labeled with letters of the Roman alphabet.

Let us describe the solution in regions I and II in more detail. In a weak magnetic field $\mu \ll 1$ the most favorable regime for beam transport in a plasma is that in which the modulation frequency satisfies $-\kappa \gg 1$ $\left(\Delta \simeq -i/\sqrt{|\kappa|}\right)$. Then it follows from (3.4.7) that the increase in the amplitude of the field is not accompanied by the significant linear phase shift of the wave relative to beam. Hence when the beam undergoes a transition to the nonlinear state, the particles which initially had phases $\cos \phi > 0$ persist as focused bunches, while those with phases $\cos \phi < 0$ are defocused. The results of numerical integration for this case are shown in Fig. 3.4.2a (point a in Fig. 3.4.1b).

As the size of mismatch $\kappa$ (for $\kappa < 0$) and the maximum growth rate

$$\Delta = -\left(1 + i\sqrt{3}\right)\left(\nu/16\right)^{1/3}, \quad |\kappa| \ll 1 \tag{3.4.18}$$

is approached, the frequency shift becomes complex and a large "linear" phase shift develops as the wave amplitude grows. Consequently, some of the beam particles leave the region where the field phases are focusing and enter the defocusing region; these are lost when nonlinear stabilization of the phase occurs. The transition to steady state near the maximum growth rate $\kappa = -0.2$ is shown in Fig. 3.4.2b (point b in Fig. 3.4.1b).

As can be seen from Fig. 3.4.2, as the mismatch $\kappa$ increases there is a qualitative change in the behavior of the nonlinear oscillations of the field amplitude. In the region $-\kappa \gg 1$ the radial oscillations of the electron disks representing cross sections of the beam at different wave phases become desynchronized because the clumps are thick, and the nonlinear oscillations of the amplitude and phase of the field, which are a global effect, damp out. As the extent of the phases of these clumps decreases for $-\kappa \ll 1$ these waves become coherent and the amplitude and phase of the field are strongly modulated.

Increasing the magnetic field (parameter $\mu$) results in partial suppression of the instability, reducing the growth rate and amplitude of the waves which still remain finite even near the instability threshold of Fig. 3.4.2c (point c in Fig. 3.4.1b). For positive values of $\kappa$ in the beam system the wave moves with linear phase velocity $v_{\mathrm{ph}} \sim \mathrm{Re}\Delta < 0$, corresponding to radial oscillations with frequency $kv_{\mathrm{ph}}$. The amplitudes of the nonlinear oscillations increases without bound with the growth rate $\Gamma$ (Fig. 3.4.3). In a weak magnetic field $\mu \ll 1$ it satisfies $\Gamma \sim kv_{\mathrm{ph}}$ and the beam is defocused by the wave field in a time $\sim (kv_{\mathrm{ph}})^{-1}$ (see

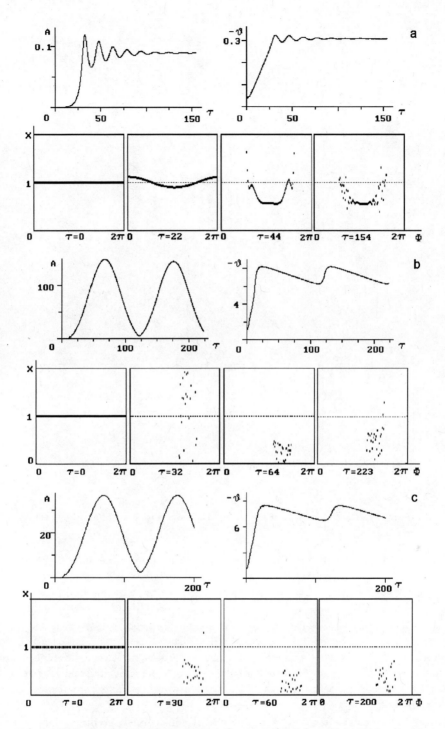

Fig. 3.4.2. Time dependence of the amplitude $A$ and the phase $\vartheta$ of the field and of the beam radius $X$ respectively at the points a, b, and c of Fig. 3.4.1b: (a) $\kappa=-10$, $\mu=0.01$; (b) $\kappa=-0.2$, $\mu=0.01$; (c) $\kappa=-0.2$, $\mu=0.05$.

Fig. 3.4.3a, point d in Fig. 3.4.1b). The growth rate $\Gamma$ falls off as a function of $\mu$ and the wave frequency $kv_{\mathrm{ph}}$ increases, so that the defocusing process slows down (see Fig. 3.4.3b, point e in Fig. 3.4.1b).

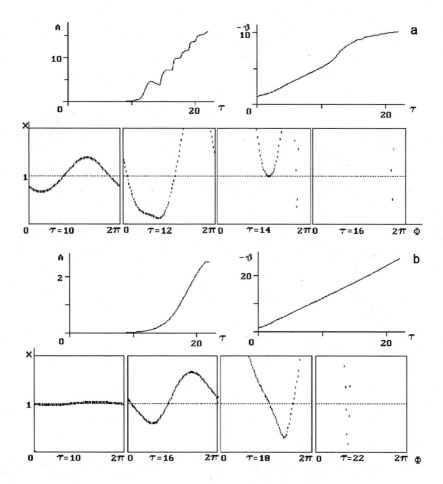

Fig. 3.4.3. Time dependence of the amplitude $A$ and the phase $\vartheta$ of the field and of the beam radius $X$ respectively at the points d and e of Fig. 3.4.1b: (a) $\kappa=0.2$, $\mu=0.01$; (b) $\kappa=2$ $\mu=0.298$.

The physical mechanism for the excitation of transverse waves in a plasma with thin electron beam in regimes I and II can be explained by analyzing the change in the wave phase velocity (here $\vartheta$ is the phase of the field), $v_{\mathrm{ph}} = \nu^{1/3} v\,\dot{\vartheta}$, in the nonlinear stage of the instability (Fig. 3.4.5).

The perturbations with phase velocities less than the beam velocity $v_{\mathrm{ph}} < 0$ are unstable. In the linear stage this inequality, which follows from Eqs. (3.4.8), is satisfied automatically. However, the subsequent behavior of $v_{\mathrm{ph}}$ in regions I and II very different. Numerical solution shows that in region II the condition $v_{\mathrm{ph}} < 0$ persists, and in the region where the present thin-beam model is applicable the wave amplitude increases without bound (see Fig. 3.4.5, trace 2).

An alternative course for the growth rate of the instability, accompanied by a shift in the wave phase velocity toward $v_{\mathrm{ph}} > 0$ when the phase velocity in the laboratory frame is greater than the beam velocity, is observed in the region I. This instability regime therefore leads only to partial breakup of the beam, with some of the particles remaining "frozen" into the focusing phases of the field (see Fig. 3.4.5, trace 1 ).

The solutions obtained in the region III (weak nonlinearity) do not need to be described in more detail, since they have already been considered in Sec .3 (Fig. 3.4.4a, point f in Fig. 3.4.1b). Note that the nature of the solution changes abruptly in the passage through

112

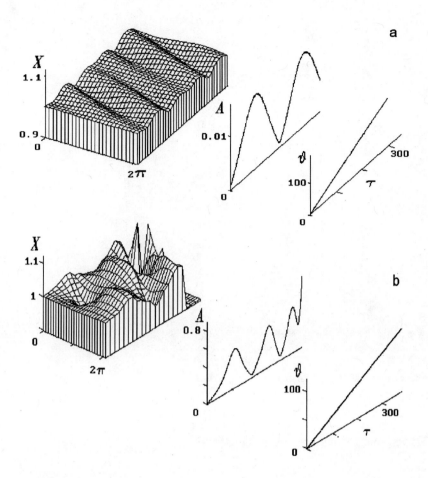

Fig. 3.4.4. Time dependence of the beam radius $X$ and of the amplitude $A$ and the phase $\vartheta$ of the field at the points f and g of Fig. 3.4.1b: (a) $\kappa = 0.005$, $\mu = 0.298$; (b) $\kappa = 0.007$, $\mu = 0.298$.

the boundary of region III (e.g., into region 2; see Fig. 3.4.4b, point g in Fig. 3.4.1b).

## §3.5 Cyclotron instability in a strong magnetic field.

As noted above, in the region $\kappa > 0$ where the effect of relatively weak magnetic field $\mu \leq 1$ on the growth of transverse waves is important, the growth of the field amplitude is not stabilized nonlinearly (Fig. 3.4.4). A similar effect is observed in a strong magnetic field, $\sqrt{\mu} \gg 1$, where the Cherenkov resonance is suppressed, and the instability for quasitransverse harmonics, $k_\perp \gg k_\parallel$, occurs at cyclotron resonance: $\omega - kv = -\omega_B$ (the anomalous Doppler effect, see Refs. [8,9]).

The solution of the dispersion relation (3.1.7) for a low-density beam is

$$
\omega = \Omega_p - \frac{\omega_p^2 \omega_{Bp}^2}{2\Omega_p^3} \frac{k_\perp^2}{k_\parallel^2}
$$

$$
+ \frac{1}{2} \left( \Omega_p + \omega_B - k_\parallel v \pm \sqrt{(\Omega_p + \omega_B - k_\parallel v)^2 - \frac{\omega_b^2 \omega_p^2}{\Omega_p \omega_B}} \right),
\tag{3.5.1}
$$

where $\omega_b$ and $\omega_p$, and $\omega_B$ and $\omega_{Bp}$ are the Langmuir frequencies and gyrofrequencies of the beam and plasma, and $\Omega_p^2 = \omega_p^2 + \omega_{Bp}^2$ is the plasma hybrid frequency.

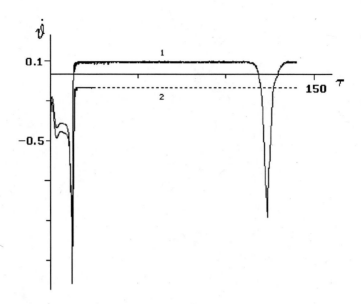

Fig. 3.4.5. Nonlinear saturation of the wave phase velocity: (1) $\kappa = -0.2$, $\mu = 0.01$ (region I) and (2) $\kappa = 0.2$, $\mu = 0.01$ (region II).

The frequency acquires an imaginary part if

$$\left(\Omega_p + \omega_B - k_\parallel v\right)^2 < 4\delta^2,$$

and the corresponding maximum growth rate is

$$\delta = \left(\frac{\omega_b^2 \omega_p^2}{4\Omega_p \omega_B}\right)^{1/2}, \quad \Omega_p - k_\parallel v = -\omega_B. \tag{3.5.2}$$

This wave mode can be excited in a plasma by a sufficiently thin beam of finite radius, since in this case we have $k_\perp \simeq 1/R_0$ (where $R_0$ is the beam radius), and the condition $k_\perp^2/k_\parallel^2 \gg 1$ is equivalent to $\left(k_\parallel R_0\right)^2 \ll 1$.

The instability in bounded beam should be accompanied by the decrease in the growth rate due to radiative damping;

$$\Gamma_{\mathrm{rad}} \simeq \frac{1}{R}\frac{d\omega}{dk_\perp} \simeq (k_\parallel R)^2 \frac{\omega_p^2 \omega_{Bp}^2}{\Omega_p^3}, \tag{3.5.3}$$

if, however, $\Gamma_{\mathrm{rad}} \ll \delta$, i.e.,

$$\left(\frac{n_b}{n_p}\right)^{1/2} \gg (k_\parallel R)^2 \left(\frac{\omega_{Bp}}{\Omega_p}\right)^{5/2}, \tag{3.5.4}$$

then the power dissipated in the beam region, $r < R_0$, is much higher than the loss due to the energy carried across the beam surface. In this case the oscillations are actually confined to the interior of the beam, and growth rate is approximately that given in (3.5.1) and (3.5.2).

Note that the inequality (3.5.4) can be found by using the substitution of (3.5.1) and (3.5.2) into general Eq. (3.1.8) under the conditions of the anomalous Doppler effect, $\omega - kv =$

114

$-\omega_B$, where the amplitude saturates only in the next order of the expansion in $(k_\parallel R)^2$ (see Ref. [8,9]).

### 1. Nonlinear anomalous Doppler mode.

In the approximation of linear plasma, $k_\parallel R \ll 1$, the system of equations (3.2.12) and (3.2.14) can be used

$$\left(\frac{\partial}{\partial t} + v\frac{\partial}{\partial z}\right)^2 R + \left[\frac{\omega_B^2}{4}\left(1 - \frac{R_0^4}{R^4}\right) + \frac{eN}{m\gamma}\right] R = 0,$$

$$\left(\frac{\partial^2}{\partial t^2} + \Omega_p^2\right) N = 2\pi e n_0 \left(\frac{\partial^2}{\partial t^2} + \omega_{Bp}^2\right)\left(1 - \frac{R_0}{R^2}\right). \tag{3.5.5}$$

Let us rewrite the first equation (3.5.5), transforming to the new variable $w = (R/R_0)^2 - 1$:

$$\ddot{w} + \omega_B^2 w - \frac{1}{2}\frac{\dot{w}^2 + \omega_B^2 w^2}{1 + w} = -\frac{2e}{m\gamma}N(1 + w). \tag{3.5.6}$$

Then, using

$$\frac{d}{dt}\frac{\dot{w}^2 + \omega_B^2 w^2}{1 + w} = \frac{4e}{m\gamma}N\dot{w}, \tag{3.5.7}$$

we can write system (3.5.5) as

$$\frac{d}{dt}(\ddot{w} + \omega_B^2 w) = -\frac{2e}{m\gamma}\left[N\dot{w} + \frac{d}{dt}N(1 + w)\right],$$

$$\left(\frac{\partial^2}{\partial t^2} + \Omega_p^2\right) N = 2\pi e n_0 \left(\frac{\partial^2}{\partial t^2} + \omega_{Bp}^2\right)\frac{w}{1 + w} \tag{3.5.8}$$

[the dot in Eqs. (3.5.6)–(3.5.8) means the derivative $d/dt = \partial/\partial t + v\partial/\partial z$].

We seek the solution of system (3.5.8) in the form of traveling waves with the time-dependent amplitudes:

$$N = \frac{m\gamma\omega_B\delta}{e}A(t)\sin\eta,$$

$$w = w_0(t) + a(t)\cos\eta + w_2(t)\sin 2\eta, \tag{3.5.9}$$

where $\eta = \Omega_p t - k_\parallel z$, and the longitudinal wave number is determined by the equation from the linear theory $k_\parallel = (\Omega_p + \omega_B)/v$. Assuming that the amplitudes are slowly varying functions of the time ($|\dot{a}| \ll \omega_B a$), and substituting (3.5.9) in (3.5.8), we find

$$\dot{a} = -\delta A(1 + w), \tag{3.5.10}$$

$$\dot{w}_0 = -\delta Aa, \qquad w_2 = -\frac{2\delta}{3\omega_B}Aa, \tag{3.5.11}$$

$$\dot{A} = -2\delta\frac{1}{a}\left(\sqrt{1 + a^2} - 1\right). \tag{3.5.12}$$

It follows from (3.5.10) and (3.5.11) that the amplitude of the second harmonic is small ($\delta \ll \omega_B$) in comparison with the first and zeroths. Integrating (3.5.10) and (3.5.11), and using the initial conditions $w_0(0) = 0 \ a(0) = 0$, we find $w_0 = \sqrt{1 + a^2} - 1$. Correspondingly,

115

the system (3.5.10) and (3.5.12) leads to a second-order nonlinear equation for the function $a(t)$:

$$\frac{d}{dt}\frac{\dot{a}}{\sqrt{1+a^2}} = 2\delta^2 \frac{1}{a}(\sqrt{1+a^2}-1). \tag{3.5.13}$$

In the linear stage of instability ($a \ll 1$) the wave amplitude increases exponentially over time with a growth rate $\delta$. The nonlinearity does not suppress the instability, since the first integral of (3.5.13) has no turning points. If $a \gg 1$ solution (3.5.13) is (where

$$a(t) = C_1 \exp(\delta^2 t^2 + C_2 t) \tag{3.5.14}$$

(where $C_1$ and $C_2$ are the integration constants), and the field amplitude linearly with the time. The function $a(t)$ varies slowly if $t \ll \omega_B/\delta^{-2}$.

The first equation in (3.5.5) has an interesting feature: Even at a small wave amplitude, the variation of the frequency

$$\omega_B^2(R) = (\omega_B^2/4)(1 - R_0^4/R^4)$$

with the field amplitude should cause the oscillations of the beam electrons to go out of phase with the plasma (a linear oscillator). It turns out, however, that the solution of the nonlinear equation is the same as that of the linear equation (with $N = 0$),

$$R^2 = R_0^2(\sqrt{1+a^2} + a\cos\eta), \tag{3.5.15}$$

and the beam remains in phase with the wave even if there is a pronounced nonlinearity, $a \gg 1$.

THe energy density of the electric field, determined from the condition $A \approx$, is on the order of

$$\frac{E_r^2}{8\pi} \simeq n_0 mc^2 \gamma \frac{\omega_B \omega_p^2}{\Omega_p^3}(k_\parallel R)^2. \tag{3.5.16}$$

At the Cherenkov resonance (see §3.3) the beam electrons in their proper reference frame are in the quasistatic electric field of the perturbation which focuses or defocuses beam electrons, depending on the phase with respect to the wave. The growth of the transverse Cherenkov mode is thus accompanied by a "ripping" of the beam surface. The longitudinal magnetic field prevents a radial displacement of the electrons and suppresses the growth of the Cherenkov mode if the gyrofrequency of the beam electrons is higher than the Cherenkov growth rate. In the case of radial stability, however, the beam electrons become oscillators, and transverse waves grow under the conditions of anomalous Doppler effect, so that the oscillation frequency of the beam electrons, converted to the reference frame of the plasma, $\Omega_\perp = k_\parallel v - \omega_B$, is equal to the frequency $\Omega_p$. As mentioned above, as the oscillations grow the beam does not go out of phase with the wave in a nonlinear fashion and – as at Cherenkov resonance – the nonlinearity does not suppress the instability according to the model of a "thin" beam. In contrast with Cherenkov mode, however, the cyclotron instability does not cause a rippling of the beam surface, since radial fluctuations of the electrons occurs uniformly over the entire length and are accompanied by a radial expansion of the plasma.

## 2. Nonlinear stabilization of the cyclotron mode.

As noted above, in the region $\kappa > 0$ where the effect of a relatively weak magnetic field $\mu \leq 1$ on the growth of transverse waves is important, the growth of the field amplitude is not stabilized nonlinearly. A similar effect is observed in a strong magnetic field $\mu \gg 1$ (under the conditions of anomalous Doppler effect [9]), where the amplitude saturates only in the next order of the expansion in $(kR)^2 \ll 1$.

Note that the growth of the waves in a fairly strong magnetic field $\mu \lesssim 1$, where the beam electron gyrofrequency is comparable with the Cherenkov growth rate, is qualitatively similar to the instability under the conditions of the anomalous Doppler shift $\Delta = -2\sqrt{\mu}$. The difficulty in treating the nonlinear problem in this region of parameters analytically is that the growth rate is equal to the frequency in order of magnitude in the beam frame ($\text{Im}\Delta \approx \text{Re}\Delta$, and the oscillations of the beam radius cannot be described in the single-mode approximation). When we pass to the strong magnetic field region $\mu \gg 1$, these oscillations are synchronized, since the cyclotron harmonic dominates.

If we take this into account, we can conveniently take $\omega = kv - \omega_B$ in the first equation of (3.4.2). In the dimensionless variables (3.4.5) we have

$$\left(\frac{\partial}{\partial \tau} - 2\sqrt{\mu}\frac{\partial}{\partial \Phi}\right)^2 X + \left[\mu\left(1 - \frac{1}{X^4}\right) + \text{Re}\left(Y\exp(i\Phi)\right)\right] X = 0. \tag{3.5.17}$$

According to §3.2, in this case the field equation (3.2.20) is change to the form

$$2i\dot{Y} - \left(\kappa + Q < X^2 >\right)Y = \left\langle \frac{\exp(-i\Phi)}{X^2} \right\rangle; \tag{3.5.18}$$

where

$$Q = \mu\nu^{1/3}\frac{\omega_p^2}{2\Omega_p^2}\frac{1+\beta^2}{1-\beta^2}(kR)^2 \ll 1$$

Equation (3.2.18) can be used to estimate the order of the maximum amplitude of the oscillations in the beam radius, since as the beam is defocused by the instability field due to the nonlinear dependence of the wave frequency on the radius the "effective mismatch" $\kappa_{\text{eff}} = \kappa + Q\langle X^2\rangle$ between the wave frequency and the upper hybrid frequency $\Omega_p$ increases, reaching the threshold value $\kappa_1(\mu)$ (see Fig. 3.4.1b). The maximum amplitude $X_m$ of in of the oscillations in the beam radius can be estimated by setting $\kappa_{\text{eff}} \simeq \kappa_1(\mu)$:

$$\langle X_m^2\rangle \simeq \frac{\kappa_1(\mu) - \kappa}{Q}. \tag{3.5.19}$$

The numerical solution for the finite values of the magnetic field in the case $Q > 0$ is shown in Fig. 3.5.1.

Retaining in the (3.5.17) the terms proportional $\mu \gg 1$, we obtain the asymptotic solution

$$X^2 = \sqrt{1+a^2} + a\cos(\Phi + \varphi) \tag{3.5.20}$$

(here $a$ and $\varphi$ are integration constants), corresponding to nonlinear beam oscillations in a strong magnetic field. In the next order to the small parameter $\mu^{-1/2}$ the solution remains the form (3.5.20), but amplitude and phase are slowly varying functions of the time.

Substituting in (3.5.17) and (3.5.18)

$$Y = \mu^{1/4}A\exp(-2i\mu^{1/4}\tilde{\tau} + i\vartheta), \quad \tau = \mu^{1/4}\tilde{\tau},$$

$$\kappa + Q = 4\mu^{1/2} + \chi\mu^{-1/4}, \quad Q = q\mu^{-1/4}$$

and evaluating the integral on the right-hand side of (3.5.18) [using (3.5.20)], we find the following system of nonlinear equations for the amplitudes and phase differences $\eta = \vartheta - \varphi$

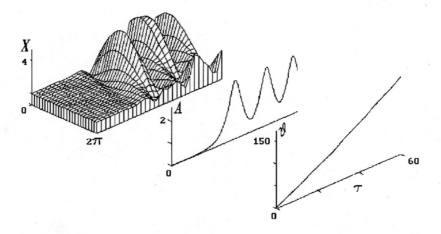

Fig. 3.5.1. Nonlinear development of the beam radius $X(\tau,\Phi)$, amplitude $a(\tau)$, and phase $\vartheta(\tau)$ of the field for $\kappa=2$, $\mu=0.298$ and $Q=0.3$.

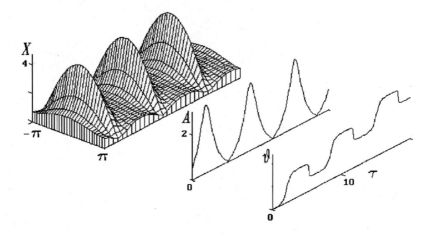

Fig. 3.5.2. Nonlinear stabilization of the instability under the conditions of the anomalous Doppler effect.

(see Ref. [9]:

$$\dot{A} = \frac{\sqrt{1+a^2}-1}{2a}\sin\eta, \quad \dot{a} = \frac{A\sqrt{1+a^2}}{2}\sin\eta,$$

$$\dot{\eta} = \frac{1}{2}\left[-\chi - q(\sqrt{1+a^2}-1) + \left(\frac{A\sqrt{1+a^2}}{a} + \frac{\sqrt{1+a^2}-1}{Aa}\right)\cos\eta\right]. \tag{3.5.21}$$

The system (3.5.21) generalizes Eqs. (3.5.10)–(3.5.12) to the case in which the beam modulation frequency is different from $\Omega_p$ ($\chi \neq 0$).

In the linear approximation $a \ll 1$ the perturbation has the growth rate

$$\tilde{\Gamma} = \frac{1}{4}\sqrt{2-\chi^2}, \quad \cos\eta = \frac{\chi}{\sqrt{2}}, \tag{3.5.22}$$

which reaches a maximum value $\tilde{\Gamma}_m = 2^{-3/2}$ at $\chi = 0$.

Using the integrals

$$A^2 = 2\ln\left(\frac{1 + \sqrt{1 + a^2}}{1 + \sqrt{1 + a_0^2}}\right),$$

$$Aa\cos\eta = \frac{q}{2}(a^2 - a_0^2) + (\chi - q)\left(\sqrt{1 + a^2} - \sqrt{1 + a_0^2}\right),$$

(3.5.23)

we can reduce the order of Eqs. (3.5.21):

$$\dot{a}^2 = \frac{1}{4}(1 + a^2)\left[2\ln\left(\frac{1 + \sqrt{1 + a^2}}{1 + \sqrt{1 + a_0^2}}\right)\right.$$

$$\left. - \frac{1}{a^2}\left(\frac{q(a^2 - a_0^2)}{2} + (\chi - q)(\sqrt{1 + a^2} - \sqrt{1 + a_0^2})\right)^2\right].$$

(3.5.24)

Equation (3.5.24) describes nonlinear radial oscillations of the beam with an amplitude that varies from $a_0$ to

$$a_m \simeq \frac{2^{3/2}}{q}\left(\ln\frac{1}{q}\right)^{1/2}, \quad a_m \gg a_0.$$

(3.5.25)

The behavior of these oscillations is shown in Fig. 3.5.2.

Note that the appearance in Eq. (3.5.24) of the parameter $\chi$ reduces the growth rate, but does not cause a qualitative change in the behavior of the nonlinear solution.

### 3. Principal results.

The investigation reported here (§3.4 and §3.5) enables us to analyze the nonlinear stage of a thin low-density relativistic electron beam in magnetized plasma when the dominant effect is the growth of transverse oscillations. The use of numerical techniques permits the behavior the nonlinear oscillations to be described over the entire range of variation of the modulation frequencies and magnetic field corresponding to the unstable region.

In a weak magnetic field $\omega_B \leq \nu^{1/3}\omega_p$ the phase velocity is close to the beam velocity, and the growth of the Cherenkov mode in the wavelength range $kv < \omega_p$ is accompanied by the development of electron clumps in the focusing phases of the field; for $kv > \omega_p$ the instability is not stabilized nonlinearity in the thin-beam model $kR \ll 1$. The difference between these instability regimes is related to the behavior of the wave phase velocity when the system goes nonlinear.

We have determined numerically and analytically the parameter range near the instability threshold for which the perturbations are stabilized at small wave amplitudes. These analytical solutions are confirmed by the numerical calculation.

Increasing the longitudinal magnetic field reduces the growth rate and slows down the growth of the oscillations in the linear stage. At the same time, however, an additional shift in the wave phase occurs relative to the beam, so that the electrons periodically enter the focusing and defocusing phase regions. Hence, as shown by the numerical calculations, the amplitudes of the field oscillations and beam radius increase without bound. Treating the nonlinear dispersion of the plasma (the dependence of the plasma dielectric function on the beam radius) results in an additional dephasing of the beam oscillations and the field, accompanied by nonlinear saturation of the amplitude. In the region of the strong magnetic field (under the conditions of the anomalous Doppler effect) an analytical solution has been found which enables us to physically interpret the numerical calculations.

## §3.6. Nonlinear dynamics of a thin electron-ion beam in a plasma.

It is shown above that in a weak magnetic field the phase velocity of the wave is closed to the beam velocity, and the growth of the Cherenkov mode in the wavelength range $kv < \omega_p$ is accompanied by the development of electron clumps in the focusing phases of the field. The injection of such a previously bunched beam into a plasma does not result in significant particle losses in the initial stage of instability at which the clumps become charge-neutralized (curve 3 in Fig. 3.3.1).

This section studies analytically and numerically the nonlinear dynamics of a thin electron-ion beam in a plasma [12]. Just as in the case of an electron beam, the initial modulation frequencies

$$\Omega_p - \omega \gg \nu^{1/3}\Omega_p, \quad \nu = \frac{\omega_b^2 \omega_p^2}{\Omega_p^4} \ll 1,$$

(corresponding to the case when the produced bunches are traveling in the focusing phases of the wave) turns out to be most favorable.

Due to the difference between the electron $m$ and ion $M$ masses, the rapid electron oscillations arise [during the time interval $\tau_e \simeq (\omega_p \nu^{1/3})^{-1}$] on the background of the unchanged ions. The following nonlinear evolution of the ion beam is accompanied by the adiabatically slow changes in the parameters of the quasi-equilibrium electron bunches. The obtained results show that the processes in the ion stage of the instability qualitatively repeat those occurring in the electron stage, so that during the time interval $\tau_i \simeq (M/m)^{1/3}\tau_e$ a complete bunching of the electron-ion beam occurs.

To describe the instability of the cylindrical electron-ion beam (moving at a constant velocity $\mathbf{v}$ in a plasma along an external magnetic field $\mathbf{B}_0$), we use the equation for the radial component of the electric field in the form

$$E_r = m\gamma\Omega_p^2\nu^{2/3}e^{-1}\text{Re}\left[A(t)\exp i\omega(t - z/v)\right] r,$$

and following §3.3-§3.5, we use a set of nonlinear equations for the radii of electron $R_e(t, z)$ and ion $R_i(t, z)$ beams and equation for the beam

$$2i\dot{A} - \kappa A = \left\langle e^{-i\Phi}\left(\frac{1}{X^2} - \frac{1}{Y^2}\right)\right\rangle, \tag{3.6.1}$$

$$\ddot{X} + \left[\mu - (\mu + \eta_e)\frac{1}{X^4} + \text{Re}\left(Ae^{i\Phi}\right)\right]X = 0, \tag{3.6.2}$$

$$\ddot{Y} - \sigma\left[\frac{\eta_i}{Y^4} + \text{Re}\left(Ae^{i\Phi}\right)\right]Y = 0, \tag{3.6.3}$$

where the following notation for dimensionless quantities is used:

$$X = \frac{R_e}{R_0}, \quad Y = \frac{R_i}{R_0}, \quad \kappa = \frac{k^2v^2 - \Omega_p^2}{\Omega_p^2\nu^{1/3}},$$

$$\mu = \frac{\omega_B^2}{4\Omega_p^2\nu^{2/3}}, \quad \eta_{e,i} = \frac{v_{Te,i}^2}{4\Omega_p^2R_0^2\nu^{2/3}}, \quad \sigma = \frac{m}{M}.$$

Here, we denote by $v_{Te,i}$ the electron and ion thermal velocities, and the point denotes the derivative with respect to $\tau = \nu^{1/3}\Omega_p t$. Also, the effect of the magnetic field on the ion motion is neglected due to a larger ion mass ($\omega_B \ll MT_i/mR_0^2$).

120

We assume the beam electron temperature to be fairly small $\eta_e \ll \mu$ to neglect the thermal electron motion. Then, the initial force-free equilibria of the electron beam can be characterized by the parameters $X = 1$, $A = 0$. Defocusing of the ion beam occurs due to the gradient in the thermal pressure (proportional to $\eta_i/Y^4$). In the electron stage of the instability, we can neglect changes in the ion beam radius ($\Delta Y \sim \sigma \ll 1$), and the dispersion relation for the small oscillations, $A \sim \exp(i\Delta\tau)$, coincides with that for the electron beam [see (3.4.7)].

$$(2\Delta + \kappa)\left(\Delta^2 - 4\mu\right) = 1$$

which has complex roots. The instability has a threshold and takes place for $\kappa < 3$.

Below, we will show that the behavior of the nonlinear system is determined by the value of the linear growth rate. That is why the solutions are classified with respect to the value of $\kappa$.

Let us consider the equation for the field (3.6.1) for large negative value of $-\kappa$. In this case, the term $2i\dot{A}$ can be omitted, the field amplitude is a real function of $\tau$, and there is practically no displacement between the wave and the beam.

In the first stage of the instability ($\tau \leq 10^2$), [3]. the changes in the ion beam radius can be neglected. The formation of the bunched structure of the electron beam proceeds in this wave: the particles located in the field in the regions with $\pi/2 \leq \Phi \leq 3\pi/2$ are focused by the wave fields, whereas, in the regions with $-\pi/2 \leq \Phi \leq \pi/2$, irreversible beam defocusing takes place. Detuning of the nonlinear oscillations in various beam cross sections (different wave phases) is accompanied by the decrease in the temporal oscillations of the field amplitude with respect to its average value (Figs. 3.6.1a and 3.6.1b).

After the electron bunching stage, within the time interval $\tau \sim 10^3$, the similar bunching of the ions develops; meanwhile, the radius of the electron bunches oscillates near its equilibrium value. This equilibrium value decreases slowly with the increase in the field (Figs. 3.6.1c and 3.6.1d). Because the focusing phases for electrons and ions are displaced in phase by $\pi$, at the end of the ion stage, we get a train of bunches with an alternating order of electron and ion bunches (Fig. 3.6.1e).

The slow ion stage of the instability can be described in terms of the analytical theory. Due to presence of a new small parameter $\sigma \ll 1$ in equation (3.6.3), we can separate the slow-scale dependence $\dot{A}/A \sim \sqrt{\sigma} \ll 1$. As a result, we can integrate the equation for the electron radius (3.6.2) (for constant field amplitude) to obtain

$$X^2(\tau, \Phi) = \sqrt{X_0^4(\Phi) + b^2} + b\cos\vartheta, \quad \vartheta = \omega_0\tau + \alpha, \quad \omega_0 = 2\sqrt{\mu}/X_0^2(\Phi), \tag{3.6.4}$$

where $\alpha$ and $b$ are the constant of integration and the parameter $X_0(\Phi)$ is defined by

$$X_0(\Phi) = \left(1 + \frac{A}{\mu}\cos\Phi\right)^{-1/4} \tag{3.6.5}$$

and corresponds to the force-free beam equilibria. Expression (3.6.5) can be found from (3.6.2) for $\ddot{X}(\Phi) = 0$ (see **Appendix 3.2**). For $A/\mu > 1$ formulas (3.6.4) and (3.6.5) determine the condition when the beam radius is finite

$$|\Phi| < \Phi_0 = \begin{cases} \pi, & A/\mu < 1, \\ \arccos(-\mu/A), & A/\mu > 1. \end{cases} \tag{3.6.6}$$

---

[3] We note that, in the case under consideration, the instability growth rate is small in comparison with its maximum value and, consequently, the time of the nonlinear evolution of the mode under consideration, is about $\tau \simeq 30\sqrt{|\kappa|}$.

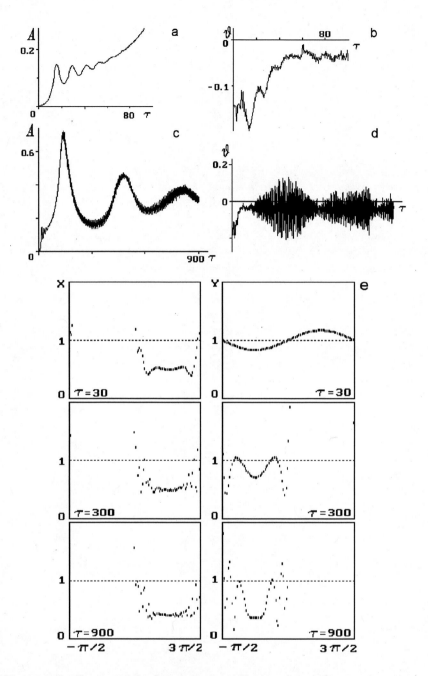

Fig. 3.6.1. Development of the electron-ion instability in a plasma when $\kappa=-10$, $\mu=0.09$. Evolution of the amplitude (a) and phase (b) of the field in the electron $\tau\lesssim10^2$ and ion [(c) and (d), for $\tau\lesssim10^3$] stages of instability. Plot (e) shows the evolution of the electron $X(\tau,\Phi)$ and ion $Y(\tau,\Phi)$ beam radii along a perturbation wavelength.

Substituting the electron radius (3.6.4) into the equation for the field (3.6.1) yields the following integral:

$$J(\tau) = \frac{1}{2\pi} \int\limits_{-\Phi_0}^{\Phi_0} e^{-i\Phi}\, \frac{d\Phi}{X^2(\Phi,\tau)}. \qquad (3.6.7)$$

122

For $\tau \gg \omega_0^{-1}$, at the transition to the ion stage of the instability, the method of the stationary phase [13] can be used to calculate the asymptotics of integral (3.6.7). Incorporating the periodic nature of the function $X^{-2}$ $\tau$ with respect to the variable $\tau$, we expand this function in a Fourier series with respect to $\tau$

$$X^{-2}(\tau, \Phi) = \sum_{n=-\infty}^{\infty} B_n(\Phi) \exp[in\omega_0(\Phi)\tau].\tag{3.6.8}$$

Substituting (3.6.8) into (3.6.7), we have the integrals of the form

$$J_n(\tau) = \int_{-\Phi_0}^{\Phi_0} B_n(\Phi) \exp[in\omega_0(\Phi)\tau - i\Phi]\, d\Phi$$

$$\simeq B_n(0) \left(\frac{\sqrt{\mu + A}}{nA\tau}\right)^{1/2}.$$

For $\tau \to \infty$, the amplitudes of harmonics with $|n| \geq 1$ are small and the main contribution to the integral (3.6.7) comes from the term with n = 0:

$$J(\tau) = \frac{1}{2\pi} \int_{-\Phi_0}^{\Phi_0} B_0(\Phi) \exp(-i\Phi)\, d\Phi, \quad B_0 = \frac{1}{2\pi} \int_0^{2\pi} \frac{d\vartheta}{X^2(\Phi, \vartheta)}.\tag{3.6.9}$$

Substituting $X^2(\Phi, \vartheta)$ given by the formula (3.6.5) into (3.6.9) and integrating yield

$$J(A) = \frac{1}{2\pi} \int_{-\Phi_0}^{\Phi_0} \left(1 - \frac{A}{\mu} \cos \Phi\right)^{1/2} \cos \Phi\, d\Phi.\tag{3.6.10}$$

Comparing (3.6.10) and the initial expression (3.6.7), we can see that the electron contribution is determined by the value of the electron equilibria radius and does not depend on the amplitude $b$ or on the phase $\vartheta$ of the wave [(3.6.4) and (3.6.5)]. In the ion stage of the instability the field amplitude is governed by the equation (3.6.1) and slowly varies with $\tau$. Calculating the value of integral (3.6.10) and substituting the result in equation (3.6.1), we get

$$\kappa A + J(A) = \frac{1}{2\pi} \int_{-\pi/2}^{3\pi/2} \frac{\cos \Phi}{Y^2}\, d\Phi,$$

$$\tag{3.6.11}$$

$$J(A) = \frac{1}{3\pi} \sqrt{\frac{8A}{\mu}} \left[(2k^2 - 1)E(k) + (1 - k^2)K(k)\right],$$

where $K(k)$ and $E(k)$ are the complete elliptic integrals with argument with argument $k = \sqrt{(1 + \mu/A)/2}$.

The ion stage of the instability can be described by equation (3.6.11) along with the equation for the electron motion

$$\ddot{Y} = \sigma \left(\frac{\eta_i}{Y^4} + A \cos \Phi\right) Y.\tag{3.6.12}$$

123

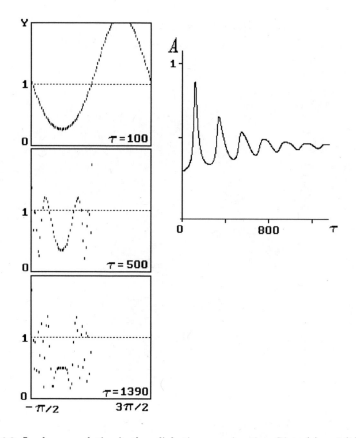

Fig. 3.6.2. Ion beam evolution in the adiabatic approximation. Plots (a) and (b) show evolutions of the field amplitude $A(\tau)$ and ion beam radius $Y(\tau, \Phi)$.

The analytical description of the initial part of this stage can be introduced by the expansion of (3.6.11) and (3.6.12) in powers of the small parameter $\sigma$. In the zeroth approximation, which corresponds to an electron equilibrium with immobile background ions, we have

$$\kappa A_0 + J(A_0) = 0, \quad Y = 1. \tag{3.6.13}$$

For small perturbations of the equilibria, $Y = 1 + y$, $|y| \ll 1$, integrating (3.6.12) yields

$$y = \frac{\sigma \tau^2}{2}(\eta_i + A_0 \cos \Phi). \tag{3.6.14}$$

Correspondingly, the characteristic time for ion-beam distortions to develop under the action of the electron-mode field is $\tau \simeq \sigma^{-1/2}$. We note that during this time the ion-beam instability with a maximum growth rate can develop. However, the effect of this mode on the process under consideration is not essential while the condition $\sigma^{1/6} < \ln(A_0/A_{oi})$ holds (where $A_{oi}$ is the perturbation amplitude).

Substituting (3.6.14) into (3.6.11) yields

$$A = A_0 \left(1 - \frac{\sigma \tau^2}{2\tilde{\kappa}}\right), \quad \tilde{\kappa} = \kappa + \frac{dJ(A_0)}{dA_0}. \tag{3.6.15}$$

Our investigation shows (see **Appendix 5.4**) that there is $\tilde{\kappa} < 0$ in the case under consideration. This corresponds to the temporal growth in the perturbation amplitude.

124

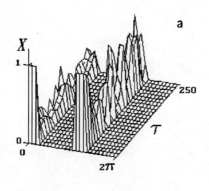

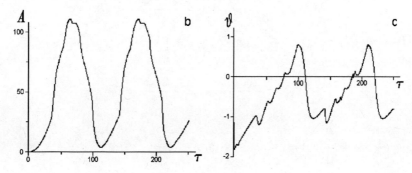

Рис.3.6.3. Evolution of the electron $X(\tau,\Phi)$ and ion $Y(\tau,\Phi)$ beam radii (a), the amplitude (b) and phase (c) of the field along a perturbation wavelength when $\kappa=-0.2$, $\mu=0.09$.

Figure 3.6.2 shows the numerical solution of (3.6.11) and (3.6.12) that are in a good agreement with the exact solution (see Fig. 3.6.1). The exact solution demonstrates small-scale oscillations of the amplitude and phase, whereas, there are no such oscillations in the case of an approximate solution. This discrepancy seems to be due to the excitation of the eingenmode of plasma oscillations, $A \sim \exp(-i\kappa\tau/2)$. This mode is a solution of the homogeneous equation, that is, the equation (3.6.1), where the beam related terms are neglected.

In the resonance case ($\kappa = -0.2$, Fig. 3.6.3), in the fast electron stage of the instability, the temporal phase deviation results in sufficient particle losses. As a result, the formed bunches of electrons are fairly short. Consequently, the slow ion stage of the instability is weakly affected by the preceding electron stage and the process of the ion-bunch formation is much alike that in the absence of an electron beam.

**Appendix 3.1**

We consider the Vlasov equation

$$\frac{\partial f}{\partial t} + v_z \frac{\partial f}{\partial z} + v_r \frac{\partial f}{\partial r} + \left(\frac{e}{m}E_r + \frac{v_\theta^2}{r}\right)\frac{\partial f}{\partial v_r} - \frac{v_r v_\theta}{r}\frac{\partial f}{\partial v_\theta} = 0 \qquad (A3.1.1)$$

to define the system of moment equations

$$n = \int f\,d\mathbf{v}, \quad nu_i = \int v_i f\,d\mathbf{v}, \quad nT_{ik} = \int (v_i - u_i)(v_k - u_k)f\,d\mathbf{v},$$

$$n\,q_{ijk} = m\int (v_i - u_i)(v_j - u_j)(v_k - u_k)f\,d\mathbf{v}, \qquad (A3.1.2)$$

125

where $i, k \equiv r, \ \theta, \ z$.

The equations for $u_r$, $u_\theta$, $T_{rr}$ and $T_{\theta\theta}$ in the cylindrical coordinates become

$$\frac{du_r}{dt} + u_r \frac{\partial u_r}{\partial r} = \frac{e}{m} E_r + \frac{u_\theta^2 + T_{\theta\theta} - T_{rr}}{r} - \frac{1}{n} \left[ \frac{\partial}{\partial z} \left( n T_{rz} \right) + \frac{\partial}{\partial r} \left( n T_{rr} \right) \right],$$

$$\frac{du_\theta}{dt} + u_r \frac{\partial u_\theta}{\partial r} = -\frac{u_r u_\theta + 2 T_{r\theta}}{r} - \frac{1}{n} \left[ \frac{\partial}{\partial z} \left( n T_{\theta z} \right) + \frac{\partial}{\partial r} \left( n T_{r\theta} \right) \right],$$

$$\frac{dT_{rr}}{dt} + u_r \frac{\partial T_{rr}}{\partial r} + 2 \left( T_{rr} \frac{\partial u_r}{\partial r} + T_{rz} \frac{\partial u_r}{\partial z} \right) - 4 \frac{u_\theta T_{r\theta}}{r}$$

$$+ \frac{1}{n} \left[ \frac{\partial}{\partial z} \left( n q_{rrz} \right) + \frac{\partial}{\partial r} \left( n q_{rrr} \right) \right] + \frac{q_{rrr} - 2 q_{rr\theta}}{r} = 0, \qquad \text{(A3.1.3)}$$

$$\frac{dT_{\theta\theta}}{dt} + u_r \frac{\partial T_{\theta\theta}}{\partial r} + 2 \left[ T_{\theta\theta} \frac{u_r}{r} + T_{\theta z} \frac{\partial u_\theta}{\partial z} + T_{r\theta} \frac{1}{r} \frac{\partial}{\partial r} \left( r u_r \right) \right] - 4 \frac{u_\theta T_{r\theta}}{r}$$

$$+ \frac{1}{n} \left[ \frac{\partial}{\partial z} \left( n q_{\theta\theta z} \right) + \frac{\partial}{\partial r} \left( n q_{\theta\theta r} \right) \right] + \frac{3 q_{r\theta\theta}}{r} = 0,$$

where $d/dt = \partial/\partial t + u_z \partial/\partial z$.

To proceed, we assume that $\partial/\partial z = 0$, $q_{ijk} = 0$, and $T_{ik} = 0$ if $i \neq k$, and we reduce Eqs. (A3.1.3) to the form

$$\frac{du_\theta}{dt} + u_r \frac{\partial u_\theta}{\partial r} = -\frac{u_r u_\theta}{r},$$

$$\frac{du_r}{dt} + u_r \frac{\partial u_r}{\partial r} = \frac{e}{m} E_r + \frac{u_\theta^2 + T_{\theta\theta} - T_{rr}}{r} - \frac{1}{n} \frac{\partial}{\partial r} \left( n T_{rr} \right), \qquad \text{(A3.1.4)}$$

$$\frac{dT_{rr}}{dt} + u_r \frac{\partial T_{rr}}{\partial r} + 2 T_{rr} \frac{\partial u_r}{\partial r} = 0, \quad \frac{dT_{\theta\theta}}{dt} + u_r \frac{\partial T_{\theta\theta}}{\partial r} + 2 T_{\theta\theta} \frac{u_r}{r} = 0.$$

In the special case $u_r = r \Omega_r$, the equations for $T_{rr}, T_{\theta\theta}$ have the same form as the continuity equation:

$$\frac{dT_{rr}}{dt} + u_r \frac{\partial T_{rr}}{\partial r} + 2 T_{rr} \Omega_r = 0, \quad \frac{dT_{\theta\theta}}{dt} + u_r \frac{\partial T_{\theta\theta}}{\partial r} + 2 T_{\theta\theta} \Omega_r = 0. \qquad \text{(A3.1.5)}$$

Note, that the equations for $T_{ik}$ in the Cartesian coordinates have the exact form:

$$\left( \partial/\partial t + u_m \nabla_m \right) T_{ik} + T_{im} \nabla_m u_k + T_{km} \nabla_m u_i + \frac{1}{n} \nabla_m \left( n q_{ikm} \right) = 0. \qquad \text{(A3.1.6)}$$

# CHAPTER 4

## STEADY INJECTION OF A MODULATED ELECTRON BEAM INTO A PLASMA

Considered in Chapters 1 thru 3, the "temporary" model of instability of electron beam in a plasma pre-supposes that the self-contained equation set solution is a monochromatic wave with the fixed wave number and a complex frequency that corresponds to the Langmuir perturbation amplitude growth vs. time steadily all over the length of the system.

An alternative problem formulation is with a fixed frequency and a complex wave number, corresponding to the steady-state injection of a modulated electron beam in the plasma semi-space [1-6]. The small longitudinal Langmuir perturbation on the plasma boundary grows exponentially with the coordinate to be stabilized by the plasma [1] or beam [2,3] nonlinearity. If the both phenomena are values of the same order, then the beam slows down synchronously with decreasing of the wave phase velocity, and the synchronism of resonance electrons with the wave remains on the non-linear stage of instability. This is why the amplitude of longitudinal oscillations increases relative to the linear plasma [4].

Thin beam steady-state injection into plasma along the constant magnetic field is accompanied by spatial enhancement of the quasi-transverse oscillations (in similarity in Chap. 3 to the "temporary" problem). Depending on the beam modulation frequency there are two different regimes of beam interaction with plasma. Unstable perturbations grow exponentially with the coordinate and are stabilized on account of beam electrons oscillating out of sync in different field phases. In the region of stability the applied electric field at the beam input into plasma provides for the radial beam focusing by the high-frequency pressure gradient [5].

This Chapter elaborates on the non-linear theory of regular oscillation enhancement in a semi-finite plasma, using a modulated electron beam. In the case of longitudinal Langmuir perturbations the plasma and beam nonlinearity is taken into account so that the papers [1] and [2] are the limiting cases of the common problem. Beam interactions with the transverse Langmuir oscillations (radial focusing or defocusing) are considered in the thin beam approximation.

### §4.1. Amplification of a regular wave by an electron beam at the nonlinear phase resonance in a plasma.

*1. Introduction.*

It is well known that, when an electron beam with sinusoidally perturbed longitudinal parameters is injected into a plasma, the amplitude of the small perturbations increases with the rate [1,2]

$$\kappa = k_0 \begin{cases} \sqrt{3}(\alpha\nu/16)^{1/3}, & |\epsilon| \ll (\nu/\alpha^2)^{1/3} \\ \sqrt{\nu/|\epsilon|}, & |\epsilon| \gg (\nu/\alpha^2)^{1/3} \end{cases} \qquad (4.1.1)$$

where $k_0 = \omega/v_0$, $\omega$ and $v_0$ are the modulation frequency and the velocity of the electron beam, $\nu = n_b/n_p$, $n_b$ and $n_p$ are the beam and plasma densities, $\alpha = v_0/v_g$, $v_g$ is the group velocity of the wave exited in the plasma, $\epsilon = 1 - \omega_p^2/\omega^2$ is the dielectric function, and $\omega_p = 4\pi e^2 n_p/m$ is the plasma frequency.

Since the equations of motion for plasma particles are nonlinear, the plasma density depends on the field amplitude, which results in a deviation of the frequency of the plasma oscillations from the resonant value $\omega \simeq \omega_p$ [1]. Additionally, the phase matching $\omega = kv_0$ between the beam electrons and the wave breaks down by virtue of the nonlinearity of the beam electron motion [2]. The relative contribution of these effects to the amplitude saturation of the field is governed by the nonlinearity parameter

$$Q = \nu \left(\frac{\alpha}{2}\right)^3. \qquad (4.1.2)$$

A beam with a low energy-flux density $Q \ll 1$ excites a wave with a constant phase velocity. In the field of this wave, the beam separates into individual bunches. Because of the periodic acceleration and deceleration of these bunches, no further energy exchange between the beam and the wave occurs over the period of the nonlinear oscillations [2]. In the opposite limit $Q \gg 1$, the motion of an electron beam is linear, and the instability is suppressed because the beam electrons are expelled (due to the high-frequency pressure gradient) from the regions in which the wave field is maximum [1].

In the range $Q \simeq 1$, both the beam and plasma nonlinearities are important. The time-dependent problem for a train of electron bunches was solved analytically in [4]. In that paper, it was shown that, when nonlinear phase velocity $v_{NL} \simeq \omega_p(E)/k_0$ of the wave and the velocity of the bunches decrease synchronously, $v(t) = v_{NL}$, the rate with which the field amplitude increases is substantially higher than in the case of a linear plasma response.

This effect is important for optimizing the parameters of the decelerating down system consisting of sequentially located resonators filled by a plasma with imposed strong longitudinal magnetic field [6]. The physical mechanism responsible for the conversion of the kinetic energy of the beam into the energy of the electromagnetic field is the plasma-beam instability, and the maximum amplitude of the electric-field of the excited plasma wave is governed by the trapping of electron bunches by this wave.

A set of nonlinear equations for the slowly varying amplitudes with allowance for the dependence of the plasma density on the field amplitude have analyzed in the paper [4]. It is shown that the maximum field amplitude increases from $E_L = 1.5mv_0\omega_p e^{-1}(\alpha\nu)^{2/3}$ (in the linear case) to $E_{NL} = 5.54E_L$ (in the nonlinear case for the parameters $Q = 1.45$ and $\epsilon = -0.8$). Accordingly, at resonance, the efficiency $\eta \sim E^2$ of the plasma-beam amplifier increases to the value $\eta_{NL}/\eta_L \simeq 30.7$.

The calculations show that 72% of the beam electrons are trapped by the wave with decreasing phase velocity. Hence, further increase in the efficiency of the system can be ensured by the use of a preliminary modulated beam in order to group all the beam electrons in the decelerating regions of the wave. For $Q = 1$ and $\epsilon = 0$, the bunches whose sizes are equal to $\Delta\psi = \pi/5k_0$ in phase space are completely trapped by the wave excited in a plasma, and the efficiency of the plasma-modulated-beam amplifier increases to $\eta_{mod}/\eta_{NL} \simeq 1.6$.

## 2. Set of averaged (slow) equations.

We write the current density of an electron beam injected in the plasma ($x > 0$) in the form [12]

$$j_b(t, x) = \frac{2\pi j_0}{\omega S} \sum_{s=-\infty}^{\infty} \delta \left[ t - t_s(x) \right], \qquad (4.1.3)$$

where the function $t_s(x)$ satisfies the equation of motion

$$\frac{d^2 t_s}{dx^2} = -\frac{e}{m} \left( \frac{dt_s}{dx} \right)^3 E \left[ x, t_s(x) \right], \qquad (4.1.4)$$

$\omega$ is the modulation frequency of the beam, $S$ is the number of bunches injected during a time equal to $2\pi/\omega$ and the quantity

$$< j_b >= \frac{\omega}{2\pi} \int_0^{2\pi/\omega} j_b \, dt = j_0$$

is the integral of motion for the space problem [unlike in the time-dependent problem, in which the charge is conserved over the spatial period (the wavelength) of the beam modulation].

A high-frequency electric field excited by the beam in the plasma is described by the equation

$$\frac{\partial}{\partial t} \left[ \hat{F} + \omega_e^2(\varphi) \right] E = -4\pi e \hat{F} j_b, \qquad (4.1.5)$$

where

$$\hat{F} = \partial^2/\partial t^2 - v_T^2 \partial^2/\partial x^2,$$

$$\omega_e^2(\varphi) = 4\pi e^2 n_e(\varphi)/m, \quad v_T^2 = T/m,$$

and $T$ is the plasma temperature.

The electrostatic potential $\varphi$, which is caused by the charge separation in the plasma, and the plasma electron density $n_e(\varphi)$ are determined by the Poisson equation and the condition for plasma equilibrium in the high-frequency electric field,

$$\frac{\partial^2 \varphi}{\partial x^2} = -4\pi e(n_e - n_i), \quad n_i = n_p \exp(e\varphi/T),$$

$$\frac{d}{dx} \left[ \frac{m}{2} \left\langle v_e^2(t, x) \right\rangle + e\varphi \right] = -\frac{T}{n_e} \frac{dn_e}{dx}, \qquad (4.1.6)$$

where $n_p$ is the unperturbed plasma density.

In a weakly nonuniform high-frequency field

$$E(t, x) = \text{Re} \left[ E(x) \exp(i\omega t - ik_0 x) \right], \quad E'(x) \ll k_0 E(x) \qquad (4.1.7)$$

plasma electrons are accelerated to the velocity

$$v_e(t, x) = \frac{1}{m\omega} \text{Re} \left[ iE(x) \exp(i\omega t - ik_0 x) \right]$$

and the second equation (4.1.6) yields

$$n_e = n_p \exp \left[ -(e\varphi + \phi)/T \right], \quad \phi = e^2 |E|^2 / 4m\omega^2. \qquad (4.1.8)$$

129

If the wavelength $k_0^{-1}$ of the unstable mode is larger than the Debye radius $v_T/\omega_p$ ($v_T/v_0 \ll 1$), the second derivative of the potential in the first equation from (4.1.6) can be neglected, and the condition that the plasma be quasi-neutral, $n_e = n_i$, yields $e\varphi = -\phi/2$.

From expression (4.1.8), we find the plasma frequency as a function of the field amplitude [7],

$$\omega_p^2(E) = \omega_p^2 \exp\left(-\frac{|E|^2}{E_T^2}\right), \quad E_T^2 = \frac{8m\omega^2 T}{e^2}, \tag{4.1.9}$$

We substitute (4.1.7) into (4.1.5) and average the resulting equation over the spatial period of the beam modulation. As a result, we obtain the following equation for a slowly varying complex field amplitude:

$$2ik_0 v_T^2 E' + \left[-\omega^2 + k_0^2 v_T^2 + \omega_p^2(|E|)\right]E = \frac{8\pi e\omega_p^2}{i\omega} < j_b \exp(ik_0 x - i\omega t) > . \tag{4.1.10}$$

We assume that electron bunches are injected into the plasma at uniform time intervals $\Delta t = 2\pi/\omega S$ and set $t_s(x) = x/v_0 + s\Delta t + \tau_s(x)$, where the term $\tau_s(x)$ describes the deviation of the velocity of electron bunches from its constant value. From (4.1.3), (4.1.4), and (4.1.10), we obtain the closed set of nonlinear equations

$$2k_0 v_T^2 E' - i\left[-\omega^2 + k_0^2 v_T^2 + \omega_p^2 \exp(-|E|^2/E_T^2)\right]E = -\frac{8\pi e\omega_p^2 j_0}{\omega S}\sum_{s=1}^{S}\exp(-i\psi_s),$$

$$\tau_s'' = -\frac{1}{mv_0^3}\left(1 + v_0\tau_s'\right)^3 \operatorname{Re} E \exp(i\psi_s), \tag{4.1.11}$$

where $\psi_s = 2\pi s/S + \omega\tau_s(x)$ and $\omega_p \equiv \omega_p(0)$.

### 3. Analytical solution.

In the dimensionless variables

$$W = \frac{eE}{mv_0\omega}, \quad W_T = \frac{eE_T}{mv_0\omega}, \quad \nu = \frac{j_0}{n_p v_0}, \quad \alpha = \frac{v_0}{v_g}, \quad \xi = k_0 x$$

equations (4.1.11) take the form

$$W' + i\frac{\alpha}{2}\left[\epsilon_p + 1 - \exp\left(-\frac{|W|^2}{W_T^2}\right)\right]W = -\frac{\alpha\nu}{S}\sum_{s=1}^{S}\exp(-i\psi_s),$$

$$\psi_s'' = -(1 + \psi_s')^3 \operatorname{Re} W \exp(i\psi_s), \tag{4.1.12}$$

where

$$\epsilon_p = 1 - \omega_L^2/\omega^2, \quad \omega_L^2 = \omega_p^2 + k^2 v_T^2, \quad v_g = v_T^2/v_0.$$

Equations (4.1.12) have the integral

$$|W|^2 = \frac{\alpha\nu}{S}\sum_{s=1}^{S}\left[1 - (1 + \psi_s')^{-2}\right], \tag{4.1.13}$$

which satisfies the initial condition $W = \psi_s' = 0$.

130

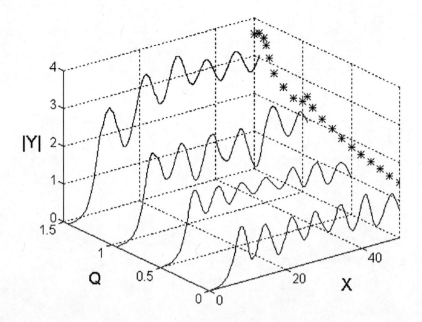

Fig. 4.1.1. Coordinate dependence of the electric-field amplitude for various values of the parameter $Q$ at $D=0$. The stars correspond to the maximum values of $|Y|$ in the computation region.

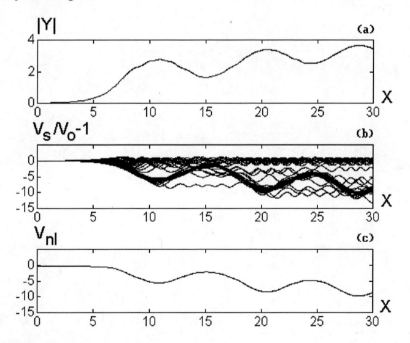

Fig. 4.1.2. (a) Electric-field amplitude, (b) beam electron velocity, and (c) nonlinear phase velocity of the wave in the beam-at-rest frame of reference for $Q=1$ and $D=0$. 60% of beam electrons are trapped by the wave.

If only one ($S = 1$) electron bunch is injected into the plasma during each period of the beam modulation, equations (4.1.12) can be solved analytically (by analogy with the time-dependent problem, §1.1).

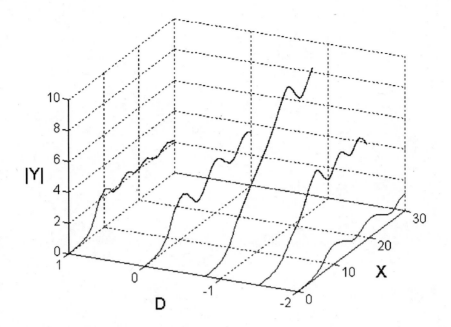

Fig. 4.1.3. Coordinate dependence of the electric-field amplitude for various values of the parameter $D$ at $Q=1.45$. The maximum value of the function $|Y|$ relalates $D=-0.8$.

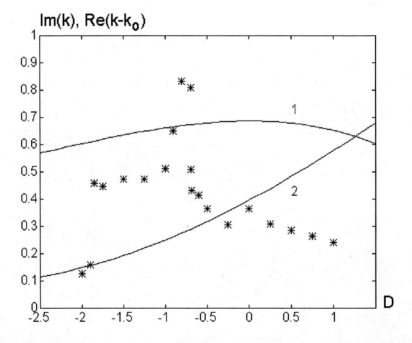

Fig. 4.1.4. (1) Imaginary and (2) real parts of the correction to the wave vector $k$ as functions of the parameter $D$ for $Q=1.45$. The stars correspond to the maximum values of $|Y|/10$ in the computation region.

We represent the electric field in the form $W = A\exp(i\vartheta)$ and separate the real and

132

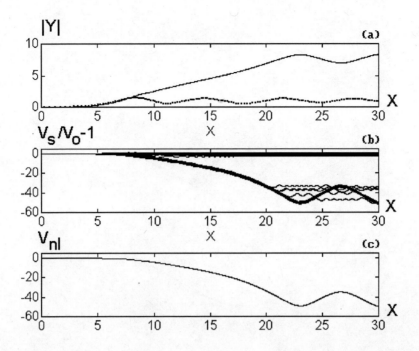

Fig. 4.1.5. (a) Electric-field amplitude in the case of a nonlinear (solid curve) and linear (dotted curve) plasma response, (b) beam electron velocity, and (c) nonlinear phase velocity of the wave in the beam-at-rest frame of reference for $Q$=1.45 and $D = -0.8$.

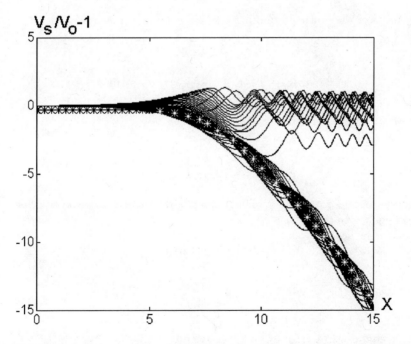

Fig. 4.1.6. Trapping of electron bunches for $Q$=1.45 and $D$=−0.8. The stars denote the nonlinear phase velocity of the wave. 72% of beam electrons are trapped by the wave.

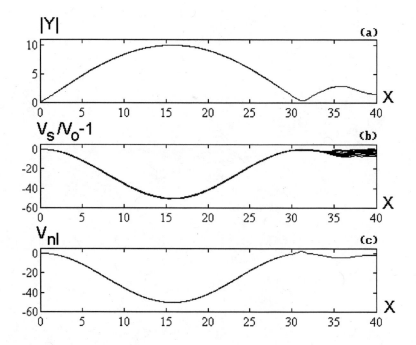

Fig. 4.1.7. (a) Electric-field amplitude in the case of a nonlinear (solid curve) and linear (dotted curve) plasma response, (b) beam electron velocity, and (c) nonlinear phase velocity of the wave in the case of a train of bunches for $Q=1$ and $D = -2^{-4/3}$. All beam electrons are trapped by the wave.

imaginary parts of the first equation in (4.1.12),

$$A' = -\alpha\nu\cos(\vartheta + \psi),$$

$$\vartheta' = -\alpha\left[\epsilon_p + 1 - \exp\left(-\frac{A^2}{W_T^2}\right)\right] + \frac{\alpha\nu}{A}\sin(\vartheta + \psi). \tag{4.1.14}$$

We rewrite integral (4.1.13) in the form

$$\psi' = \left(1 - \frac{A^2}{\alpha\nu}\right)^{-1/2} - 1 \tag{4.1.15}$$

and pass over the phase $\eta = \vartheta + \psi$ of the bunch with respect to the wave in order to lower the order of equations (4.1.14),

$$A' = -\alpha\nu\cos\eta,$$

$$\eta' = \frac{1}{\sqrt{1 - \dfrac{A^2}{\alpha\nu}}} - 1 - \alpha\left[\epsilon_p + 1 - \exp\left(-\frac{A^2}{W_T^2}\right)\right] + \frac{\alpha\nu}{A}\sin\eta. \tag{4.1.16}$$

We integrate equations (4.1.16) to find "phase" integral

$$2\alpha\nu A\sin\eta = 2\alpha\nu\left[\left(1 - \frac{A^2}{\alpha\nu}\right)^{1/2} - 1\right] + A^2 + \frac{\alpha}{2}\left[(\epsilon_p + 1)A^2 - W_T^2\left(1 - \exp\left(-\frac{A^2}{W_T^2}\right)\right)\right]. \tag{4.1.17}$$

We assume that the inequalities

$$\frac{A^2}{\alpha\nu} \ll 1, \quad \frac{A^2}{W_T^2} \ll 1$$

hold and expand the right-hand side of integral (4.1.17) in powers of $A^2$,

$$A' = -\frac{\kappa}{2}E_b\cos\eta,$$

$$\sin\eta = 2\delta\frac{A}{E_b} - (1-Q)\left(\frac{A}{E_b}\right)^3 - 2\kappa\left(1+\frac{2}{3\alpha}Q^2\right)\left(\frac{A}{E_b}\right)^5, \tag{4.1.18}$$

$$\kappa = (\alpha\nu)^{1/3}, \; E_b = 2\kappa^2, \; \delta = \frac{\alpha\epsilon_p}{\kappa}, \; Q = \frac{\nu\alpha^3}{8}.$$

Passing over from (4.1.17) to (4.1.18), we set $\omega \simeq \omega_p$ and $W_T^2 \simeq 8/\alpha$.

According to (4.1.18), the beam ($Q < 1$) or plasma ($Q > 1$) nonlinearities prevent the linear increase in the field amplitude when the phase shift of the bunch with respect to the wave reaches the value $|\eta_m| = \pi/2$. At the nonlinear phase resonance ($Q \to 1$), when the bunch velocity and the phase velocity of the wave decrease synchronously, and the phase shift is governed by the higher-order terms in expansion (4.1.18), the field amplitude increases. When $\delta = 0$, the asymptotic solutions to the algebraic equation (4.1.18) have the form

$$\frac{A_m}{E_b} = \begin{cases} |1-Q|^{-1/3}, & |1-Q| \gg (2\kappa)^{3/5} \\ (2\kappa)^{-1/5}, & |1-Q| \ll (2\kappa)^{3/5}. \end{cases} \tag{4.1.19}$$

In the dimensionless variables, we have $E_m = (mv_0\omega_p/e)A_m$, and the ratio between the energy-flux densities of the field and of the beam is

$$\frac{v_g E_m^2}{4\pi n_b m v_0^3} = \begin{cases} 4\kappa(1-Q)^{-2/3}, & |1-Q| \gg (2\kappa)^{3/5} \\ 2^{9/5}\nu^{2/15}, & |1-Q| \ll (2\kappa)^{3/5}. \end{cases} \tag{4.1.20}$$

We have used the relationships

$$Q = 1, \; \alpha = 2/\nu^{1/3}, \; \kappa = 2^{1/3}\nu^{2/9}, \; A_m = 2^{7/2}\nu^{2/5}.$$

to eliminate the parameter $\alpha$ from the second asymptotic expression in (4.1.20).

When electron bunches are injected into the plasma at uniform time intervals, the density of the field energy can be higher than that of the beam energy [2]. However, from the conditions for the applicability of asymptotic equations (4.1.12), it follows that the ratio (4.1.20) between the energy-flux densities of the field and of the beam should, as before, be small.

### 4. Numerical integration.

When the number of electron bunches injected into the plasma during each period of the beam modulation is large ($S \gg 1$), we can use nonlinear equations (4.1.12) to describe the injection of a continuous electron beam into the plasma with the required accuracy [8]. Below, the results of numerical integration of these equations by a macroparticle method (the number of macroparticles was $S = 50$).

135

In terms of the slowly varying variables

$$Y = W/\kappa^2, \ X = \kappa\xi, \ Q = \alpha^3\nu/8, \ D = \alpha\epsilon_p/\kappa,$$

correct to small terms of order $\kappa \ll 1$, equations (4.1.12) can be written in the form

$$Y' + (i/2)\left(D + Q|Y|^2\right)Y = -S^{-1}\sum_{s=1}^{S}\exp(-i\psi_s), \tag{4.1.21}$$

$$\psi_s'' = -\operatorname{Re}Y\exp(i\psi_s).$$

According to expressions (4.1.18), the higher-order terms of expansion are important only for bunches whose size is zero in phase space when there is a singularity at $Q = 1$.

In the linear approximation, we have

$$\psi_s = 2s/S + \psi_{s1}, \ \psi_{so} \gg |\psi_{s1}|, \ |D| \gg Q|Y|^2$$

and equations (4.1.21) yield the equation for the field amplitude

$$2Y''' + iDY'' + iY = 0, \tag{4.1.22}$$

which is a familiar hydrodynamic equation. We must use the solution to equation (4.1.22),

$$Y = Y_0\exp(-i\Delta X), \tag{4.1.23}$$

where

$$\Delta = \frac{\delta}{6} + \frac{1}{2}(u + v) + \frac{i\sqrt{3}}{2}(u - v),$$

$$u = (-q + \sqrt{d})^{1/3}, \qquad u = (-q - \sqrt{d})^{1/3},$$

$$q = (D/6)^3 - 1/4, \quad d = [1 - (D/3)^3]/16$$

in order to choose the following, physically correct, initial conditions for the equations of motion:

$$\psi_{s1} = \operatorname{Re}\left[(Y/\Delta^2)\exp(i\psi_{so})\right], \psi_{s1}' = -\operatorname{Re}\left[(iY/\Delta)\exp(i\psi_{so})\right]. \tag{4.1.24}$$

Numerical solutions of equations (4.1.21) for the resonant harmonic with $D = 0$ are shown in Fig. 4.1.1. The field amplitude increases with the parameter $Q$ from $|Y_L| = 1.5$ (in the linear case) to $|Y_{NL}| = 3.64$ at $Q = 1.45$ when 60% of beam electrons are trapped by the wave with a decreasing phase velocity (Fig. 4.1.2). A further increase in $Q$ leads to a decrease in the field amplitude.

The calculations show that the conditions for the trapping of beam electrons by the wave depend on the ratio between the modulation frequency and the Langmuir frequency (Fig. 4.1.3).

For the fixed values $Q = 1.45$, the maximum field amplitude over the computation region ($|Y_{NL}| = 8.31$) corresponds to $D = -0.8$. For these parameters, 72% of beam electrons are trapped by the wave. Figure 4.1.4 displays the maximum field amplitude as a function of the parameter $D$.

In the frame of reference in which the beam particles are at rest, the nonlinear phase velocity of the wave is

$$V_{NL} = \vartheta', \quad \vartheta = -i\ln\left(Y/|Y|\right).$$

Using equation (4.1.21), we can rewrite this velocity as

$$V_{NL} = -\frac{1}{2}\left(D + Q|Y|^2\right) - \frac{1}{|Y|^2 S}\,\mathrm{Im}\sum_{s=1}^{S} Y^* \exp(-i\psi_s). \qquad (4.1.25)$$

The calculations show that the phase velocity of the wave and the velocity of the trapped bunch are essentially the same (Fig. 4.1.5 and 4.1.6). For resonant harmonics with $|D| \ll 1$, formula (4.1.25) can be approximated with a high accuracy by the expression

$$V_{NL} = -2^{-4/3}(1 + \sqrt{3}|Y|^2). \qquad (4.1.26)$$

A further increase in the field amplitude can be ensured by using a preliminary modulated beam in order to optimize the conditions for the trapping of beam electrons by the wave and to reduce the number of untrapped electrons. Figure 4.1.7 presents the calculated results for a train of bunches with the width $\Delta\psi = \pi/5$. We can see that all beam electrons are trapped by the wave, and the field amplitude reaches the value $|Y_{NL}| = 10.53$ at $D = -2^{-4/3}$ and $Q = 1$.

In this case, we choose the initial perturbation in the form

$$Y(0) = S^{-1}\sum_{s=1}^{S} \exp(-i\psi_s) \qquad (4.1.27)$$

in order to avoid a discontinuity in the phase velocity (4.1.25) when the field amplitude is small.

By analogy with formula (4.1.20), we can define the efficiency of the plasma-modulated-beam amplifier as the ratio between the energy-flux densities of the field and of the beam [5]

$$\eta_{NL} = \frac{v_g E_m^2}{4\pi n_b m v_0^3} = \kappa|Y(X)|^2. \qquad (4.1.28)$$

When the beam is preliminary modulated, we have $\max(|Y(X)|^2) = 110.87$, which exceeds the corresponding value in the linear case by a factor of approximately 49. However, we must take into account the fact that the conditions for the applicability of equations (4.1.21), $\eta_{NL} \ll 1$, should be satisfied.

### §4.2. Amplification of a transverse wave by a thin electron beam in a plasma.

It is shown above that nonlinear saturation of the amplitude of Langmuir oscillations excited in a plasma by a single-velocity beam of electrons of limited density arises as the result of beam electron capture by the wave, the disintegration of the beam into bunches, and as a result of the appearance of nonlinear oscillations on the part of the captured particles in potential wells. Such a model of plasma-beam interaction is attained in an infinitely powerful magnetic field for beams whose radius $R$ is greater than the plasma length $\lambda = 2\pi v_0/\omega_p$ ($v_0$ is the velocity of the beam and $\omega_p$ is the frequency of the plasma) thus enabling us to determine the amplitude of the longitudinal monochromatic wave generated within the plasma by the modulated electron beam.

Another formulation of the problem of interest from the standpoint of the transportation of the electron beams within the plasma is that model of a "thin" beam ($R \ll \lambda$) dealt with in Chap. 3. Such beams, moving along a finite magnetic field, build up quasitransverse Langmuir oscillations within the plasma, and since the transverse field component $E_r \simeq$

$k_\perp \varphi \simeq \varphi/R$ is considerably in excess of the longitudinal component $E_z \simeq \varphi/\lambda$ ($\varphi$ is the self-consistent potential), then the principal nonlinear effect is the radial defocusing of the beam to $R_m \simeq \lambda$. In this case, the effect of the longitudinal phase modulation of the beam makes itself apparent in the subsequent approximation of the small parameter $R/\lambda \ll$, and the beam is propagated in the plasma without significant deceleration.

The nonlinear quasitransverse oscillations have been examined in Chap. 3 within the framework of the "time" model in which the perturbation has the form of a monochromatic wave with an amplitude that is depend on time, and the development of instability is accompanied by the appearance of nonlinear oscillations in the radius of the beam, these oscillations uniform over the entare length of the beam.

In the present study we have investigated the process involved in the steady injection of a thin beam into a plasma half-space ($z > 0$), along an external magnetic field $\mathbf{B}_0$ [5]. It is assumed at the boundary of the plasma the continuously injected beam is simulated at a frequency $\omega$, with the modulation signal suppressing the development of instability at all of the remaining frequencies, and that a monochromatic wave with an amplitude changing slowly in space excited within the plasma. The buildup of the quasitransverse oscillations in such a system is describes by the self-consistent system of nonlinear partial differential equations (3.2.12) and (3.2.14) (see §3.2).

In the case of a cold beam, $T_0 = 0$F, the linear perturbations $\sim \exp(i\omega t - ikz)$ satisfy the dispersion relation

$$k = \frac{1}{v}\left(\omega \pm \sqrt{\frac{\omega_b^2}{\epsilon} + \omega_B^2}\right), \qquad (4.2.1)$$

$$\epsilon = 1 - \frac{\omega_p^2}{\omega^2 - \omega_{Bp}^2}, \qquad \omega_b^2 = \frac{4\pi e^2 n_b}{m\gamma}, \qquad \omega_p^2 = \frac{4\pi e^2 n_p}{m},$$

$$\omega_{Bp} = \frac{eB_0}{mc}, \qquad \omega_B = \frac{\omega_B}{\gamma},$$

following from Eqs. (3.2.12) and (3.2.14) and defining the relationship between the perturbation wavelength $\lambda = (\text{Re}\,k)^{-1}$ and the spatial instability increment $\alpha = \text{Im}\,k$ as a function of the beam modulation frequency.

If the beam modulation frequency $\omega$ at the inlet to the half-space satisfies inequalities

$$\epsilon < 0, \qquad \frac{\omega_b}{\epsilon} + \omega_B^2 < 0. \qquad (4.2.2)$$

it follows from (4.2.1) that we have spatial amplification of the oscillations within the system. We will assume that the modulation frequency is close to the transverse plasma-oscillation frequency $|\omega - \Omega_p| \ll \Omega_p$, with the fundamental resonance field harmonic:

$$N = \frac{\omega_b^2}{|\epsilon|} \frac{m\gamma}{e} \, \text{Re}\,[A(z)\exp i\Phi], \qquad \Phi = \omega t - kz. \qquad (4.2.3)$$

After we substitute (4.2.3) into (3.2.12) and (3.2.14), and having averaged over the oscillation period, we have [see Eqs. (3.4.4)]

$$A = \frac{1}{2\pi}\int_0^{2\pi} \frac{\exp(-i\Phi)}{X^2} \, d\Phi,$$

$$\frac{d^2 X}{d\zeta^2} + \left(\mu - \frac{\eta}{X^4} + \text{Re}[A(z)\exp i\Phi]\right) X = 0, \qquad (4.2.4)$$

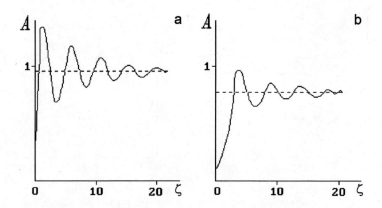

Fig. 4.2.1. Electric-field amplitude as a function of coordinate for (a) $\mu=0.1$, $\eta=\mu$ and (b) $\mu=0$, $\eta=0.1$.

where the dimensionless variables are

$$X = \frac{R}{R_0}, \quad \zeta = \frac{\omega_b z}{v\sqrt{|\epsilon|}}, \quad \mu = \frac{\omega_B^2 |\epsilon|}{4\omega_b^2}, \quad \eta = \frac{|\epsilon|}{\omega_b^2}\left(\frac{\omega_B^2}{4} + \frac{T_0}{m\gamma R_0^2}\right), \quad (4.2.5)$$

and we have put $k = \omega/v$.

In approximation of small nonlinearity, the analytical solution of (4.2.4) can be found near the stability threshold $|\delta| \ll 1$, $\delta^2 = 1 - 4\mu \ll 1$ (assuming the beam to be cold: $v_T = 0$ and $\eta = \mu$). Substituting into (4.2.4)

$$X = 1 + x_0(\zeta) + \mathrm{Re}[x_1(\zeta)\exp(i\Phi) + x_2(\zeta)\exp(2i\Phi)]. \quad (4.2.6)$$

(the higher harmonics $|x_n| \sim |x_1|^n$ for $n \geq 3$ can be neglected), we obtain the following, after standart computations:

$$x_0 = \frac{5}{4}\,x_1 x_1^*, \quad x_2 = \frac{5}{4}\,x_1^2, \quad A = x_1 - \frac{33}{8}\,x_1^2 x_1^*, \quad (4.2.7)$$

where $x_1$ satisfies the equation

$$x_1'' - \left(\delta^2 + \frac{3}{2}\,|x_1|^2\right) = 0. \quad (4.2.8)$$

Solution (4.2.8), changing in the case of $|x_1| \ll 1$ to the increasing branch of the oscillations $x_1(\zeta) = x_1(0)\exp(\delta\zeta)$, has the form

$$x_1(\zeta) = x_1(0)\,\frac{\mathrm{sh}(\delta\zeta_0)}{\mathrm{sh}[\delta(\zeta_0 - \zeta)]}, \quad \zeta_0 = \frac{1}{\delta}\,\mathrm{arsh}\left[\frac{4\delta}{\sqrt{3}x_1(0)}\right], \quad (4.2.9)$$

where $x_1(0)$ is the perturbation amplitude in the $\zeta = 0$ plane.

As is demonstrated by our study, the nonlinear addition to the increment is positive and the reverse effect of the oscillations on the beam is accompanied by an increase in the increment, i.e., the instability is explosive in nature. The plasma-beam system is therefore unstabile, even in the region of linear-theory stability $\delta^2 < 0$, provided that the amplitude of the initial perturbation is sufficiently large:

$$[x_1']^2(0) + \delta^2 x_1^2(0) - \frac{3}{4}x_1^4(0) > \frac{1}{3}|\delta|^4.$$

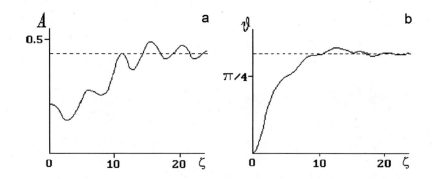

Fig. 4.2.2. (a) Amplitude and (b) phase of the field as functions of coordinate in a finite magnetic field beyond the threshold of linear-theory instability, $\mu=0.26$, and $\eta=\mu$.

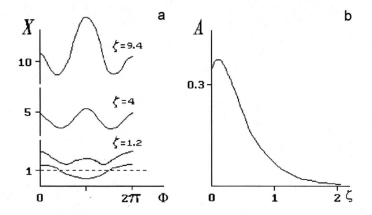

Fig. 4.2.3. (a) Beam radius modulation shape at various distances from the injection plane and (b) the field amplitude as functions of coordinate for the thermal beam, $\eta=1$, in the absence of a magnetic field $\mu=0$.

The latter contention is confirmed by the numerical results presented in Ref. [5]. Under the conditions of cold beam ($\eta = 0$) injection into a nonmagnetized plasma ($\mu = 0$), the spatial amplification of the oscillations is accompanied by the beam phase modulation and an exponential increase in the amplitude of the field as functions of coordinate. In the phase-focus region, a constriction is formed within the cross section of the beam, and this constriction, at the distance $z_{\min}$ on the order of the reciprocal of the linear-theory increment, reaches to the axis of the beam. Since the beam density increases without limit at this point, this leads to disruption of the conditions of linearity in the equations of plasma motion $n_b/n_p|\epsilon| \ll 1$. If we take into consideration that the relationship between beam density and the radius of the beam is expressed by $n_b = n_b(0)R_0^2/R^2$, we find the minimum value of the radius $R_{\min} \simeq R_0 \left(n_b/n_p|\epsilon|\right)^{1/2}$, corresponding to this constriction.

If we take into consideration the finite magnetic field which blocks the radial collapse of the electron bunches, we eliminate the point foci and can continue numerical calculation in the region of high values of $z > z_{\min}$. In this case, system of equations (4.2.4) depends only on a single parameter $\mu$. The results of numerical integration carried out for the case $\mu = 0.1$ are shown in Fig. 4.2.1a. As follows from this figure, at small distances from the injection plane the form of the function $A(\zeta)$ qualitatively reproduces the shape of the curves fo the nonmagnetized plasma. However, beginning from $\zeta \simeq 2$, considerating of the magnetic field becomes fundamentally important and leads to the appearance of phase

140

modulation, associated with the appearance of radial oscillations in the electron disks in the focus phases, as a result of the combined effect of the electric field and the radial projection of the Lorentz force. Consequently, the increase in the amplitude of the field as functions of the coordinate is slowed down and is replaced by the appearance of attenuating oscillations near the constant asymptotic value. At the same time, that portion of the electron beam injected into the defocusing phases of the wave makes virtually no contribution to the change in the field amplitude.

We should take note of the fact, although the found functions $A(\zeta)$ qualitatively replicate the earlier-derived relationships for the longitudinal oscillations (see Fig. 4.1.1), each of these effects exhibits a diverse physical nature. Indeed, if in the case of amplitude oscillation, governed by the nonlinear capture of electrons by the longitudinal field and the mixing of phases. Then in the latter case the appearance of attenuating oscillations is associated with the radial oscillations of the electron disks.

As shown by numerical calculations for large values of $\mu$, the increase in the field amplitude is accompanied by suppression of the radial electron motion and a reduction in the linear increment. As a result, the buildup process for the oscillations is decelerated [5]. With a further increase in the magnetic field and on egress beyond the threshold of linear-theory stability $\mu > 1/4$, we now have the above-noted spatial amplification of the oscillations in the nonlinear regime. The results of numerical integration for $\mu = 0.26$ can be found in Fig. 4.2.2.

The final thermal scattering, just as in the case of the longitudinal magnetic field, blocks the radial collapse of the bunches and eliminates the singularity in Eqs. (4.2.1b). In the case $\eta \ll 1$, the thermal scattering has only a slight effect on the reduction of the increment relative to the cold beam, but it proves to be significant in the region of strong electron bunch focusing, bloking the formation of the point foci.

With increase in $\eta$, the role of the transverse thermal scattering of the bunches increases, thus leading to a deceleration in the growth of the field amplitude. Figure 4.2.3 shows the case of a very large scattering $\eta \simeq 1$, where we have an irreversible dispersion of the beam, uniformly as a function of the coordinate.

### §4.3. High-frequency radial beam focusing.

We will set the beam modulation frequency $\omega$ at the inlet to the half-space so as to satisfy the condition

$$\epsilon(0) > 0, \qquad (4.3.1)$$

when the oscillations are stable [see formula (4.2.1)]. In this case, the phase velocity $v_{\text{ph}} = \omega/k$ of the wave does not coincide with the velocity $v$ of the beam and the field in the beam system becomes high-frequency. Since the field amplitude increases from the center of the beam to the boundary in proportion to $r$, then, as demonstrated in [7], we are dealing with spatial radial focusing of the beam under the action of the high-frequency ponderomotive force. In this case, the external modulating field (on entry of the beam into the plasma) plays a dual role: first of all, it suppresses the development of the beam instability and, secondly, it accomplishes the radial focusing of the beam.

Since under the conditions of the experiment the plasma is usually magnetized, we have some interest in investigating the effect of external magnetic field on the process of radial beam focusing, where that beam is injected into the plasma along the magnetic lines of force, as well as to evaluate the comparative effectiveness of the high-frequency and magnetic compression of the beam.

Taking the foregoing into consideration, we seek the solution of (4.2.4) in the form [4].

$$X = x_0(\zeta) + \mathrm{Re}[x_1(\zeta)\exp(i\omega t)], \qquad (4.3.2)$$

assuming that the beam is cold, $\eta = \mu$. Substituting (4.3.2) into Eqs. (3.2.12) and (3.2.14), and following standard transformations, we obtain

$$\left(\frac{d}{d\zeta} + i\sqrt{\epsilon}\,\frac{\omega}{\omega_b}\right)^2 x_1 + \frac{x_1}{x_0^2} = 0,$$

$$\frac{d^2 x_0}{d\zeta^2} + \mu\left(1 - \frac{1}{x_0^4}\right)x_0 + \frac{1}{2}\frac{|x_1|^2}{x_0^3} = 0. \qquad (4.3.3)$$

In the derivation of (4.3.3) we made the simplifying assumption that $\mu \ll 1$ and the corresponding terms have been dropped. Since in this section $\epsilon > 0$, it is essential that we change the sign in the right-hand side of the formula for field (4.2.4).

Assuming $x_0$ to be a slowly changing function of $z$ ($L \gg v\sqrt{\epsilon}/\omega_b$, $L$ is the characteristic length of the change in $x_0$), by means of the WKB method we find the solution for the first of equations in (4.3.3) [9]:

$$x_1(\zeta) = a\sqrt{x_0}\exp\left\{-i\left[\sqrt{\epsilon}\,\frac{\omega}{\omega_b}\zeta \pm \int_0^\zeta \frac{d\zeta'}{x_0(\zeta')}\right]\right\}, \qquad (4.3.4)$$

where the constant $a$ can be expressed in terms of the field modulation amplitude at the plasma boundary $z = 0$:

$$a = \frac{e\epsilon E_r(0)}{m\gamma\omega_b^2 R_0}.$$

Substituting (4.3.4) into the second of the equations in (4.3.3), we derive an equation for the beam radius, namely:

$$x_0'' + \mu\left(1 - \frac{1}{x_0^4}\right)x_0 + \frac{1}{2}\frac{a^2}{x_0^2} = 0. \qquad (4.3.5)$$

In the absence of the external magnetic field, $\mu = 0$, solution (4.3.5) with boundary conditions $x_0(0) = 1$ and $x_0'(0) = 0$ has the form

$$\sqrt{x_0(1 - x_0)} + \mathrm{Arccos}\sqrt{x_0} = a\zeta. \qquad (4.3.6)$$

At the points $\zeta_n = (n + 1/2)(\pi/a)$, where $x_0(0) = 0$ and the beam density becomes infinite, and the approximation $x_0 \gg a^2$ of Eq. (4.3.5) is disrupted, so that it becomes necessary to provide for a larger number of harmonics. However, the model proves to be correct for the case in which $\mu > 0$, when consideration of the pressure of the magnetic field eliminates the point foci.

In the presence of the magnetic field $\mu > 0$, the first integral in (4.3.5), satisfying boundary conditions $x_0(0) = 1$ and $x_0'(0) = 0$ has the form

$$x_0'^2 = \frac{1}{(\Lambda x_0)^2}(1 - x_0)(x_0 - x_{\min})\left[\left(x_0 + \frac{1 + x_{\min}}{2}\right)^2 + b^2\right], \qquad (4.3.7)$$

---

[4] In this case it is assumed that in the absence of spatial amplification condition $|x_1| \ll 1$ is conserved at any distance from the injection plane, so that the higher harmonics $|x_n| \sim |x_1|^n$ can be dispensed with.

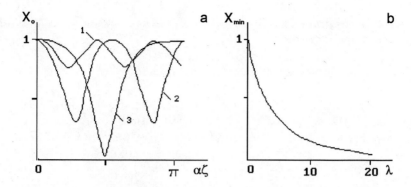

Fig. 4.3.1. (a) Beam radius $x_0$ as a function of the spatial coordinate $\zeta$ in a finite magnetic field for (1) $\Lambda = 1$, (2) $\Lambda = 4$, and (3) $\Lambda = 25$;
(b) Minimum mean radius $x_{min}$ as a function of $\Lambda$.

where minimum radius of the focused beam is determined from the formula

$$x_{min} = \sqrt[3]{\frac{8}{27} + \frac{\Lambda}{6} + \sqrt{D}} + \sqrt[3]{\frac{8}{27} + \frac{\Lambda}{6} - \sqrt{D}} - \frac{1}{3}, \tag{4.3.8}$$

with the remaining notation

$$b^2 = \frac{1}{x_{min}} - \frac{(1 + x_{min})^2}{4} > 0, \quad D = \frac{\Lambda}{27}\left(\Lambda^2 - \frac{13}{4}\Lambda + 8\right) > 0,$$

$$\Lambda = \frac{a^2}{\mu} = \frac{4\omega_b^2 a^2}{\epsilon \omega_B^2}.$$

Having integrated (4.3.7) with the initial conditions $x_0(0) = 1$, we derive the equation determining, in implicit form, the function $x_0(\zeta)$:

$$x_0(\zeta) = \frac{C + x_{min}A + (C - x_{min}A)\cos\varphi}{C + A + (C - A)\cos\varphi},$$

$$\zeta = -\frac{B}{2}\sqrt{\frac{A^5}{C}}F(\varphi, k) + \frac{B}{2x_{min}}\sqrt{\frac{A}{C}}(x_{min}^2 + 2x_{min} - 1 + AC)\Pi(\varphi, n, k) \tag{4.3.9}$$

$$+ \frac{A^2 B}{\sqrt{x_{min}}}\left[\arctan\left(\frac{1 + x_{min}^2 - AC}{4\sqrt{x_{min}A^3 C}}\frac{\sin 2\varphi}{\Delta}\right) + \arctan\left(\frac{B^2}{2}\sqrt{\frac{x_{min}}{A^3 C}}\frac{\sin\varphi}{\Delta}\right)\right],$$

where we have introduced the notation

$$\Delta = \sqrt{1 - k^2\sin^2\varphi}, \quad A = \sqrt{1 + x_{min}}, \quad B = \sqrt{1 - x_{min}},$$

$$C = \sqrt{1 - x_{min} + 2x_{min}^2}, \quad n = \frac{x_{min}B^4}{2A^2(1 + x_{min}^2 + AC)},$$

$$k^2 = \frac{1}{2}\left(1 - \frac{2 + x_{min} + 4x_{min}^2 + x_{min}^3}{2A^3 C}\right),$$

$F(\varphi, k)$ and $\Pi(\varphi, n, k)$ are incomplete elliptical integrals of the first and third kind [10].

The condition for the applicability of solution (4.3.9) is determined by the following inequality:

$$\omega_b a^2 \ll \omega_B\sqrt{\epsilon} \ll \omega_b. \tag{4.3.10}$$

Figure. 4.3.1a shows the beam radius $x_0$ as a function of $a\zeta$ for various values of the parameter $\Lambda$, as defined in (4.3.9). With $\Lambda = 25$ (curve 3) the solution is close to that obtained earlier in (4.3.6) for the case $\mu = 0$. Figure 4.3.4b illustrates the relationship between $x_{min}$ and $\Lambda$.

Thus, the high-frequency focusing of the beam predominates when $\Lambda \gg 1$, which corresponds to a weak magnetic field

$$B_0 \ll B_c = c\sqrt{\epsilon}\, E_r(0)/\omega_b R_0.$$

If $B_0 \sim B_c$ ($\Lambda \simeq 1$), the effects of the high-frequency and magnetic compression of the beam turn out to be of the same order of magnitude, while with $B_0 \gg B_c$ ($\Lambda \ll 1$) the beam is focused by the external magnetic field. In the latter case, the solution of (4.3.10) has the form

$$x_0 = 1 - \frac{\Lambda}{8}\left(1 - \cos\frac{\omega_B \zeta}{v}\right) \tag{4.3.11}$$

and describes small oscillations in the radius of the beam about the equilibrium value of $R_0$.

The derived formula (4.3.11) can easily be generalized to the case of a stronger magnetic field, when the second inequality in (4.3.10) is violated:

$$x_0 = 1 + \frac{a^2}{2}\left(\frac{3}{2} - \frac{1}{\epsilon}\frac{\omega_b^2}{\omega_B^2}\right)\left(1 - \cos\frac{\omega_B \zeta}{v}\right). \tag{4.3.12}$$

An interesting solution of the system of equation (4.3.5) is found when the external magnetic field offsets the effect of the wave field and as a result the amplitude of the wave $A = A_0$ and the form of the beam modulation $X(\Phi)$ do not change as the distance $\zeta$ increases from the injection plane.

Assuming in the second of the equations in (4.2.4) that $X'' = 0$, we find the relationship between the radius of the beam and the parameter $\Phi$:

$$\frac{R(\Phi)}{R_0} = X(\Phi) = \frac{1}{\sqrt[4]{1 - a\cos\Phi}}, \quad a = \frac{A_0}{\mu} \tag{4.3.13}$$

(the beam is assumed to be cold, $v_T = 0$, so that $\eta = \mu$).

Correspondingly, from the first of the equations in (4.2.4) we obtain

$$a = -\frac{1}{2\pi\mu}\int_0^{2\pi}\sqrt{1 - a\cos\Phi}\,\cos\Phi\,d\Phi. \tag{4.3.14}$$

The integral in right-hand side is referred to the complete elliptic integrals and, as a result, from (4.3.14) we have the equation for amplitude of the field under conditions of radial electron-beam equilibrium within the plasma:

$$\mu - \frac{2\sqrt{1 + a}}{3\pi a^2}\left[(a - 1)K\left(\sqrt{\frac{2a}{1 + a}}\right) + E\left(\sqrt{\frac{2a}{1 + a}}\right)\right] = 0. \tag{4.3.15}$$

Here $K(k)$ and $E(k)$ are the complete elliptical integrals of the first and second kinds [10].

Using the asymptotic expansions $K(k)$ and $E(k)$ for small amplitudes $a \ll 1$, we find

$$\mu = \frac{1}{4} + \frac{1}{128}a^2. \tag{4.3.16}$$

It is easy to see that when $a = 0$ the value of the parameter $\mu$ corresponds to the linear-theory stability threshold [see (4.2.2)].

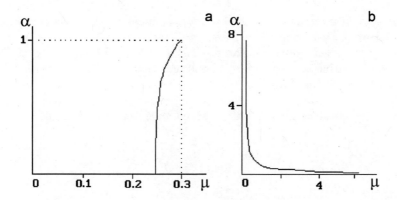

Fig. 4.3.2. Equilibrium beam modulation amplitude as a function of $\mu$: (a) continuous beam; (b) bunched beam.

Figure 4.3.2a shows the function $a(\mu)$ for the large-amplitude region. As we can see from the graph, a solution exists when $a < 1$. With $a \geq 1$ a singularity arises in Eq. (4.3.13) and the radius of the beam is undefined at the points $\cos\Phi = a^{-1}$.

From the physical point of view, this regime of interaction between the beam and the plasma can be interpreted as follows. The presence of external beam modulation leads to the formation of the plasma wave propagating synchronously with the beam at the velocity $v_{ph} = v$. Therefore, the electrons in the beam system are in fixed phases of the static field and are focused or defocused by a force that is linear with respect to field amplitude. Simultaneously, the lateral projection of the Lorentz force prevents the dispersion or collapse of the electrons. It is clear that this effect can make itself felt only in the case of a specific relationship between the beam modulation amplitude and the magnitude of the magnetic field.

As was demonstrated in the previous section, the radial focusing and defocusing of the modulated electron beam by the instability field, on injection of the beam into the plasma, leads to the formation of a sequence of electron bunches, found in the focus phases of the stationary electrostatic wave. However, if the beam injected into the plasma is initially formed into bunches, then these bunches will be in radial equilibrium, by analogy with the earlier-considered case of a continuous beam. Assuming the beam electron density to be equal to

$$n_b = n_0 \begin{cases} X^{-2}(\Phi), & \pi/2 \leq \Phi \leq 3\pi/2, \\ 0, & 0 < \Phi \leq \pi/2, \ 3\pi/2 < \Phi < 2\pi, \end{cases} \tag{4.3.17}$$

and the offsetting charge to be uniformly distributed over the beam, by analogy with Eqs. (4.3.13) and (4.3.15) we derive the equation for the amplitude of the stationary-wave field under conditions of equilibrium for the cold electron bunches in the plasma:

$$\mu + \frac{2}{3\pi a} - \frac{2\sqrt{1+a}}{3\pi a^2}$$
$$\times \left[ (a-1)F\left(\frac{\pi}{4}, \sqrt{\frac{2a}{1+a)}}\right) + E\left(\frac{\pi}{4}, \sqrt{\frac{2a}{1+a)}}\right) \right] = 0, \tag{4.3.18}$$

where $F(\varphi, k)$ and $E(\varphi, k)$ are the incomplete elliptical integrals of the first and second kinds [10]. The radius of the equilibrium bunches is determined by the formula (4.3.13).

The graph of the function $a(\mu)$ can be seen in Fig. 4.3.2b. As we can see from the figure, unlike the case of the continuous beam a solution exists for any values of the parameter $\mu$.

145

This is associated with the fact that the bunches are found in the focus phases of the wave and the radial projection of the Lorentz force, increasing rapidly on compression of the bunches, can prevent their collapse regardless of the magnitude of the magnetic field.

The studies conducted above into the interaction between a thin electron beam $(k_\parallel R \ll 1)$ and the plasma, based on system of equations (4.2.1), makes no provision for the deceleration of the beam and the capture of electrons by the longitudinal electric field $E_z$. The frequency of the phase oscillations which arise in this case is on the order of

$$\Omega_{\mathrm{ph}}^2 \simeq \frac{ek_\parallel E_z}{m\gamma^3} \approx \frac{eE_r}{m\gamma^3} k_\parallel^2 R. \tag{4.3.19}$$

Having made the substitution $E_r = Nr$ from formula (4.2.4) we obtain

$$\Omega_{\mathrm{ph}}^2 \simeq \frac{(k_\parallel R)^2}{\gamma^2} \frac{\omega_b^2}{|\epsilon|}. \tag{4.3.20}$$

It follows from formula (4.3.20) that the phase oscillations arise at a distance of

$$L_{\mathrm{ph}} \simeq \frac{v}{\Omega_{\mathrm{ph}}} \simeq \frac{\gamma}{k_\parallel R} \frac{v}{\omega_b} \sqrt{|\epsilon|}, \tag{4.3.21}$$

from the injection plane, and here this quantity exceeds the length $L_\perp \simeq v\sqrt{|\epsilon|}\,/\omega_b$ of instability development by a factor of $\gamma/k_\parallel R \gg 1$.

# CHAPTER 5

## INTERACTION OF ELECTRON AND ION BEAMS WITH ELECTROMAGNETIC WAVE IN A MAGNETIC FIELD

The oscillator motion with the super–speed of light in the slow-down medium is accompanied by electromagnetic radiation [11]. In the case of charged particle beam under the constant magnetic field this radiation is coherent, while small electromagnetic perturbations grow exponentially under the conditions of the anomalous Doppler effect [2-4]. The physical mechanism that limits the growth of field amplitude is the beam phase locking getting out of sync with the wave on account of the longitudinal beam slowdown [5-7]. In a strong electromagnetic field the retarding medium non-linearity comes through, accompanied by increasing or decreasing of the wave phase velocity [8]. In the latter case, the phase resonance remains on the nonlinear instability stage, and the radiation enegry density increases relative to the linear medium (in similarity to the Langmuir oscillations considered in Chapter 4). The nonlinear "pinching" of the phase resonance under the conditions of electron-ion beam instability has to do with appearance of a strong electrostatic field during electrons slowdown relative to ions [9].

The slow electromagnetic wave that is needed for swinging oscillations under the conditions of the anomalous Doppler effect exists in the plasma with relativistic ion beam [10]. Increasing of the phase velocity of the unstable mode wave builds up the conditions for charged particle acceleration in plasmas [11].

An effective physical mechanism of charge acceleration in the magnetic field by radiation pressure is autoresonance acceleration [12-16]. The resonance remains during relativistic increasing of the electron mass, and in the absence of the radiation drag force [17] and Coulomb collisions [18], the energy of the particle in applied field grows unlimitedly. In the case of the finite density beam the problem outgrows the limits of the single particle approach, and the accelerated particle energy gets limited by the beam spatial charge [19-21].

This Chapter elaborates on the nonlinear theory of charged beam interactions with a slow regular electromagnetic wave under the conditions of the anomalous and normal Doppler effects, including the effects of coherent radiation excitation by beam and charged particle acceleration by radiation pressure.

### §5.1. Instability of the relativistic electron beam in the presence of the anomalous Doppler effect.

We consider the excitation of an electromagnetic wave by a relativistic electron beam mowing down a constant external magnetic field $\mathbf{B}_0$ with the velocity $\mathbf{v}_0$ which is

147

greater than the velocity of light in the slowing down system with an effective refractive index $n_0$, where $v_0 > c/n_0$.

The self-consistent electromagnetic field **E** and **B** satisfy the system of Maxwell equations

$$\text{curl}\,\mathbf{E} = -\frac{1}{c}\frac{\partial \mathbf{B}}{\partial t}, \quad \text{curl}\,\mathbf{B} = \frac{1}{c}\frac{\partial}{\partial t}\hat{\epsilon}\mathbf{E} + \frac{4\pi e}{c}n_b\mathbf{v}, \tag{5.1.1}$$

and the beam density $n$ and the velocity **v** are described by the following system of hydrodynamic equations

$$\frac{\partial n_b}{\partial t} + \text{div}(\mathbf{v}n_b) = 0,$$

$$\frac{\partial \mathbf{p}}{\partial t} + (\mathbf{v}\nabla)\mathbf{p} = e\mathbf{E}/\frac{e}{c}[\mathbf{v}, \mathbf{B} + \mathbf{B}_0], \tag{5.1.2}$$

where $\mathbf{p} = m\mathbf{v}(1 - v^2/c^2)^{-1/2}$ is the momentum of the beam, and $\hat{\epsilon}$ is a retarding system dielectric function matrix.

The solution of Eqs. (5.1.1) and (5.1.2) will be sought in the form of cilcularly polarized waves with the slowly-varying amplitudes and phases propagating along the magnetic field:

$$E_x + iE_y = E(t)\exp[i\Phi - i\varphi(t)], \quad v_x + iv_y = v(t)\exp[i\Phi - i\vartheta(t)],$$

$$\Phi = \omega t - kz, \quad \dot{\varphi} \ll \omega\varphi, \quad \dot{\vartheta} \ll \omega\vartheta. \tag{5.1.3}$$

We assume, moreover, that

$$n_b(t, z) = n_{bo}, \quad v_z(t, z) = v_{\parallel}(t)$$

(the continuity equation is then automatically satisfied).

The small perturbation $\exp i(\pm\omega't \mp kz)$ satisfy the dispersion relation [2]

$$\left(\frac{ck}{\omega'}\right)^2 = n_0^2(\omega') - \frac{\omega_b^2(\omega' - kv_0)}{\omega'^2(\omega' - kv_0 \pm \omega_B^*)}, \tag{5.1.4}$$

where

$$\omega_b^2 = \frac{4\pi n_b e^2}{m\gamma_0}, \quad \omega_{Be}^* = \frac{|e|B_0}{mc\gamma_0}, \quad \gamma_0 = \frac{1}{\sqrt{1 - \beta_0^2}}, \quad \beta_0 = \frac{v_0}{c},$$

$k$ is the fixed wave number, and $\omega'$ is the complex frequency of the wave, corresponding to the unstable system. In the derivation of (5.1.4) we have used the relation [22]

$$\hat{\epsilon}\exp(i\omega't) = \epsilon(\omega')\exp(i\omega't),$$

at which $\epsilon(\omega') = n_0^2(\omega')$ is the effective permittivity of the retarding system.

The conditions of the anomalous Doppler effect [2-7] exists in the frequency range $n_0(\omega) > \beta_0^{-1}$ which corresponds to the plus sign "+" in equation (5.1.4). In approximation of small beam density, the solution of (5.1.4) can be found in the form

$$\omega' = \omega - i\delta^*, \quad \delta^* \ll \omega,$$

$$ck = \omega n_0(\omega), \quad \omega - kv_0 = -\omega_B^* \tag{5.1.5}$$

at which the resonant frequency and the growth rate of instability are

$$\omega = \frac{\omega_B^*}{\beta_0 n_0 - 1}, \quad \delta^* = \left(\frac{\omega_b^2 \omega_B^*}{2\omega Q}\right)^{1/2}, \quad Q = \frac{d}{d\omega^2}\left(\omega^2 n_0^2\right). \tag{5.1.6}$$

148

Substituting (5.1.3) into Eqs. (5.1.1) and (5.1.2), we obtain the following set of nonlinear equations in total derivatives with respect to time (see **Appendix 5.1**):

$$\frac{d}{dt}\gamma\beta_\perp = \frac{e}{mc}(1 - \beta_\| n_0)E\cos\eta,$$

$$\frac{d}{dt}\gamma\beta_\| = \frac{e}{mc}n_0\beta_\perp E\cos\eta, \quad \frac{d}{dt}(\hat{n}^2 E) = -2\pi e n_b c\beta_\perp\cos\eta, \qquad (5.1.7)$$

$$\frac{d\eta}{dt} = \frac{\omega_B}{\gamma} + \omega(1 - \beta_\| n_0) + \frac{\omega}{2\hat{n}^2}\gamma(n_0^2 - n^2) - \mathrm{tg}\,\eta\,\frac{d}{dt}\ln\left(\hat{n}^2\,\gamma\beta_\perp E\right),$$

where

$$\omega_B = \gamma_0\omega_B^*, \quad \eta = \vartheta - \varphi, \quad \gamma = (1 - \beta_\perp^2 - \beta_\|^2)^{-1/2},$$

$$\beta_{\perp,\|} = v_{\perp,\|}/c, \quad \hat{n}^2 = d(\omega^2 n^2)/d\omega^2,$$

and $n(\omega, E^2)$ is a function of the field amplitude.

In terms of the same variables, the change in the beam energy with time is described by the equation

$$\frac{d\gamma}{dt} = \frac{e}{mc}\beta_\perp E\cos\eta. \qquad (5.1.8)$$

Subtructing Eq. (5.1.8) from first Eq. (5.1.7) term by term, we obtain

$$\frac{d}{dt}(\beta_\| n_0 - 1)\gamma = \frac{e}{mc}(n_0^2 - 1)\beta_\perp E\cos\eta. \qquad (5.1.9)$$

### 1. Linear retarding medium.

In this section, we will assume that for small beam densities the main nonlinear effect limiting the increase in oscillation amplitude is the loss of synchronism between the particles and the field as a result of beam deceleration [5,6], and hence the refractive index of retarding system does not depend on the field amplitude, $n \simeq n_0$.

In the dimensionless variables

$$a = \beta_\perp\gamma, \quad b = (\beta_\| n_0 - 1)\gamma, \quad \mathcal{E} = E/B_0,$$

$$q^2 = 2\pi e^2 n_b/m\omega_B^2 Q, \quad \Omega = \omega/\omega_B, \quad d\tau = \omega_B dt/\gamma$$

equations (5.1.7) take the form

$$a' = -b\mathcal{E}\cos\eta, \quad b' = (n_0^2 - 1)a\mathcal{E}\cos\eta,$$

$$\mathcal{E}' = -q^2 a\cos\eta, \quad \eta' = 1 - \Omega b - \mathrm{tg}\,\eta\,(\ln\mathcal{E}a)'. \qquad (5.1.10)$$

As is seen from system (5.1.10), the main source of its nonlinearity is the Lorentz force in the equation for the longitudinal motions of beam particles. In the linear approximation, ignoring this term, we obtain for $\Omega b_0 = 1$ (the anomalous Doppler effect condition)

$$\mathcal{E}'' - \delta^2\mathcal{E} = 0, \quad \eta = 0, \qquad (5.1.11)$$

i.e., the wave amplitude increases exponentially over time $\tau$ with a growth rate $\delta = q/\sqrt{\Omega}$.

Integrating (5.1.10) subject to the initial conditions

$$\eta_0 = a_0 = 0, \quad b_0 = \Omega^{-1}, \quad \mathcal{E}(0) = \mathcal{E}_0,$$

149

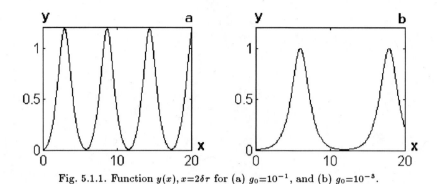

Fig. 5.1.1. Function $y(x)$, $x=2\delta\tau$ for (a) $g_0=10^{-1}$, and (b) $g_0=10^{-3}$.

we obtain the following integrals:

$$(n_0^2 - 1)a^2 + b^2 = \Omega^{-2},$$

$$\mathcal{E}^2 = \mathcal{E}_0^2 + \frac{2q^2}{n_0^2 - 1}\left(\frac{1}{\Omega} - b\right), \quad \sin\eta = -\frac{(1 - \Omega b)^2}{2a\mathcal{E}\Omega(n_0^2 - 1)}. \qquad (5.1.12)$$

These relationships enable us to reduce the set of equations given by Eq. (5.1.10) to a first-order equation.

In view of Eq. (5.1.12), we substitute

$$\sqrt{n_0^2 - 1}\,\Omega a = \sin\psi, \quad \Omega b = \cos\psi.$$

The result of this is the following equation for the function $\psi$:

$$\psi' = -\sqrt{n_0^2 - 1}\,\mathcal{E}\cos\eta. \qquad (5.1.13)$$

Expressing $\eta$ in right-hand side of Eq. (5.1.13) in terms of $\psi$ by means of Eqs. (5.1.12), we get

$$\psi'^2 = \mathcal{E}_0^2(n_0^2 - 1) + 2\delta^2(1 - \cos\psi) - \frac{(1 - \cos\psi)^4}{4\sin^2\psi}. \qquad (5.1.14)$$

According to Eq. (5.1.13), the maximum value of the function $\psi(\tau)$ is obtained from the condition $\psi'(\psi_m) = 0$ corresponds to $\eta = \pm\pi/2$. Since $\psi_m \ll 1$ when

$$\mathcal{E}_0^2(n_0^2 - 1) \ll 1, \quad \delta^2 \ll 1, \qquad (5.1.15)$$

we can reduce Eq. (5.1.14) to the form:

$$\frac{dy}{d\tau} = 2\delta\sqrt{y(g_0 + y - y^3)},$$

$$y = \left(\frac{\psi}{\psi_m}\right)^2, \quad \psi_m = \sqrt{8\delta}, \quad g_0 = \left(\frac{1}{2\delta}\right)^3\mathcal{E}_0^2(n_0^2 - 1). \qquad (5.1.16)$$

Numerical solutions of equation (5.1.15) are represented in Fig. 5.1.1. The distance between the spikes of function $y(\tau)$ increases if the amplitude of the external field, $g_0$, decreases. In the case of $g_0 \to 0$, a periodic function transforms to soliton,

$$y = \text{ch}^{-1}\left(\text{arech}\,y_0^{-1} - 2\delta\tau\right) \qquad (5.1.17)$$

($y_0$ is the small initial perturbation).

150

The increase in the field amplitude is accompanied by the deceleration of beam particles. This corresponds to the temporal growth in the transverse velocity amplitude. It follows from (5.1.12) and (5.1.16) that the maximum values of the momentum amplitude $a_m$ and the field amplitude $\mathcal{E}_m$ are

$$a_m = \frac{1}{\Omega}\sqrt{\frac{8\delta}{n_0^2 - 1)}}, \quad \mathcal{E}_m = \sqrt{\frac{8\delta^3}{n_0^2 - 1}} \tag{5.1.18}$$

(if the inequality $\mathcal{E}_m \gg \mathcal{E}_0$ holds).

An anomalous Doppler effect condition (5.1.6) can be fulfilled only for $n_0 > \beta_0^{-1}$, and hence, $\sqrt{n_0^2 - 1} > (\beta_0\gamma_0)^{-1}$. Since for low-density beam, $\psi_m \ll 1$, only a small part of beam energy transforms into the oscillation energy, the following inequalities hold: $\gamma \simeq \gamma_0$ and $\tau \simeq \omega_B^* t$.

### 2. Nonlinear retarding medium.

The variation of the refractive index of the retarding medium with the field amplitude, $n(\omega, \mathcal{E})$, leads to an additional term in (5.1.7), and this term along with $1 - \Omega b$, leads to a phase shift $\eta$ between the transverse beam velocity and the field. To modify the system of integrals (5.1.12), we will use the following differential relations

$$\mathcal{E}\, d(\hat{n}^2 \mathcal{E})' = -Q q^2\, d\gamma,$$
$$d(\hat{n}^2 \mathcal{E} a \sin \eta) = \Omega[(n_0^2 - 1)(\gamma_0 - \gamma)\hat{n}^2 + (\gamma/2)(n_0^2 - n^2)]\, d\gamma \tag{5.1.19}$$

[which is evident from Eqs. (5.1.7)].

In the quite general case, the refractive index can be approximated by

$$n^2(\omega, \mathcal{E}) = n_0^2(\omega) + \alpha(\omega)\mathcal{E}^2 + \upsilon(\omega)\mathcal{E}^4. \tag{5.1.20}$$

Substituting $n^2(\omega, \mathcal{E})$ given by the formula (5.1.20) into (5.1.19) and integrating yield

$$\gamma = \gamma_0 - \frac{\mathcal{E}^2}{2q^2}\left(1 + \frac{3\hat{\alpha}}{2Q}\mathcal{E}^2 + \frac{5\hat{\upsilon}}{3Q}\mathcal{E}^4\right), \quad \sin\eta = -\frac{B}{C}\mathcal{E}^2 - \frac{A}{C}\mathcal{E}^4, \tag{5.1.21}$$

where

$$A = \frac{1}{3Q}\left[5\hat{\alpha}\left(n_0^2 - 1\right) + \alpha - 2q^2\gamma_0\upsilon\right], \quad B = n_0^2 - 1 - \frac{\alpha}{Q}q^2\gamma_0, \quad C = 8\delta^3.$$

In the derivation of phase integral (5.1.21) we made the simplifying assumption that $|B| \ll 1$ and the corresponding terms $\sim B\mathcal{E}^4$ have been dropped. Besides, we have introduced the following notation:

$$\hat{n}^2 = Q + \hat{\alpha}\mathcal{E}^2 + \hat{\upsilon}\mathcal{E}^4,$$

at which $Q$ is defined by Eq. (5.1.6).

The maximum field amplitude is governed by an algebraic equation which follows from (5.1.21) with $\sin\eta = \mp 1$:

$$\mathcal{E}_m^2 = \frac{1}{2A}\left(-B + \sqrt{B^2 \pm 4AC}\right). \tag{5.1.22}$$

151

The asymptotic solutions to the algebraic equation (5.1.22) have the form

$$\mathcal{E}_m^2 \simeq \begin{cases} C/|B|, & AC \ll B^2, \\ \sqrt{C/|A|}, & AC \gg B^2. \end{cases} \qquad (5.1.23)$$

The soliton solutions of Eqs. (5.1.19) and (5.1.21) (in the case $\mathcal{E}_0 \to 0$), corresponding to the formulas (5.1.23), become

$$\mathcal{E}^2 \simeq \begin{cases} (C/|B|)\mathrm{sech}(2\delta\tau), & AC \ll B^2, \\ \sqrt{(C/|A|)}\mathrm{sech}(4\delta\tau), & AC \gg B^2. \end{cases} \qquad (5.1.24)$$

The first asymptotics (5.1.23) may be used in the case $\alpha < 0$ $(B > 0)$, at which the phase velocity of the wave as a function of the field amplitude increases in comparison with a linear medium (5.1.18). In the opposite case $\alpha > 0$, the phase velocity decrease is accompanied by an increase in the maximum field amplitude. The second asymptotics (5.1.23) corresponds to the nonlinear phase resonance $(B \to 0)$, when the beam velocity and the phase velocity of the wave decrease synchronously, and the phase shift is governed by the higher-order terms in expansion (5.1.20). Correspondingly, the rate with which the field amplitude increases, $\mathcal{E}_m^2 \sim \delta^{3/2}$, is substantially higher than in the case of a linear medium response.

The nonlinearity both of the equations of motion of the beam and of the oscillatory properties of the retarding medium are values of the same order if the beam energy becomes

$$n_b m c^2 \gamma_0 \sim (n_0^2 - 1)\, \frac{Q^2}{2\pi\alpha}\, B_0^2. \qquad (5.1.25)$$

(the first and second terms in the coefficient $B$ turn out to be of the same order of magnitude).

Let us now consider the physical mechanism causing the effect of nonlinear stabilization of the instability under the conditions of the anomalous Doppler effect

$$\omega = k v_0 - \omega_B^*, \quad k = \omega n_0/c.$$

In the linear stage of instability $(\eta = 0)$, the beam particle energy increases in the electric field $\mathbf{E}$ of the wave transverse to $\mathbf{B}_0$, and the magnetic field $\mathbf{B}$ of the wave converts the transverse motion to a longitudinal motion and decelerates the particle along the $\mathbf{B}_0$ field as a result of the Loretz force $\mathbf{F} = (e/c)(\mathbf{v} \times \mathbf{B})\mathbf{B}_0/B_0$. This process, accompanied by the transformation of the energy of the longitudinal motion into wave energy, stops when the system reaches maximum of the wave amplitude, since at that time the difference in phases between the transverse velocity and the electrical field vectors, due to the detuning of the anomalous Doppler resonance as the result of the longitudinal retardation, reaches the value $\eta_m = \pi/2$. After this the particle reaches a retarding phase, loses transverse energy, and accelerates along the magnetic field to reach its initial velocity.

The model of an unbounded beam considered above is physically correct if the slow wave guide is filled with a sufficiently dense plasma, $n_p \gg n_b$, which neutralizes the Coulomb charge of the beam. We can take into account a plasma contribution in a dispersion with the help of the substitution

$$N^2 = n_0^2 - \frac{\omega_p^2}{\omega(\omega + \omega_{Be})}, \quad \omega_p^2 = \frac{4\pi e^2 n_p}{m} \qquad (5.1.26)$$

152

in the dispersion relation (5.1.6). A plasma addition is small in comparison with the growth rate of anomalous Doppler instability if the following inequalities hold:

$$\omega_b^2 \ll \omega_p^2 \ll \omega_B^{*2}\beta_0 n_0^3/(\beta_0 n_0 - 1)^2, \tag{5.1.27}$$

where $\omega_b^2$ and $\omega_p^2$ are the Langmuir frequencies of the beam and plasma, respectively.

## §5.2. Instability of an electron-ion beam in a retarding medium.

As noted above, the Coulomb charge of the beam can be neutralized by means of a sufficiently dense plasma. An alternative method for achieving charge and current neutrality in the system, which is realized for a small density of the background plasma (or its absence), is the accompanying ion beam [9]. The longitudinal polarization electric field, which appears upon electron shift with respect to ions, impedes the electron-beam retardation and diminishes the role of the high-frequency nonlinearity. The time of synchronous interaction between electron beam and the wave increases, and, therefore, the radiation-energy density also increases.

In the linear stage of instability, we can neglect changes in the ion beam velocity, and the dispersion relation for the small oscillations coincides with that for the electron beam (5.1.6):

$$\omega = \frac{\omega_{Be}}{\beta_0 n_0 - 1}, \quad \delta = \sqrt{\frac{\omega_b^2}{2n_0^2}\left(\beta_0 n_0 - 1\right)}, \tag{5.2.1}$$

where $v_0 = \beta_0 c$ is the initial velocity of the beam, $\omega_{Be} = |e|B_0/mc$, is the gyrofrequency of the beam electrons, and $n_0$ is the refractive index of the retarding medium (without dispersion).

The nonlinear stage of instability of the electron-ion beam is described by the following system of hydrodynamic equations

$$\dot{E} = -\frac{2\pi e n_b}{n_0^2}v_\perp \cos\eta, \quad \dot{v}_\perp = \frac{e}{m}\left(1 - \frac{v_\| n_0}{c}\right)E\cos\eta,$$

$$\dot{v}_\| = \frac{en_0}{mc}v_\perp E\cos\eta + \frac{e}{m}E_\|, \quad \dot{v} = \frac{\omega n_0}{c}(v_0 - v_\|) - \left(\frac{\dot{E}}{E} + \frac{\dot{v}_\perp}{v_\perp}\right)\mathrm{tg}\,\eta, \tag{5.2.2}$$

The longitudinal polarization (low-frequency) electric field $E_\|$ satisfies the Poisson equation, which, with allowance for the equation of longitudinal motion of ion beam, is presented in the form

$$\dot{E}_\| = -\frac{4\pi e n_b}{n_0^2}(v_\| - V), \quad \dot{V} = -\frac{e}{M}E_\|, \tag{5.2.3}$$

where $V$ is the velocity of the ion beam.

Integrating (5.2.2) and (5.2.3), we find the following system of integrals:

$$n_b\left[m(v_\perp^2 + v_\|^2) + MV^2\right] + \frac{n_0}{2\pi}\left(E^2 + \frac{E_\|^2}{2}\right) = n_b(m + M)v_0^2,$$

$$n_b(mv_\| + MV) + \frac{n_0^3 E^2}{4\pi c} = n_b(m + M)v_0, \tag{5.2.4}$$

$$\sin\eta = -\frac{\omega}{2\omega_{Be}}\frac{B_0}{mcv_\perp E}\left[m(v_0 - v_\|)^2 + M(v_0 - V)^2 + \frac{E_\|^2}{4\pi n_b}\right].$$

153

To proceed, let us differentiate first equation (5.2.3) with respect to $t$ and use the relations (5.2.2) to transform it to the form

$$\frac{d^2 E_\parallel}{dt^2} + \frac{\omega_b^2}{n_0^2}(1+\mu)E_\parallel = \frac{en_0}{mc}\frac{dE^2}{dt}, \tag{5.2.5}$$

where $\omega_b^2 = 4\pi e^2 n_b/m$ and $\mu = m/M$. Introducing the vector potential $E_\parallel = -\dot{A}_\parallel/c$, we can write (5.2.5) as

$$\ddot{A}_\parallel + \frac{\omega_b^2}{n_0^2}(1+\mu)A_\parallel = -\frac{en_0}{m}E^2. \tag{5.2.6}$$

Then it follows from second formula (5.2.3) that

$$M(V - v_0) = (e/c)A_\parallel. \tag{5.2.7}$$

The above relation allows us to eliminate the ion velocity from the formulas (5.2.4).

In the dimensionless variables

$$\beta_\perp = v_\perp/c, \quad \mathcal{E} = E/B_0, \quad A = eA_\parallel/mc^2,$$
$$\tau = \omega_{Be}t, \quad \Omega = \omega/\omega_{Be}, \quad q^2 = \omega_b^2/2\omega_{Be}^2.$$

equations (5.2.2) and (5.2.6) take the form

$$\mathcal{E}' = -\frac{q^2}{n_0^2}\beta_\perp \cos\eta, \quad A'' + \frac{2q^2}{n_0^2}(1+\mu)A = -n_0\mathcal{E}^2, \tag{5.2.8}$$

where the prime means the derivative with respect to $\tau$, and the functions $\beta_\perp(\mathcal{E})$ and $\eta(\mathcal{E})$ are defined by Eqs. (5.2.4):

$$\beta_\perp^2 = \frac{n_0^2}{q^2\Omega}\mathcal{E}^2 + \frac{2}{\Omega}\beta_\perp\mathcal{E}\sin\eta,$$
$$-\frac{2}{\Omega}\beta_\perp\mathcal{E}\sin\eta = \left(\frac{n_0^3}{2q^2}\mathcal{E}^2 + A\right)^2 + \mu A^2 + \frac{1}{2}\left(\frac{n_0 A'}{q}\right)^2. \tag{5.2.9}$$

Last term in the first relation (5.2.9) (which is small term of order $q \ll 1$) can be dropped.

In the new variables

$$u = \frac{n_0^5\Omega^{3/2}}{8q^3}\mathcal{E}^2, \quad w = \frac{n_0^2\Omega^{3/2}}{4q}A_\parallel, \quad x = 2\delta\tau$$

($\delta = q/n_0\Omega^{1/2}$ is the dimensionless growth rate), the equations (5.2.8) and (5.2.9) can be represented in the form

$$u'^2 = u^2 - [(u+w)^2 + \mu w^2 + (2/\Omega)w'^2]^2,$$
$$w'' = -(\Omega/2)[(1+\mu)w + u]. \tag{5.2.10}$$

At the limit case of

$$(2/\Omega)w'^2 \ll \mu, \quad (2/\Omega)w'' \ll \mu, \quad u + w \simeq -\mu u,$$

the system (5.2.10) reduces to the following nonlinear equation

$$u' = u\sqrt{1 - \mu^2 u^2}, \tag{5.2.11}$$

154

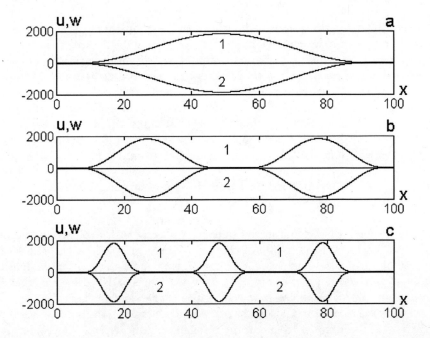

Fig. 5.2.1. Functions $u$ (1) and $w$ (2) for different initial values of the parameter $\Omega/2$ (a) 10; (b) 50, and (c) 400. The ratio between the masses of the electron and ion is equal to $\mu=1/1840$.

which coincides with (5.1.15) in the case $\mu = 1$.

The appearance in Eq. (5.1.15) of the parameter $\mu \ll 1$ reduces the role of the high-frequency nonlinearity. Respectively, the maximum of the field amplitude increases and reaches the value $u_m = \mu^{-1}$, i.e.,

$$\mathcal{E}_m^2 = \frac{1}{\mu} \frac{8q^3}{n_0^5 \Omega^{3/2}}. \qquad (5.2.12)$$

Note that the excitation of an electromagnetic wave by a nonrelativistic, $\beta_0 \ll 1$, electron-ion beam under the conditions of the anomalous Doppler effect, $\beta_0 n_0 - 1 > 0$, corresponds to $n_0 \gg 1$.

Figure 5.2.1 shows the numerical solution of (5.2.10) that are in a good agreement with the analytical solution (5.2.11) in the limit case of $\Omega \gg 1$. We have used the equation

$$u'' = u - 2(u + w)[(u + w)^2 + \mu w^2 + (2/\Omega)w'^2], \qquad (5.2.13)$$

to fulfill the numerical integration. The accuracy of calculations was monitored on the basis of integral (5.2.10).

Let us now consider in more detail the physical mechanism causing the effect of nonlinear stabilization of the instability of electron-ion beam in a retarding medium which a refractive index satisfy the inequality $n_0 > 1/\beta_0$. Under the conditions of the anomalous Doppler effect the growth rate of the low-frequency ion instability

$$\omega_i = \mu\omega, \quad \delta_i = \sqrt{\mu}\,\delta$$

is much less than the growth rate of the high-frequency electron instability (5.2.1) due to the difference between the electron and ion masses, $\delta \gg \delta_i$. Therefore, the rapid electron oscillations arise on the background of the ions moving with a constant velocity.

A deceleration of the electron beam relatively ions accompanies by the charge separation in the frame of reference in which the ions are at rest. Arising the polarization field is so large that electron retardation due to radiation occurs in synchronism with ions (when the ion Langmuir frequency $\omega_{bi} = \sqrt{\mu}\omega_b/n_0$ much exceed the electron growth rate, $\sqrt{\beta_0 n_0 - 1} \ll \sqrt{\mu}$), since in the considered case the electron beam is "frozen" into the ion one. In other words, a peculiar mass anisotropy emerges, since electrons act as particles with mass $m$ as they travel across the magnetic field, while their effective mass equals $M$ as they travel along the field. As a result, due to the large ion mass, the soliton amplitude becomes greater by a factor of $(M/m)^{1/2}$ than that in a case of an electron beam.

### §5.3. Collective slowing of a relativistic electron beam in a nonlinear dielectric medium.

Let us consider the soliton solutions in a plasma with a highly modulated relativistic electron beam. When a beam of charged particles moves in a gas whose dielectric constant is sufficiently close to unity, $n_0(\omega) - 1 \ll 1$, the condition of the anomalous Doppler effect, $\beta_0 n_0(\omega) > 1$ ($\beta_0 = v_0/c$, $v_0$ is the beam velocity), is satisfied only in the case of a relativistic beam and in a narrow frequency range. Therefore in describing the medium one can use a two-level approximation [23,24], with the assumption that the instability frequency $\omega$ [see (5.1.6)] is closed to one of the resonance frequencies of the medium, say $\Omega_R$: $|\omega - \Omega_R| \ll \Omega_R$.

Substituting the refractive index

$$n_0^2 = 1 + \frac{\omega_g^2}{\Omega_R^2 - \omega^2}, \quad \omega_g^2 = \frac{8\pi N_0 d_0^2 \Omega_R}{\hbar}. \tag{5.3.1}$$

into formulas (5.1.5), we find the frequency $\omega$ and the growth rate $\delta$ of the beam-gas instability:

$$\omega^2 \simeq \Omega_R^2 - \frac{\omega_g^2}{n_0^2 - 1}, \quad \delta = \left(\frac{\omega_b^2 \omega_B^*}{2\omega Q}\right)^{1/2},$$

$$Q = 1 + (n_0^2 - 1)\frac{\Omega_R^2}{\omega_g^2}, \quad n_0 \simeq \frac{1}{\beta_0}\left(1 + \frac{\omega_B^*}{\Omega_R}\right), \quad \omega_g \ll \Omega_R. \tag{5.3.2}$$

The propagation of an electromagnetic wave in a dielectric (gaseous) medium is accompanied by the appearance of a ponderomotive force [22]

$$\mathbf{f} = -T\nabla N + \frac{n^2 - 1}{8\pi}\left(\nabla \mathbf{E}^2 + \frac{1}{c}\frac{\partial}{\partial t}(\mathbf{E} \times \mathbf{B})\right), \tag{5.3.3}$$

($T$ is the temperature, $N$ is the density of the medium, $n^2 = 1 + 4\pi N\alpha(\mathbf{E})$, and $\mathbf{E}$ and $\mathbf{B}$ are the electric and magnetic field). The function $\alpha(\mathbf{E})$ gives the connection between the dipole moment $\mathbf{d}$ of the individual molecule and the electric field $\mathbf{E}$, $\mathbf{d} = \alpha(\mathbf{E})\mathbf{E}$, and is given by the equation (**Appendix 6.1**)

$$\ddot{\mathbf{d}} + \Omega_R^2 \mathbf{d} = \frac{2d_0}{\hbar}\sqrt{\Omega_R^2 d_0^2 - \Omega_R^2 \mathbf{d}^2 - \dot{\mathbf{d}}^2}\,\mathbf{E}, \tag{5.3.4}$$

($d_0$ is a dipole-transition constant characterizing the properties of the medium).

In the case of a standing wave

$$\mathbf{E} = \mathbf{E}(\mathbf{r})\cos\omega t, \tag{5.3.5}$$

156

$$d_0^2 - \mathbf{d}^2 \gg (\Delta/\Omega_R)\mathbf{d}^2, \quad \Delta = (\Omega_R^2 - \omega^2)/\Omega_R$$

it follows from Eq. (5.3.4) that [24]

$$\alpha(\mathbf{E}) = \frac{d_0\kappa}{\sqrt{1 + \kappa^2 E^2}}, \quad \kappa = \frac{2d_0}{\hbar\Delta}. \tag{5.3.6}$$

and from the condition $< \mathbf{f} >= 0$ we get

$$N = N_0 \exp\left[\frac{d_0}{2\kappa T}\left(\sqrt{1 + \kappa^2 E^2} - 1\right)\right], \tag{5.3.7}$$

($N_0$ is the unperturbed density of the gas; the average is taken over a time of one period $T = 2\pi/\omega$, and $< \mathbf{E}^2 >= E^2/2$; the derivative with respect to time in Eq. (5.3.3) is dropped).

From Eqs. (5.3.6) and (5.3.7) it follows that the nonlinear dielectric constant $n^2(\omega, E)$ of the gaseous medium is given by

$$n^2(\omega, E) = 1 + \frac{4\pi\kappa d_0 N_0}{\sqrt{1 + \kappa^2 E^2}} \exp\left[\frac{d_0}{2\kappa T}\left(\sqrt{1 + \kappa^2 E^2} - 1\right)\right]. \tag{5.3.8}$$

The above expression generalizes (5.3.1) to the case of $E > 0$ at which the nonlinear properties of the retarding medium must be taken into account (see Refs. [25,24,8]).

The energy density of unstable electromagnetic oscillations can be found from general equation (5.1.22). Let us expand the right side of Eq. (5.3.8) in series with respect to the field amplitude

$$n^2(\omega, E) = 1 + 4\pi\kappa d_0 N_0\left[1 + \frac{(\kappa E)^2}{2}(\mu - 1) + \frac{3(\kappa E)^4}{8}\left(1 - \mu + \frac{\mu^2}{3}\right)\right], \tag{5.3.9}$$

where $\mu = d_0/2\kappa T$.

We intrduce the dimensionless field amplitude

$$\mathcal{E} = E/B_0, \quad \chi = \kappa B_0 \quad \chi\mathcal{E} = \kappa E.$$

Comparing the above expression and (5.1.20), we obtain

$$n_0^2 = 1 + 4\pi\kappa N_0 d_0, \quad \alpha = \frac{\chi^2}{2}(n_0^2 - 1)(\mu - 1),$$

$$v = \frac{3\chi^4}{8}(n_0^2 - 1)\left(1 - \mu + \frac{\mu^2}{3}\right). \tag{5.3.10}$$

Substituting the coefficients given by the formula (5.3.10) into (5.1.21), yield

$$A\mathcal{E}_m^4 + B\mathcal{E}_m^2 \mp C = 0, \tag{5.3.11}$$

where

$$A = \frac{n_0^2 - 1}{Q}\left[\chi^2\left(\frac{1}{6}(\mu - 1) - \frac{1}{4}q^2\chi^2\gamma_0(1 - \mu + \frac{\mu^2}{3})\right) + \frac{5}{6}\frac{\partial}{\partial\omega^2}\left(\omega^2\chi^2(n_0^2 - 1)(\mu - 1)\right)\right],$$

$$B = (n_0^2 - 1)\left(1 - \frac{1}{2Q}q^2\chi^2\gamma_0(\mu - 1)\right), \quad C = \frac{8q^3}{\Omega^{3/2}}.$$

For $\mu < 1$ the nonlinear dependence of the polarization of the medium on the field amplitude is descisive and leads to a decrease of the function $n(\omega, \mathcal{E})$ with the increasing $\mathcal{E}^2$. Because the phase velocity of the wave increases and and there is an additional (as compared with the case of a linear medium) mismatching in phase between the beam and the field. Omitting the small term $A\mathcal{E}^4$ in formula (5.3.11), we find

$$\mathcal{E}_m^2 \simeq \mathcal{E}_{mL}^2/|G|,$$
$$\mathcal{E}_{mL}^2 = 8\delta^3/(n_0^2 - 1), \quad G = 1 + q^2\chi^2\gamma_0(1 - \mu)/2Q. \tag{5.3.12}$$

The particles in the beam get out of resonance with the wave more quickly ($G > 1$), and the maximum energy density is smaller.

The situation is different when $\mu > 1$, ($G < 1$), and the nonlinearity caused by electrostriction is the dominant effect. Molecules of the gas are pulled into the region of high field intensity, so that the index of refraction there is increased. The phase velocity decreases along with the deceleration of the beam, so that synchronism between the beam and the wave is preserved longer than in the case of a linear medium.

This effect is particularly pronounced if the condition $G \to 0$ in (5.1.12) holds; in this case there is nonlinear resonance between the beam and the field (analogous to that considered in for a plasma wave). Omitting the term $B\mathcal{E}_m^2$ in formula (5.3.11) and assuming $G \simeq 0$, we find

$$\mathcal{E}_m^2 \simeq \sqrt{\frac{C}{|A|}}, \quad q^2\gamma_0 \simeq \frac{2Q}{\chi^2(\mu - 1)}. \tag{5.3.13}$$

After the energy density of the field has reached its maximum value, as given by (5.3.12), and the longitudinal velocity of the beam has reached its minimum value, the inverse process begins–that of acceleration of the beam in the field of the wave, back to its original energy. The process of exchange of energy between the beam and field thereafter repeats periodically, so that the system of the beam and the retarding medium exhibits oscillations with a period of the order of the reciprocal of the increment given by the linear theory.

## §5.4. Excitation of a nonlinear low-frequency electromagnetic wave by a relativistic ion beam in a plasma.

If the retarding medium is an electron–ion plasma, the condition for a decrease in the phase velocity (the anomalous Doppler effect)

$$\omega[1 - \beta_0 n_0(\omega)] = -\omega_{Bi}^*, \quad \omega_{Bi}^* = eB_0/Mc\gamma_0, \tag{5.4.1}$$

which is a consequence of $n_0(\omega) > 1$ is met in the frequency range $\omega \ll |\omega_{Be}|$ ($\omega_{Be} = -eB_0/mc$ is the gyrofrequency of the beam electrons) and condition (5.4.1) holds for the ion beam [10].

We describe the propagation of a relativistic beam in a plasma with nonlinear oscillatory properties by means of the Maxwell equations,

$$\mathrm{curl}\,\mathbf{E} = -\frac{1}{c}\frac{\partial \mathbf{B}}{\partial t},$$
$$\mathrm{curl}\,\mathbf{B} = \frac{1}{c}\frac{\partial \mathbf{E}}{\partial t} + \frac{4\pi}{c}e(n_b\mathbf{u} + n_p\mathbf{V} - n_p\mathbf{v}) \tag{5.4.2}$$

and the equations of motion of the beam ions and the plasma electrons and ions:

$$M\left[\frac{\partial}{\partial t}\gamma\mathbf{u} + (\mathbf{u}\nabla)\gamma\mathbf{u}\right] = e\mathbf{E} + \frac{e}{c}[\mathbf{u}, \mathbf{B} + \mathbf{B}_0], \tag{5.4.3}$$

158

$$m \left[ \frac{\partial \mathbf{v}}{\partial t} + (\mathbf{v}\nabla)\mathbf{v} \right] = -e\mathbf{E} - \frac{e}{c}[\mathbf{v}, \mathbf{B} + \mathbf{B}_0], \tag{5.4.4}$$

$$M \left[ \frac{\partial \mathbf{V}}{\partial t} + (\mathbf{V}\nabla)\mathbf{V} \right] = e\mathbf{E} + \frac{e}{c}[\mathbf{V}, \mathbf{B} + \mathbf{B}_0], \tag{5.4.5}$$

where $\mathbf{u}$ is the beam velocity, $\gamma = (1 - u^2/c^2)^{-1/2}$, $\mathbf{v}$, $\mathbf{V}$, $m$, and $M$ are the velocities and the masses of the plasma electrons and ions.

For waves propagating along the external magnetic field,

$$\begin{aligned}
B_x - iB_y &= n_0(E_x + iE_y) = n_0 E(t) \exp[i\Phi - i\varphi(t)], \\
u_x + iu_y &= n_0 u_\perp(t) \exp[i\Phi - i\vartheta(t)],
\end{aligned} \tag{5.4.6}$$

[$n_0$ is the linear refractive index of the plasma; $\Phi = \omega t - kz$], the plasma is "incompressible", and the continuity equations for each particle species are satisfied automatically.

In the linear approximation, system (5.4.2)–(5.4.5) reduces to the following dispersion relation for the complex frequency $\omega'$:

$$\left( \frac{ck}{\omega'} \right)^2 = n_0^2(\omega') - \frac{\omega_b^2(\omega' - ku_0)}{\omega'^2(\omega' - ku_0 + \omega_{Bi}^*)}, \tag{5.4.7}$$

($k$ is the fixed wave number, $u_0$ is the initial velocity of the ion beam, and $\omega_b^2 = 4\pi n_b e^2/M\gamma_0$). The refractive index of the plasma is given by [26]

$$n_0^2(\omega) = 1 - \frac{\omega_e^2}{\omega(\omega + \omega_{Be})} - \frac{\omega_i^2}{\omega(\omega + \omega_{Bi})}, \tag{5.4.8}$$

where $\omega_e^2 = 4\pi n_p e^2/m$, $\omega_i^2 = (m/M)\omega_e^2$, and $\omega_{Bi} = \gamma \omega_{Bi}^*$.

### 1. Alfven wave.

At frequencies $\omega \ll \omega_{Bi}$ the refractive index in (5.4.8) is

$$n_0^2 = 1 + \frac{c^2}{c_A^2}, \tag{5.4.9}$$

where $c_A = B_0/\sqrt{4\pi n_p M}$ is the Alfven velocity. Substituting

$$\omega' = \omega - i\delta, \quad \delta \ll \omega, \quad c^2/c_A^2 \gg 1,$$

into (5.4.7), we find the following equations for the frequency and growth rate:

$$\omega = \omega_{Bi}^* \left( \frac{u_0}{c_A} - 1 \right)^{-1}, \quad \delta = \omega_b \frac{c_A}{c} \left( \frac{\omega_{Bi}^*}{2\omega} \right)^{1/2}. \tag{5.4.10}$$

Correspondingly, the condition under which the growth rate is small in comparison with the frequency is

$$\frac{n_b \gamma_0}{n_p} \left( \frac{\omega_{Bi}^*}{\omega} \right)^3 \ll 1.$$

To determine the nonlinear refractive index of the plasma we must find the current density which appears on the right side of Eq. (5.4.2). Since we must find nonlinear solutions of the equations of motion (5.4.4) and (5.4.5) for this purpose, but since this is not possible in the general case, we restrict the analysis to low frequencies, for which we can make use of the small parameters $\omega/\omega_{Bi}$ and $\omega/\omega_{Be}$.

At frequencies $\omega \ll \omega_{Bi}$, corresponding to Eqs. (5.4.9) and (5.4.10) of the linear theory, we find the following results from Eqs. (5.4.4) and (5.4.5) in the drift approximation, within terms $\sim m/M$

$$\mathbf{v}_\perp = c\frac{[\mathbf{E}\mathbf{B}_0]}{B_0^2}\left(1 - \frac{B^2}{B_0^2}\right),$$

$$\mathbf{V}_\perp = \mathbf{v}_\perp + \frac{Mc^2}{e}\left[\frac{\partial}{\partial t}\left(1 - \frac{B^2}{B_0^2}\right)\mathbf{E} + V_z\frac{\partial \mathbf{E}}{\partial z}\right], \quad v_z = V_z = c\frac{[\mathbf{E}\mathbf{B}_0]_z}{B_0^2}. \tag{5.4.11}$$

It follows from (5.4.11) that the transverse current caused by the plasma particles is

$$\mathbf{j}_\perp = \frac{c^2}{4\pi c_A^2}\left[\frac{\partial}{\partial t}\left(1 - \frac{B^2}{B_0^2}\right)\mathbf{E} + V_z\frac{\partial \mathbf{E}}{\partial z}\right]. \tag{5.4.12}$$

Substituting the equation for the current in (5.4.12) into (5.4.2), we find an equation for the nonlinear refractive index of the plasma:

$$n^2(\mathcal{E}) = n_0^2 - 2n_0^4\mathcal{E}^2, \quad n_0^2 \simeq c^2/c_A^2 \gg 1. \tag{5.4.13}$$

The energy density of unstable electromagnetic oscillations can be found from general equation (5.1.22). Since $\alpha = -2n_0^4 < 0$, so that the coefficients $B$ in this equation does not vanish, one can use the first asymptotics (5.1.23) at which $q = 2\pi e^2 n_p/M\omega_{Bi}^2$ and $\Omega = \omega/\omega_{Bi}$. As a result, we get

$$\mathcal{E}_m^2 = \frac{\mathcal{E}_{mL}^2}{1 + \gamma_0 n_b/n_p}, \quad \mathcal{E}_{mL}^2 = \left(2\frac{n_b}{n_p}\frac{\omega_{Bi}}{\omega}\right)^{3/2}\frac{c_A^2}{c^2} \tag{5.4.14}$$

and, hence, the nonlinear dependence of the plasma on the field amplitude leads to a decrease of the maximum field amplitude (as compared with the case of a linear plasma) because an additional mismatching in phase between the beam and the field occurs. Note that in the considered case of a low-density beam, the contribution of the plasma [the second term in the denominator (5.4.14)] is small in comparison with the beam contribution.

### 2. Helicon wave.

The plasma nonlinearity becomes progressively more important as the working frequency $\omega$ increases. At frequencies

$$\omega_{Bi} \ll \omega \ll |\omega_{Be}| \tag{5.4.15}$$

the refractive index of the plasma is $n_0^2 = 1 + \omega_e^2/\omega|\omega_{Be}|$, so that the resonance condition (5.4.1) is fulfilled for a frequency

$$\frac{\omega_{Bi}}{\omega} = \frac{\gamma_0^2}{2}\frac{u_0^2}{c_A^2} - \gamma_0 - \sqrt{\left(\frac{\gamma_0^2}{2}\frac{u_0^2}{c_A^2} - \gamma_0\right)^2 - 1} \tag{5.4.16}$$

which satisfy the first inequality (5.4.15).

Let us assume $(u_0/c_A)^2 \gamma_0 \gg 1$ and expand the right side of Eq. (5.4.16) in series

$$\omega \simeq (\gamma_0 u_0/c_A)^2 \omega_{Bi}, \quad n_0 \simeq 1/\beta_0, \quad \beta_0 = u_0/c,$$
$$Q \simeq (1 + \beta_0^2)/2\beta_0^2. \tag{5.4.17}$$

In this approximation, Eq. (5.1.6) reduces to the following relation for the growth rate

$$\delta = \left( \frac{1}{1 + \beta_0^2} \frac{n_b}{n_p} \right)^{1/2} \frac{\omega_{Bi}}{\gamma_0^2}. \tag{5.4.18}$$

Above expression differs from (5.1.6) by the replacement $\omega_B \to \omega_{Bi}$, and the parameter $Q$ takes account a dispersion of the plasma.

The retarding properties of the plasma are due to the electrons, and the nonlinear current is

$$\mathbf{j}_\perp = -ecn_p \left( 1 - \frac{B^2}{B_0^2} \right) \frac{[\mathbf{E B_0}]}{B_0^2} \tag{5.4.19}$$

Substituting the current density from (5.4.19) into (5.4.2), we find the nonlinear refractive index of the plasma to be

$$n^2(\mathcal{E}) = n_0^2 - n_0^2(n_0^2 - 1)\mathcal{E}^2, \tag{5.4.20}$$

where $n_0$ is the linear refractive index.

Comparing Eqs. (5.4.20) and (5.1.20), we find $\alpha = -n_0^2(n_0^2 - 1)$, and the maximum field amplitude is, according to (5.1.23) and (5.4.17),

$$\mathcal{E}_m^2 = \mathcal{E}_{mL}^2 \left[ 1 + \frac{2\gamma_0}{(1 + \beta_0^2)^2} \frac{u_0^2}{c_A^2} \frac{n_b}{n_p} \right]^{-1}, \quad \mathcal{E}_{mL}^2 = \left( \frac{4}{1 + \beta_0^2} \frac{n_b}{n_p} \right)^{3/2} \frac{\beta_0^2}{\gamma_0}. \tag{5.4.21}$$

The second term in denominator in (5.4.21), which is governed by the plasma nonlinearity, is comparable to unity at a beam density

$$n_b \simeq n_p(c_A/u_0)^2 \gamma_0^{-1}. \tag{5.4.22}$$

It is easy to see that this situation does not contradict inequality $\delta \ll \omega$, which sets the applicability limits for this approximation.

As mentioned above, the retarding properties of the plasma are governed by the appearance of an electron drift across the external magnetic field $\mathbf{B}_0$, at a velocity equal to

$$\mathbf{v}_L = c \frac{[\mathbf{E B_0}]}{B_0^2}. \tag{}$$

in the linear approximation.

The magnetic field of the wave makes the electron velocity lower than $\mathbf{v}_L$ and thus reduces the refractive index of the plasma. In the resultant magnetic field, $\mathbf{B}_\Sigma = \mathbf{B} + \mathbf{B}_0$, the electrons acquire a velocity

$$\mathbf{v}_{NL} = c \frac{[\mathbf{E B_\Sigma}]}{B_\Sigma^2}, \tag{5.4.23}$$

which is directed across $\mathbf{B}_\Sigma$. The angle through which the velocity vector is rotated $\theta$ (the angle between $\mathbf{v}_L$ and $\mathbf{v}_{NL}$) is governed by the manifestation of the longitudinal drift velocity $v_z = cB_0^{-2}[\mathbf{E B}]_z$ as the result of the acceleration of the plasma electrons under the influence of the longitudinal component of the Lorentz force. This angle is $\sin^{-1} E/b_0$. Thus

$$\mathbf{v}_\perp = |\mathbf{v}_{NL}| \cos\theta \simeq |\mathbf{v}_{NL}| \left( 1 - \frac{B^2}{B_0^2} \right), \tag{5.4.24}$$

161

which agrees with Eq. (5.4.11), derived in a formal manner.

A decrease in the refractive index of the plasma with increasing field amplitude leads to an increase in the wave phase velocity $v_{ph} = c/n(\mathcal{E})$ and thus to a phase change beyond that of the linear plasma. This effect clearly leads to a decrease in the maximum field amplitude (5.4.21).

At low frequencies $\omega \ll \omega_{Bi}$, the instability is accompanied by the drift of the plasma as a whole across $\mathbf{B}_0$, and the drift current is governed by ions. The nonlinearity is manifested in the appearance of a longitudinal drift and is of the same physical nature as the nonlinearity discussed above.

### §5.5. Amplitude-modulated wave in a plasma with relativistic electron beam.

A condition for a decrease in the phase velocity (the anomalous Doppler effect) does not satisfy in a plasma with an electron beam. Therefore, the beam electron can exchange energy with the electromagnetic wave under the conditions of the normal Doppler effect [27].

In the latter case, the dispersion equation (5.1.4) has the form

$$\left(\frac{ck}{\omega}\right)^2 = n_0^2(\omega) - \frac{\omega_b^2(\omega - kv_0)}{\omega^2(\omega - kv_0 - \omega_B^*)}, \tag{5.5.1}$$

where the refractive index of the plasma is

$$n_0^2(\omega) = 1 - \frac{\omega_p^2}{\omega(\omega - \omega_B)}, \tag{5.5.2}$$

$$\omega_b^2 = \frac{4\pi n_b e^2}{m\gamma_0}, \quad \omega_p^2 = \frac{4\pi n_p e^2}{m}, \quad \omega_B = \frac{|e|B_0}{mc},$$

$\omega_B^* = \omega_B/\gamma_0$ is the gyrofrequency of the beam electrons, and $\gamma_0$ is the relativistic factor.

If the resonance conditions for a plasma and a beam are satisfied

$$ck = \omega n_0(\omega), \quad \omega - kv_0 = \omega_B^*, \tag{5.5.3}$$

the frequency of small oscillations may be represented in the form

$$\omega = \omega_R + \Delta\omega, \quad \Delta\omega \ll \omega,$$

$$\omega_R = \frac{\omega_B^*}{1 - \beta_0 n_0(\omega_R)}, \quad \Delta\omega = \left[\frac{\omega_b^2 \omega_B^*}{2\omega Q(\omega_R)}\right]^{1/2}, \tag{5.5.4}$$

where $Q(\omega) = d(\omega^2 n_0^2)/d\omega^2$ takes into account a dispersion, $\beta_0 = v_0/c$.

The solution (5.5.4) corresponds to a circularly polarized wave which complex amplitude slowly varies with respect to time. In the case of the finite field amplitude the problem outgrows the limits of the linear theory, and the system of nonlinear equations must be used. According to [27], we consider a relativistic electron beam rotating across the external magnetic field with the gyrofrequency $\omega_B^*$ which corresponds to the following refractive index of the plasma,

$$n_0^2(\omega_B^*) = 1 + \frac{\omega_p^2}{\omega_B^2}\frac{\gamma_0}{\gamma_0 - 1}, \quad Q = 1 + \frac{\gamma_0(n_0^2 - 1)}{2(\gamma_0 - 1)}. \tag{5.5.5}$$

162

The corresponding system of nonlinear equations can be found from Eqs. (5.1.10) by means of the substitution $\Omega \to -\Omega$:

$$a' = -b\mathcal{E}\cos\eta, \quad b' = (n_0^2 - 1)a\mathcal{E}\cos\eta,$$
$$\mathcal{E}' = -q^2 a\cos\eta, \quad \eta' = 1 + \Omega b - \operatorname{tg}\eta(\ln\mathcal{E}a)', \tag{5.5.6}$$

where

$$a = \beta_\perp\gamma, \quad b = (\beta_\parallel n_0 - 1)\gamma, \quad \mathcal{E} = E/B_0,$$
$$q^2 = \omega_b^2/2\omega_B^2 Q, \quad \Omega = \omega/\omega_B, \quad d\tau = \omega_B dt/\gamma.$$

Integrating (5.5.6) subject to the initial conditions

$$a(0) = a_0, \quad \mathcal{E}(0) = \mathcal{E}_0, \quad b(0) = -\Omega^{-1},$$

we get the integrals of motion

$$\left(n_0^2 - 1\right)a^2 + b^2 = \left(n_0^2 - 1\right)a_0^2 + \Omega^{-2},$$
$$\mathcal{E}^2 = \mathcal{E}_0^2 - \frac{2q^2}{n_0^2 - 1}\left(\frac{1}{\Omega} + b\right), \quad \sin\eta = \frac{(1 + \Omega b)^2}{2a\mathcal{E}\Omega(n_0^2 - 1)}, \tag{5.5.7}$$

which allow us to reduce the order of equations (5.5.6)

$$\mathcal{E}'^2 = \delta^4 \Omega^2 a_0^2 - (\mathcal{E}^2 - \mathcal{E}_0^2)\left[\delta^2 + \frac{n_0^2 - 1}{2}(\mathcal{E}^2 - \mathcal{E}_0^2)\right] - \left[\frac{n_0^2 - 1}{8\delta^2\mathcal{E}}(\mathcal{E}^2 - \mathcal{E}_0^2)^2\right]^2, \tag{5.5.8}$$

where $\delta = \Delta\omega/\omega_B^* = q/\sqrt{\Omega}$ is the frequency of small oscillations with respect to time, $\tau$.

In the partial case $\mathcal{E}_0 = 0$, the equation (5.5.8) reduces to the form

$$w'^2 = \delta^2\left(1 - w^2 - gw^6\right), \tag{5.5.9}$$

where $w = \mathcal{E}/\delta\Omega a_0$ and $g = \left[\Omega^2 a_0^2\left(n_0^2 - 1\right)/8\delta\right]^2$. In going from (5.5.8) to (5.5.9), the small term of order $\delta \ll 1$ is dropped.

Accordingly to (5.5.9), the field amplitude changes periodically with time in the limits $0 \le w \le w_m$, so that the maximum field amplitude is

$$w_m = \begin{cases} 1, & g \ll 1 \\ g^{-1/6}, & g \gg 1. \end{cases} \tag{5.5.10}$$

For $g \ll 1$ when the phase velocity is closed to a speed of light, the nonlinear electromagnetic oscillations are asymptotically closed to monochromatic one. In the inverse case $g \gg 1$ a nonlinear saturation of the field amplitude arises as the result of mismatching in phase between the wave and the transverse velocity of a beam.

In the non-relativistic approximation, $\gamma_0 - 1 \simeq \beta_{\perp o}^2/2 \ll 1$, Eqs. (5.5.5) reduces to the following relation

$$n_0^2(\omega_B^*) = 1 + \frac{\omega_p^2}{\omega_B^2}\frac{2}{\beta_{\perp o}^2}, \quad Q = 1 + \frac{n_0^2 - 1}{\beta_{\perp o}^2}, \tag{5.5.11}$$

and we must put $a_0 \simeq \beta_{\perp o}$ and $\Omega \simeq 1$ in Eq. (5.5.8).

It follows from the momentum conservation law (5.5.7) that an increase in the amplitude of the wave propagating along the external magnetic field leads to an acceleration of the rotating electrons in the opposite direction. Note that the physical mechanism responsible for the conversion of the transverse electrons motion to a longitudinal motion should cause the loss of the beam electrons in the magnetic traps.

Note that the stability of a system of phased oscillators, i.e., oscillators with fixed phase in velocity space is investigated in Ref. [28]. Such a system can be obtained when a transverse electromagnetic wave propagates along a magnetic field. In this case the problem of stability of a system of phased oscillators is identical with the problem of stability of a wave propagating in a plasma along the magnetic field.

## §5.6. Autoresonance acceleration of plasma electrons by the field of a plane wave.

Charged particle can be accelerated to high energies in the field of a plane wave traveling along the external magnetic field $\mathbf{B}_0$ at the speed of light when the conditions of the normal Doppler effect are satisfy [13-16]

$$\omega^* = \omega_B^*, \quad \omega^* = \omega(1 - \beta_\parallel), \quad \omega_B^* = \omega_B/\gamma \qquad (5.6.1)$$

[$\omega^*$ is the frequency of the wave, $\omega_B^*$ is the gyrofrequency of particle, $\gamma = (1 - \beta^2)^{-1/2}$ is the relativistic factor, $\beta^2 = \mathbf{v}^2/c^2$, and $\mathbf{v}$ is the velocity of particle].

In this method of acceleration, the particles energy increases in the electric field $\mathbf{E}$ of the wave transverse to $\mathbf{B}_0$, and the nonzero magnetic field $\mathbf{B}$ of the wave converts the transverse motion to a longitudinal motion and accelerates the particle along the $\mathbf{B}_0$ field as a result of the Loretz force

$$F_\parallel = \frac{e}{c} \frac{\mathbf{B}_0}{B_0} [\mathbf{v}\mathbf{B}]. \qquad (5.6.2)$$

The increase in the longitudinal velocity $v_\parallel$ in turn decreases the effective wave frequency $\omega^* = \omega(1 - \beta_\parallel)$ and thereby enhances the relativistic mass increase and the decrease in the gyrofrequency $\omega_B^* = \omega_B/\gamma$.

The trapped particles, for which

$$(1 - \beta_\parallel)\gamma = (1 - \beta_0)\gamma_0 = b_0, \qquad (5.6.3)$$

therefore remain in resonance with the wave at relativistic energies, and the energy of the charged particles in the wave field increases without limit if there is no magnetic bremsstrahlung [17].

In the case of the finite density beam ($n_b > 0$) the problem outgrows the limits of the single particle approach ($4\pi n_b mc^2\gamma \ll E_0^2$) [13-16], and the accelerated particle energy gets limited by the beam spatial charge [19-21].

We assume that the solution of Eqs. (5.1.1) and (5.1.2) will be sought in the form of circularly polarized waves with the slowly-varying amplitudes and phases propagating in a plasma along the external magnetic field at the speed of light

$$B_x - iB_y = E_x + iE_y = E(t) \exp\left[-i\Phi - i\varphi(t)\right],$$

$$v_x + iv_y = v(t) \exp\left[-i\Phi - i\vartheta(t)\right], \qquad (5.6.4)$$

$$\Phi = \omega(t - z/c), \quad n_b(t, z) = n_b, \quad v_z(t, z) = v_\parallel(t).$$

The corresponding system of nonlinear equations can be found from Eqs. (5.1.7) by means of the substitution $n_0 = 1$ and $\omega \to -\omega$:

$$\frac{d}{dt}\gamma v_\perp = \frac{e}{m}\left(1 - \frac{v_\parallel}{c}\right) E \cos\eta, \qquad \frac{dE}{dt} = -2\pi e n_b v_\perp \cos\eta,$$

$$\frac{d}{dt}\gamma v_\parallel = \frac{e}{m}\left(\frac{v_\perp}{c} E \cos\eta + E_\parallel\right), \qquad \frac{dE_\parallel}{dt} = -4\pi e n_b v_\parallel, \qquad (5.6.5)$$

$$\frac{d\eta}{dt} = \frac{\omega_B}{\gamma} - \omega\left(1 - \frac{v_\parallel}{c}\right) - \operatorname{tg}\eta \frac{d}{dt}\ln\left(\gamma v_\perp E\right).$$

164

In comparison with Eq. (5.1.7), the above system takes into account the longitudinal polarization electric field, $E_\parallel$, which appears upon electron shift with respect to immobile background ions.

Introducing the vector potential $E_\parallel = -\dot{A}_\parallel/c$ and integrating (5.6.5) subject to the initial conditions $A_\parallel(0) = \dot{A}_\parallel(0) = 0$, we find the following system of integrals:

$$n_b mc^2(\gamma - \gamma_0) + \frac{E^2 - E_0^2}{4\pi} + \frac{\dot{A}_\parallel^2}{8\pi c^2} = 0, \quad n_b m(v_\parallel \gamma - v_0\gamma_0) + \frac{E^2 - E_0^2}{4\pi c} + \frac{eA_\parallel}{c} = 0,$$

$$Ev_\perp\gamma\sin\eta = -\frac{\omega_B}{e}\left(\frac{E^2}{4\pi n_b} + \frac{m\omega}{2\omega_B}\gamma^2 v_\perp^2\right) + C,$$

$$(5.6.6)$$

where $v_0$ and $mc^2\gamma_0$ are the initial longitudinal velocity and energy of electrons, and $C$ the constant of integration which satisfy the condition $\eta = 0$.

Let us use the integrals (5.6.6) to represent the equation $\dot{E}_\parallel = -4\pi e n_b v_\parallel$ in the form

$$\gamma\ddot{A}_\parallel + \omega_b^2 A - \frac{e\dot{A}_\parallel^2}{mc^2} = 4\pi e n_b c^2(\gamma - \gamma_0), \qquad (5.6.7)$$

where $\omega_b^2 = 4\pi e^2 n_b/m$ is the Langmuir frequency of the beam.

In the dimensionless variables

$$\mathcal{E} = E/B_0, \quad A = eA_\parallel/mc^2, \quad d\tau = \omega_B dt, \quad \beta = \mathbf{v}/c,$$

$$\Omega = \omega/\omega_B, \quad q^2 = \omega_b^2/\omega_B^2 \ll 1,$$

the formulas (5.6.6) become

$$\mathcal{E}^2 + q^2\gamma + \dot{A}^2/2 = \mathcal{E}_0^2 + q^2\gamma_0, \quad \mathcal{E}^2 + q^2(\gamma\beta + A) = \mathcal{E}_0^2 + q^2\gamma_0\beta_0,$$

$$\mathcal{E}\beta_\perp\gamma\sin\eta = -\frac{\mathcal{E}^2}{q^2} - \frac{\Omega}{2}\beta_\perp^2\gamma^2 + \mathcal{C}.$$

$$(5.6.8)$$

Let us transform the right-hand side of third formula (5.6.8). Eliminating the term $\mathcal{E}^2/q^2$ by means of an energy conservation law [first formula (5.6.8)], and using the expressions

$$\beta_\perp^2\gamma^2 = 2\gamma b - b^2 - 1, \quad b = \gamma(1 - \beta_\parallel), \quad b_0 = 1/\Omega, \qquad (5.6.9)$$

we get the relation

$$2\mathcal{E}\beta_\perp\gamma\sin\eta = \left(1 - \frac{b}{b_0}\right)(2\gamma - b - b_0) + \frac{\dot{A}^2}{q^2}, \qquad (5.6.10)$$

at which the parameter $b$ is determined by the difference between the energy and the momentum integrals:

$$b = b_0 + A - \dot{A}^2/2q^2. \qquad (5.6.11)$$

With the help of Eqs. (5.6.8)–(5.6.11), the set (5.6.5) reduces to the following equations

$$\frac{d\gamma}{d\tau} = \beta_\perp\mathcal{E}\cos\eta - \frac{d}{d\tau}\frac{\dot{A}^2}{2q^2}, \quad \gamma\frac{d^2A}{d\tau^2} - \left(\frac{dA}{d\tau}\right)^2 + q^2 A = q^2(\gamma - \gamma_0), \qquad (5.6.12)$$

165

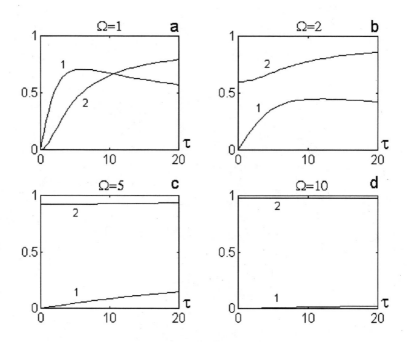

Fig. 5.6.1. Transverse $\beta_\perp$ (curve 1) and longitudinal $\beta_\parallel$ (curve 2) velocities of electron in the single particle model $q=0$ for $\beta_\perp(0)=0$ and $\gamma_0=(1+\Omega^2)/2\Omega$.

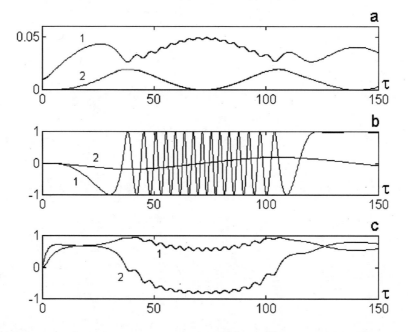

Fig. 5.6.2. Initial stage of self-consistent acceleration ($q=0.1$): (a) energy density of electrons $q^2\gamma$ (1) and longitudinal field $\mathcal{E}_\parallel^2/2$ (2); (b) beam phase $\sin\eta$ (1) longitudinal field $\mathcal{E}_\parallel$ (2); (c) transverse $\beta_\perp$ (1) and longitudinal $\beta_\parallel$ (2) velocities of electrons.

which describe a connected (transverse and longitudinal) nonlinear oscillations.

We use the new variables

$$X = \beta_\perp \gamma \mathcal{E} \cos\eta, \quad Y = \beta_\perp \gamma \mathcal{E} \sin\eta$$

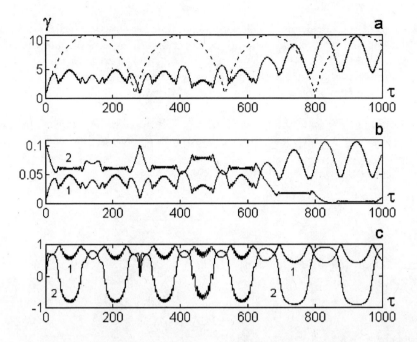

Fig. 5.6.3. Transformation of electromagnetic energy into the energy of plasma electrons (next stage of self-consistent acceleration): (a) energy of plasma electrons $\gamma$; (b) energy density of plasma electrons $q^2\gamma$ (1) and transverse field $\mathcal{E}^2$ (2); (c) transverse $\beta_\perp$ (1) and longitudinal $\beta_\parallel$ (2) velocities of electrons. Dashed curve corresponds to the solution of Eq. (5.6.17).

allowing to eliminate $\eta$ from Eqs. (5.6.5):

$$\frac{d\mathcal{E}^2}{d\tau} = -\frac{q^2}{\gamma}X, \quad \frac{dY}{d\tau} = \frac{1}{\gamma}(1 - \Omega b)X,$$

$$\frac{dX}{d\tau} = \frac{1}{\gamma}\left[-(1-\Omega b)Y + b\mathcal{E}^2 + \frac{q^2}{\gamma}(1 + b^2 - 2b\gamma)\right], \qquad (5.6.13)$$

$$\frac{db}{d\tau} = -\frac{b}{\gamma}\mathcal{E}_\parallel, \quad \frac{d\mathcal{E}_\parallel}{d\tau} = -q^2\left(1 - \frac{b}{\gamma}\right), \quad \frac{d\gamma}{d\tau} = \frac{1}{\gamma}X + \left(1 - \frac{b}{\gamma}\right)\mathcal{E}_\parallel.$$

An initial stage of acceleration which corresponds to "autoresonance" initial conditions:

$$\mathcal{E} = \mathcal{E}_0, \quad b_0 = \Omega^{-1}, \quad \eta = 0 \qquad (5.6.14)$$

coincides with the single particle model [13,14]. The results of numerical integration of Eqs. (5.6.13) for $\mathcal{E}_0^2 = 0.1$ and for different initial values of the parameter $\Omega$ is represented in Fig. 5.6.1.

Figure 5.6.2 illustrates the solution of the self-consistent problem ($\mathcal{E}_0^2 = 0.1$, $q = 0.1$ and $\Omega = \gamma_0 = 1$). The appearance of electrostatic field $\mathcal{E}_\parallel$, which is caused by the acceleration of plasma electrons relatively ions, accompanies by the mismatching in phase $\eta$ between the wave and the transverse velocity of a beam [see Eq. (5.6.10)]. As the result, a nonlinear saturation of the electron energy arises when $\sin\eta \simeq -1$ (Fig. 5.6.2a and Fig. 5.6.2b). After this the particle reaches a retarding phase and loses energy. The first energy minimum coincides with a maximum of the longitudinal field.

Next stage of self-consistent acceleration characterizes by the excitation of nonlinear phase oscillations and by two scales energy modulation arising in a plasma with an electromagnetic wave (Fig. 5.6.2). Accordingly to Fig. 5.6.3, the maximum amplitude (averaged over the nonlinear oscillation period) increases with respect to time so that this process accompanies by the transformation of the energy of the wave into a kinetic energy of plasma electrons.

A qualitative behavior of the above nonlinear solution can be analyzed in the absence of the electrostatic field which lead to an appearance of phase oscillations. Neglecting this effect, we put

$$A = \dot{A} = \eta \equiv 0, \quad b \equiv b_0$$

and rewrite Eq. (5.6.12) in the form

$$\dot{\gamma} = \beta_\perp \sqrt{\mathcal{E}_0^2 + q^2(\gamma_0 - \gamma)}. \tag{5.6.15}$$

Taking into account (5.6.9), we introduce the new variable $w$ with the help of relations (see **Appendix 5.2**)

$$\beta_\perp = \frac{2b_0}{\sqrt{1 + b_0^2}} w, \quad \gamma = \frac{1 + b_0^2}{2b_0}(1 + w^2)$$

and we transform (5.6.15) to the form [19]

$$\dot{w} = \frac{Q}{1 + w^2} \sqrt{w_m^2 - w^2}, \tag{5.6.16}$$

$$Q = q\sqrt{2} \frac{b_0^{3/2}}{1 + b_0^2}, \quad w_m^2 = w_0^2 + \frac{2b_0}{1 + b_0^2} \frac{\mathcal{E}_0^2}{q_0^2},$$

where $w_0$ is determined by the initial energy of electrons [19]. Integrating (5.6.16), we find

$$\left(1 + \frac{w_m^2}{2}\right) \arcsin\left(\frac{w}{w_m}\right) - \frac{w}{2}\sqrt{w_m^2 - w^2} = Q\tau. \tag{5.6.17}$$

Accordingly to (5.6.17), the electron reaches a maximum energy $w_m$ at the moment

$$\tau_m = \frac{\pi}{2Q}\left(1 + \frac{w_m^2}{2}\right), \tag{5.6.18}$$

and after decreases along the magnetic field to reach its initial energy. This solution is represented in Fig. 5.6.3a by a dashed curve.

### §5.7. Influence of Coulomb collisions on autoresonant acceleration in a plasma.

As distinct from §5.6, where all plasma electrons are locked into autoresonant acceleration, we will assume that only a small part of fast particles is trapped by a wave propagating in a plasma along the external magnetic field $\mathbf{B}_0$ at the speed of light:

$$\frac{\omega^*}{\omega_B^*} = \frac{\omega}{\omega_B}\left(1 - \beta_\parallel\right)\gamma = 1 \tag{5.7.1}$$

We then have a two-component system in which a relatively small number of fast electrons, $n_b \ll n_p$, pass through a cold dense plasma. At low energies $v \ll c$, Coulomb collisions of the resonant particles with the background plasma do not limit the energy increase transverse to magnetic field. However, for $v \simeq c$ the relative decrease in the gyrofrequency takes the electrons out of resonance with the wave and acceleration is terminated [18].

The kinetic equation for the accelerated component of the plasma takes the form

$$\frac{\partial f}{\partial t} + v_z \frac{\partial f}{\partial z} + e \left( \mathbf{E} + \frac{1}{c}[\mathbf{v}, \mathbf{B} + \mathbf{B}_0] \right) \frac{\partial f}{\partial \mathbf{p}} = I(f), \qquad (5.7.2)$$

where $I(f)$ is the Landau relativistic collision integral, $v_z$ is the projection of the velocity on $\mathbf{B}_0$, and $\mathbf{p}$ is the electron momentum.

The inclusion of Coulomb scattering of the accelerated beam by the plasma particles makes the kinetic equation considerably more complicated, since the distribution function generally depends in a complicated way on the integrals of motion and evolves both spatially and temporally as a result of collisions (beam heating). However, we will show below that the longitudinal motion of the particles dominates and that the beam does not heat up significantly during acceleration. We may therefore use the relativistic hydrodynamic equations to truncate the system of moment equations and assume an accelerated particle distribution function in the form

$$f = \frac{N}{4\pi m^2 c T K_2(mc^2/T)} \exp \left[ -\frac{\Gamma}{T} \left( c\sqrt{mc^2 + p^2} - \mathbf{pU} \right) \right], \qquad (5.7.3)$$

where $K_2$ is a modified Bessel function of the second kind; $\Gamma = (1 - U^2/c^2)^{-1/2}$, $N$, $T$, $\mathbf{U}$ are the density, temperature, and average velocity of the beam and depend on $t$ and $z$.

The system of equations has the following form [29]

$$\frac{\partial}{\partial t}(N\Gamma) + \frac{\partial}{\partial z}(N\Gamma U_z) = 0,$$

$$\left( \frac{\partial}{\partial t} + U_z \frac{\partial}{\partial z} \right)(mG\Gamma \mathbf{U}) = e \left( \mathbf{E} + \frac{1}{c}\, [\mathbf{U}, \mathbf{B} + \mathbf{B}_0] \right)$$

$$- \nu_c m\mathbf{U} - \frac{1}{N\Gamma} \frac{\partial N T}{\partial z} \frac{\mathbf{B}_0}{B_0}, \qquad (5.7.4)$$

$$N \left( \frac{\partial}{\partial t} + U_z \frac{\partial}{\partial z} \right)(mc^2 G - T) - T \left( \frac{\partial}{\partial t} + U_z \frac{\partial}{\partial z} \right) N$$

$$= 2\nu_c N mc^2 \left( G - \frac{4T}{mc^2} \right),$$

where $G = K_3\left(mc^2/T\right)/K_2(mc^2/T)$, $\nu_c = 4\pi e^4 n_p \Lambda/m^2 c^3$, $\Lambda$ is the Coulomb logarithm, and $n_p$ is the density of the background plasma.

We can make the simplifying approximation $\Gamma \gg 1$ when calculating the terms proportional to $\nu_c$ arising from the collision integral, since we are interested in high energies. In addition, we assume that the background plasma consists of singly charged particles with a distribution function of the form $f_{e,i} = n_p \delta(\mathbf{p})$. Concerning the last assumption we note that the results of calculating the collision terms will remain unchanged if we the assume the electrons in the background plasma obey a Maxwellian distribution with temperature $T_e \ll mc^2$ and linearize the collision integral [30].

Let us switch to the variation $\psi = \omega(t - z/c)$,

$$B_x - iB_y = E_x + iE_y = E_0 \exp(-i\psi).$$

169

We assume that the beam is sufficiently cool [5]

$$G \simeq 1 + \frac{5T}{2mc^2}, \quad T \ll mc^2,$$

and rewrite system (5.7.4) in terms of the new dimensionless variables

$$U = \frac{\Gamma}{c}(U_x + iU_y)\exp(i\psi), \quad b = \Gamma\left(1 - \frac{U_z}{c}\right),$$

$$\mathcal{E} = \frac{eE_0}{mc\omega}, \quad \mathcal{T} = \frac{T}{mc^2}, \quad \Omega = \frac{\omega}{\omega_B}, \quad \nu = \frac{\nu_c}{\omega}. \tag{5.7.5}$$

to get the system

$$\dot{U} = \mathcal{E} + i\left(1 - \frac{1}{\Omega b}\right)U - \nu\frac{U}{b}, \tag{5.7.6}$$

$$\dot{b} = -\nu\left(1 + \frac{4}{3b}\right)\left(1 - \frac{5\mathcal{T}}{3\Gamma b}\right)^{-1}, \tag{5.7.7}$$

$$\dot{\mathcal{T}} + \frac{2\dot{b}}{3b}\mathcal{T} = \frac{4\nu}{3}\frac{\Gamma}{b}, \tag{5.7.8}$$

$$\Gamma = \frac{1}{2b}(1 + b^2 + |U|^2), \quad N = N_0\frac{b_0}{b}. \tag{5.7.9}$$

The dots denote derivatives with respect to $\psi$.

If there are no collisions, so that $\nu = 0$, we have the resonance solution

$$U = \mathcal{E}\psi, \quad b = b_0 = \Omega^{-1}, \quad \mathcal{T} = 0, \tag{5.7.10}$$

and the transverse momentum of the beam trapped by the wave grows without limit.

However, if $\nu > 0$ the phase of the beam particles shifts during acceleration (the function $U$ becomes complex) and beam-wave resonance breaks down. In this case we can find a simple solution of system (5.7.6)–(5.7.9) if we assume that

$$\nu \ll \nu\psi \ll 1, \quad b_0 \simeq 1, \tag{5.7.11}$$

so that we can neglect the frictional term in Eq. (5.7.6) and omit the thermal correction to unity in Eq. (5.7.7). We then find from Eq. (5.7.7) that

$$b - b_0 - \frac{4}{3}\ln\left(\frac{4 + 3b}{4 + 3b_0}\right) = -\nu\psi. \tag{5.7.12}$$

Since the right-hand side of (5.7.12) is small by (5.7.11), we can expand the left-hand side in powers of $1 - b/b_0 \ll 1$. Retaining the leading term of the expansion, we find that

$$b = b_0 - \nu\psi\left(1 + \frac{4}{3b_0}\right). \tag{5.7.13}$$

---

[5]  Using the asymptotic expansions [31]

$$K_n(x) \simeq \left(\frac{\pi}{2x}\right)^{1/2}e^{-z}\left(1 + \frac{4n^2 - 1}{8x}\right), \quad x \gg 1.$$

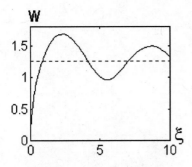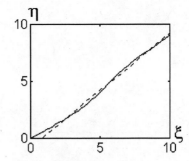

Fig. 5.7.1. Transverse momentum $W = (\sqrt{\delta}/\mathcal{E})|U|$ and phase $\eta$ of the beam as a function of $\xi = \delta\psi^2/2.$ . The dashed lines indicates the asymptotic values (5.7.16).

Upon substituting (5.7.13) into Eq. (5.7.6) and recalling (5.7.10), we have

$$\dot{U} = \mathcal{E} - i\delta\psi U, \quad \delta = \frac{\nu}{b_0}\left(1 + \frac{4}{3b_0}\right) \ll 1. \tag{5.7.14}$$

The solution of this equation satisfying zero initial conditions can be expressed in terms of the Fresnel sine and cosine integrals [31]

$$C(\xi) = \sqrt{\frac{2}{\pi}} \int_0^{\sqrt{\xi}} \cos t^2 \, dt, \quad S(\xi) = \sqrt{\frac{2}{\pi}} \int_0^{\sqrt{\xi}} \sin t^2 \, dt.$$

If we write $U$ in the form
$$U = U_1 + U_2 = |U| \exp(-i\eta),$$

we have

$$U_1 = \mathcal{E}\sqrt{\frac{\pi}{\delta}}\left[C(\xi)\cos\xi + S(\xi)\sin\xi\right], \quad U_2 = \mathcal{E}\sqrt{\frac{\pi}{\delta}}\left[S(\xi)\cos\xi - C(\xi)\sin\xi\right], \tag{5.7.15}$$

where $\xi = \delta\xi^2/2$.

For small $\xi \ll 1$, Eqs. (5.7.15) reduces to (5.7.10). For large $\xi \gg 1$ we can use the asymptotics values of the Fresnel integrals [31] and obtain

$$C(\xi) \simeq S(\xi) = 1/2, \quad U_\infty(\xi) \simeq \sqrt{\frac{\pi}{2\delta}}\mathcal{E}\exp i\left(\frac{\pi}{4} - \xi\right). \tag{5.7.16}$$

Figures 5.7.1 illustrate the approach of the solution (5.7.15) to the asymptotic values (5.7.16). We see that the characteristic time for the energy of the accelerated beam to increase and for beam phase to fall out of resonance with the wave is $\xi \simeq 1$.

By Eqs. (5.7.8), (5.7.9), the energy and temperature of the beam corresponding to (5.7.16) are equal to

$$\Gamma_\infty = \frac{1}{2b_0}\left(1 + b_0^2 + \frac{\pi\mathcal{E}^2}{2\delta}\right), \quad \mathcal{T}_\infty \simeq \sqrt{\delta}\Gamma_\infty\left(1 + \frac{4}{3b_0}\right)^{-1}. \tag{5.7.17}$$

Inequality (5.7.11), which determines the validity of Eqs.(5.7.16) and (5.7.17), is satisfied if

$$\sqrt{\xi\delta} \ll 1 + 4/3b_0 \tag{5.7.18}$$

171

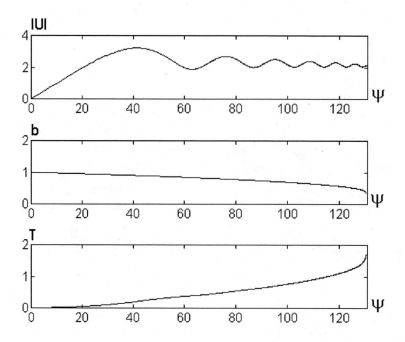

Fig. 5.7.2. Numerical solution of Eqs. (5.7.6)–(5.7.9) corresponding to the initial conditions (5.7.20). For $\mathcal{T} > 1$ a singularity in Eq. (5.7.7) arises.

and since $\xi \simeq 1$, the constraint (5.7.18) is not significant.

We observe that by Eq. (5.7.17) the longitudinal energy of the beam exceeds the beam temperature,

$$\mathcal{T}_\infty / \Gamma_\infty \simeq \sqrt{\delta} \ll 1 \tag{5.7.19}$$

so that collisional dephasing of the beam relative to the wave is primarily responsible for limiting the energy increase. We also remark that the condition $\mathrm{T} \ll 1$ assumed above imposes the additional constraint $\mathcal{E}^2 / \sqrt{\delta} \ll 1$.

Numerical solution of set (5.7.6)–(5.7.9) which satisfy the conditions

$$\mathcal{E} = 10^{-1}, \quad \nu = 10^{-3}, \quad b_0 = 1, \quad U_0 = \mathcal{T}_0 = 0, \tag{5.7.20}$$

is represented in Fig. 5.7.2. It follows from this figure that the temporal growth in the transverse momentum amplitude accompanies by the decrease in the parameter $b < 1$ takes the electrons out of resonance with the wave and acceleration is terminated [18]. Because of the beam temperature increases, a singularity in Eq. (5.7.7) in the range $\mathcal{T} > 1$ arises.

The influence of the radiative-damping force on autoresonant acceleration is considered in **Appendix 5.3**

## §5.8. Acceleration of charged particle in the field of a plane wave with increasing phase velocity.

In the absence of the external magnetic field a synchronism between the accelerated charge particle and the wave, $v_\| = v_{\mathrm{ph}}$, may be realized by means of increasing in the phase velocity $v_{\mathrm{ph}}(z)$ along the acceleration length [11]. At the same time autophasing occurs, since during acceleration the phase spread of the particles in an accelerated bunch decreases (by analogy with longitudinal wave [32]).

172

The coordinate of particle $\mathbf{r}(t)$ satisfy the equation

$$m\frac{d\mathbf{v}}{dt} = e\mathbf{E} + \frac{e}{c}\,[\mathbf{v}, \mathbf{B}], \quad \frac{d\mathbf{r}}{dt} = \mathbf{v}, \tag{5.8.1}$$

where $\mathbf{E}$ and $\mathbf{B} = \sqrt{\epsilon}\,[\mathbf{n}, \mathbf{E}]$ are the electric and magnetic field of the wave, $\epsilon(z)$ is the effective dielectric constant of the retarding medium,

$$E_x = B_y = E_0 \cos\psi,$$

$$\psi = \psi_0 + \omega\left(t - \int_0^{z(t)} \frac{dz'}{v_{\rm ph}(z')}\right), \tag{5.8.2}$$

and the phase velocity is determined by formula $v_{\rm ph} = c/\sqrt{\epsilon}$.

Let us use the relation

$$[\mathbf{v}, [\mathbf{n}, \mathbf{E}]] = \mathbf{n}(\mathbf{vE}) - \mathbf{E}(\mathbf{nv}),$$

to represent the vector equation (5.8.1) in the projections

$$m\frac{dv_x}{dt} = eE_0\left(1 - \frac{v_z}{v_{\rm ph}}\right)\cos\psi, \quad \frac{dv_y}{dt} = 0,$$

$$m\frac{dv_z}{dt} = eE_0\,\frac{v_x}{v_{\rm ph}}\cos\psi \tag{5.8.3}$$

(in the quite general case, we can put $v_y = 0$).

For a synchronous particle, $z = z_s(t)$ and $v_z = \dot{z}_s(t)$, moving in a fixed phase $\psi_s$ with a constant transverse velocity $v_x = v_{\perp s}$ (as a result of the combined effect of the electric field and the transverse projection of the Lorentz force), the equation of motion has the form

$$m\frac{dv_z}{dt} = eE_0\,\frac{v_{\perp s}}{v_s}\cos\psi_s. \tag{5.8.4}$$

Integrating (5.8.4), we find an implied relation for the function $v_s(t)$

$$v_s(t) = \sqrt{v_s^2(0) + \frac{2eE_0}{m}v_{\perp s}\cos\psi_s\, t}\,. \tag{5.8.5}$$

For small deviations from the trajectory of a synchronous particle

$$\psi = \psi_s + \varphi, \quad |\varphi| \ll \psi_s, \quad v_x = v_{\perp s} + q, \quad q \ll v_{\perp s},$$

upon linearizing Eq. (5.8.3) we obtain the equation

$$\ddot{\varphi} + \frac{\dot{v}_s}{v_s}\dot{\varphi} + \frac{eE_0}{mv_s^2}\left(\frac{eE_0}{v_s^2}\cos^2\psi_s - v_{\perp s}\sin\psi_s\right)\varphi = 0 \tag{5.8.6}$$

(the transverse velocity perturbation is equal to $q = \varphi\cos\psi_s$).

It follows from Eq. (5.8.6) that the region of stable phases is given by the inequality $0 > \psi_s > -\pi/2$,

$$\Omega_{\rm ph}^2 = \frac{eE_0}{mv_s^2}\left(\frac{eE_0}{v_s^2}\cos^2\psi_s - v_{\perp s}\sin\psi_s\right) > 0.$$

173

The small oscillations decrease exponentially for $\dot{v}_s > 0$ and, hence, a particle energy asymptotically reaches a synchronous particle energy.

In this method of acceleration, the particles energy growth decreases in the region of relativistic energies, $1 - \beta_s^2 \ll 1$, at which the transverse velocity decreases

$$v_{\perp s}^* = cp^* \sqrt{\frac{1 - \beta_s^2}{1 + p^{*2}}}, \quad p^* = \frac{p_\perp}{mc}. \tag{5.8.7}$$

(the transverse momentum $p_{\perp s}$ remains constant).

A synchronous particle velocity $\beta_s$ time-dependence is determined by the relation

$$\frac{\beta_s^2(t)}{1 - \beta_s^2(t)} - \frac{\beta_s^2(0)}{1 - \beta_s^2(0)} + \frac{2eE_0}{mc} \frac{p^*}{1 + p^{*2}} t, \tag{5.8.8}$$

and the frequency of relativistic phase oscillations becomes $\Omega_{\mathrm{ph}}^* \sim \Omega_{\mathrm{ph}}(1 - \beta_s^2)$ [6].

### Appendix 5.1

Eliminating the magnetic field from first equation (5.1.1), we rewrite the Maxwell equations in the form

$$\operatorname{grad} \operatorname{div} \mathbf{E} - \Delta E = -\frac{1}{c^2} \frac{\partial}{\partial t} \left( \frac{\partial D}{\partial t} + \frac{4\pi e}{c} n_b \mathbf{v} \right), \tag{A5.1.1}$$

where $\mathbf{D} = \hat{\epsilon} \mathbf{E}$, and the operator $\hat{\epsilon}$ models the effect of the plasma. If a charge of "incompressible" beam

$$\frac{\partial n_b}{\partial t} = -n_b \operatorname{div} \mathbf{v} = 0, \quad \operatorname{div} \mathbf{E} = 0 \tag{A5.1.2}$$

is neutralized by ions $\operatorname{div} \mathbf{E} = 0$, the equation (A5.1.1) takes the form

$$\Delta \mathbf{E} - \frac{1}{c^2} \frac{\partial^2 \mathbf{D}}{\partial t^2} = \frac{4\pi e n_b}{c^2} \frac{\partial \mathbf{v}}{\partial t} \tag{A5.1.3}$$

In the considered case of a low-density beam, a solution of above equation can be represented in the form

$$\mathbf{E}(t, z) = \mathbf{E}(t) \exp(i\omega t - ikz), \quad \dot{\mathbf{E}} \ll \omega \mathbf{E}. \tag{A5.1.4}$$

To transform the left-hand side of Eq. (A5.1.3), we will use the relation [22]

$$\frac{\partial^2 \mathbf{D}}{\partial t^2} \simeq \left[ -\omega^2 \epsilon(\omega) \mathbf{E} + i \frac{\partial}{\partial t} \frac{\partial(\omega^2 \epsilon)}{\partial \omega} \mathbf{E} \right] \exp(i\omega t). \tag{A5.1.5}$$

Substituting (A5.1.5) into (A5.1.3), we get the equation for a complex field amplitude slowly varying with respect to time

$$\frac{2i\omega}{c^2} \frac{d}{dt} Q\mathbf{E} + \left( k^2 - \frac{\omega^2 \epsilon}{c^2} \right) \mathbf{E} = -\frac{4\pi e n_b}{c^2} i\omega \mathbf{v}, \tag{A5.1.6}$$

where $Q = \partial(\omega^2 \epsilon)/\partial \omega^2$ takes into account a dispersion of retarding medium.

---

[6] The frequency of phase oscillations in the case of a longitudinal wave is equal to $\Omega_{\mathrm{ph}}^* \sim \Omega_{\mathrm{ph}}(1 - \beta_s^2)^{3/4}$ [32].

We represent the electric field in the form

$$E_x + iE_y = E \exp(-i\varphi), \quad v_x + iv_y = E \exp(-i\vartheta) \tag{A5.1.7}$$

and separate the real and imaginary parts in Eq. (A5.1.6)

$$\frac{d}{dt}QE = -\frac{2\pi e n_b}{\omega} v_\perp \cos(\vartheta - \varphi),$$

$$\frac{d\varphi}{dt} = -\frac{\omega}{2Q}\left(\frac{c^2 k^2}{\omega^2} - \epsilon\right) - \frac{2\pi e n_b}{\omega Q E}\sin(\vartheta - \varphi). \tag{A5.1.8}$$

Eliminating the magnetic field of the wave $\mathbf{B} = (c/\omega)[\mathbf{kE}]$ from equation (5.1.2) and using the relation

$$[\mathbf{v}, \mathbf{B}] = (c/\omega)\{\mathbf{k}(\mathbf{vE}) - \mathbf{E}(\mathbf{kv}),$$

we obtain the following system of equations

$$\left[\frac{d}{dt} + i(\omega - kv_\parallel)\right]\gamma(v_x + iv_y)$$

$$= \frac{e}{m\omega}(\omega - kv_\parallel)(E_x + iE_y) - -i\omega_B(v_x + iv_y), \tag{A5.1.9}$$

$$\frac{d}{dt}\gamma v_\parallel = \frac{ek}{m\omega}(v_x E_x + v_y E_y),$$

where $\omega_B = eB_0/mc$ and $\gamma = (1 - \mathbf{v}^2/c^2)^{-1/2}$.

Substituting (A5.1.7) in (A5.1.9) and separating the real and imaginary parts, we find

$$\frac{d}{dt}(\gamma v_\perp) = \frac{e}{m}\left(1 - \frac{kv_\parallel}{\omega}\right)E\cos(\vartheta - \varphi),$$

$$\frac{d\vartheta}{dt} = (\omega - kv_\parallel) + \frac{\omega_B}{\gamma} - \frac{e}{m}\left(1 - \frac{kv_\parallel}{\omega}\right)\frac{E}{\gamma v_\perp}\sin(\vartheta - \varphi). \tag{A5.1.10}$$

For going from Eqs. (A5.1.8) and (A5.1.10) to Eqs. (5.1.7) and (5.1.18) we must put $\eta = \vartheta - \varphi$, $k = \omega n_0/$, $n = \sqrt{\epsilon}$ and $Q = d(\omega^2 n^2)/d\omega^2$.

**Appendix 5.2**

We can use the following integral of motion

$$(1 - \beta_\parallel)(1 - \beta_\perp^2 - \beta_\parallel^2)^{-1/2} = b_0, \tag{A5.2.1}$$

to define the longitudinal velocity $\beta_\parallel$ as a function of the transverse velocity $\beta_\perp$,

$$\beta_\parallel = \frac{1}{1 + \mu}\left(1 - \mu\sqrt{1 - \frac{1 + \mu}{\mu}\beta_\perp^2}\right), \quad \mu = b_0^2. \tag{A5.2.2}$$

Using (A5.2.2), we will introduce the new variable

$$\beta_\perp = \sqrt{\frac{\mu}{1 + \mu}}\sin\varphi, \quad \beta_\parallel = \frac{1}{1 + \mu}(1 - \mu\cos\varphi), \tag{A5.2.3}$$

175

which allows to represent the energy and momentum in the form

$$\gamma = \frac{\sqrt{\mu}}{1-\beta_{\parallel}} = \frac{1+\mu}{\sqrt{\mu}}\frac{1}{1+\cos\varphi}, \quad \beta_{\perp}\gamma = \sqrt{1+\mu}\frac{\sin\varphi}{1+\cos\varphi}. \tag{A5.2.4}$$

Substituting (A5.2.3) and (A5.2.4) in first equation (5.6.5) (for $\eta = 0$)

$$\frac{d}{d\tau}\gamma\beta_{\perp} = (1-\beta_{\parallel})\mathcal{E}, \tag{A5.2.5}$$

we find

$$\frac{d\varphi}{d\tau} = \frac{\mu}{(1+\mu)^{3/2}}(1+\cos\varphi)^2\mathcal{E}. \tag{A5.2.6}$$

The electric field $\mathcal{E}$ as a function of $\varphi$ can be found from an energy conservation law (5.6.8) (for $\dot{A} = 0$)

$$\mathcal{E} = \sqrt{\mathcal{E}_0^2 + q^2\left(\gamma_0 - \frac{1+\mu}{\sqrt{\mu}}\frac{1}{1+\cos\varphi}\right)}. \tag{A5.2.7}$$

Using the relation

$$1+\cos\varphi = 2/\left[1+tg^2(\varphi/2)\right], \quad w = tg(\varphi/2),$$

one can represent (A5.2.6) and (A5.2.7) in the form

$$\frac{dw}{d\tau} = \frac{2\mu}{(1+\mu)^{3/2}}\sqrt{\mathcal{E}_0^2 + q^2\left[\gamma_0 - \frac{1+\mu}{2\sqrt{\mu}}(1+w^2)\right]} \tag{A5.2.8}$$

which is identical with (5.6.16).

## Appendix 5.3

We consider the influence of the radiative-damping force [33]

$$\mathbf{f} = \frac{2e^4}{3m^2c^5}\left(F_{kl}u^l\right)\left(F^{km}u_m\right)\mathbf{v} \tag{A5.3.1}$$

[in the ultrarelativistic approximation $1-v^2/c^2 \ll 1$] on autoresonant acceleration of charged particle in the field of a plane wave traveling along a magnetic field $\mathbf{B}_0$. Assuming $B_0 \gg B_x$, $B_y$ and $\left(F_{kl}u^l\right)\left(F^{km}u_m\right) \approx -B_0^2\gamma^2\beta_{\perp}^2$, we find

$$\mathbf{f} = \frac{2e^4B_0^2}{3m^2c^5}\gamma^2\beta_{\perp}^2\,\mathbf{v}. \tag{A5.3.2}$$

The above problem may be solved by setting $\nu = \nu_0|U|^2$, $\nu_0 = 2e^4B_0^2/3m^3c^5\omega$ in Eqs. (5.7.6), (5.7.7). In the case of cold beam ($G = 1$, $T = 0$), we get

$$\begin{aligned} \dot{U} &= \mathcal{E} + i\left(1 - 1/\Omega b\right)U - \nu_0\,|U|^2U/b, \\ \dot{b} &= -\nu_0\,|U|^2, \end{aligned} \tag{A5.3.3}$$

where the dot means the derivative with respect to $\psi = \omega(t - z/c)$.

Substituting $U = a\exp\left(-i\vartheta\right)$, we obtain the set of equations for the amplitude and phase of the transverse momentum [12]

$$\begin{aligned} \dot{a} &= \mathcal{E}\cos\vartheta - \nu_0\,a^3/b, \quad \dot{b} = -\nu_0\,a^2, \\ \dot{\vartheta} &= 1/\Omega b - 1 - (\mathcal{E}/a)\sin\vartheta. \end{aligned} \tag{A5.3.4}$$

In the absence of the radiation drag force, $\nu_0 = 0$, the resonance remains during relativistic increasing of the electron mass and the energy of the resonance particle, $\Omega b = 1$, in applied field grows unlimitedly

$$a = \mathcal{E}\psi, \quad \vartheta = 0. \tag{A5.3.5}$$

In the case of $\nu_0 > 0$, the accelerated particle energy gets limited by the charged particle radiation [17]. This value can be estimated with the help of physical mechanism responsible for the conversion of the monotonically increasing with respect to $\psi$ electrons motion to an oscillation motion. To proceed, we assume that the relation $a \approx \mathcal{E}\psi$ still remains when the appearance of radiation drag force accompanies by the mismatching in phase $\vartheta \ll 1$ between the wave and the transverse velocity of an electron. Then it follows from second and third equation (A5.3.4) that

$$b = \frac{1}{\Omega} - \frac{1}{3}\nu_0 \mathcal{E}^2 \psi^3, \quad \vartheta = \frac{1}{15}\nu_0 \Omega \mathcal{E}^2 \psi^4. \tag{A5.3.6}$$

As the result, a nonlinear saturation of the electron transverse momentum arises when $\vartheta_m \simeq 1$:

$$a_m = \left(15\mathcal{E}^2/\nu_0 \Omega\right)^{1/4}, \quad 1 - \Omega b_m = 3^{-1}\left(15\nu_0 \Omega \mathcal{E}^2\right)^{1/4}. \tag{A5.3.7}$$

After this the particle reaches a retarding phase and loses energy.

The particle energy is determined by Eq. (5.7.9): $\gamma \approx \Omega a^2/2$ $(\gamma \gg \gamma_0)$. Substituting $a_m$ from Eq. (A5.3.7), we find

$$W_m \simeq 2mc^2 \mathcal{E}\sqrt{\frac{\Omega}{\nu_0}} = 2eE\lambda_0 \sqrt{\frac{\lambda_0}{r_0}}, \tag{A5.3.8}$$

where $r_0 = 2e^2/3mc^2$ and $\lambda_0 = c/\omega_B$ are the classical radius of electron and the wavelength, relatively. Note that the condition for the applicability of above theory is determined by the following inequality: $\left(\nu_0 \Omega \mathcal{E}^2\right)^{1/4} \ll 1$.

### Appendix 5.4

We introduce the following notation:

$$F(B) = \frac{1}{2\pi} \int_{-\Phi_0}^{\Phi_0} \sqrt{1 + B\cos\Phi}\,\cos\Phi\,d\Phi, \tag{A5.4.1}$$

$$G(B) = \frac{F(B)}{B} - \frac{dF}{dB}, \tag{}$$

where $\Phi_0$ is given by (3.6.6), and

$$\lambda = \mu|\kappa|, \quad B = \frac{A}{\mu}.$$

The condition providing the equilibria can be written in the form

$$\lambda = \frac{F(B)}{B}. \tag{A5.4.2}$$

177

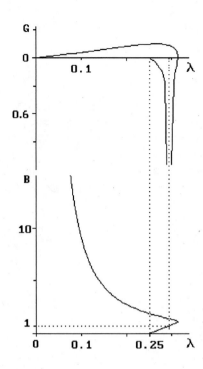

The sign of the function $G(B)$ coincides with that of the field increment in the ion stage of the instability. Figure A5.4.1 shows the dependencies $B(\lambda)$ and $G(\lambda)$ given implicitly by formulas (A5.4.1) and (A5.4.2). The dependencies in Fig. A5.4.1 show that, for $\lambda < 1/4G$ and $> 0$, the presence of electrons does not substantially affect the onset of the ion stage of the instability. For $\lambda > 1/4$, we have a two-valued function $B(\lambda)$: the branch corresponding to the lager values of $B$ relates to $G > 0$, and that for the smaller values relates to $G < 0$. However, this specific situation cannot be a result of the onset of the electron-beam instability because, for $-\kappa \gg 1$, only the values $\lambda < 1/4$ correspond to the unstable solutions of dispersion relation (3.4.7).

Fig. A5.4.1

## INSTABILITY OF AN ELECTRON BEAM AND CHERENKOV RADIATION IN A NONLINEAR GAS WITH AN INVERTED POPULATION LEVELS

The ultimate model of dense matter is the gas of oscillators [1]. To describe the nonlinear properties of a medium under the conditions of the resonance $|\omega - \Omega| \ll \Omega$ one can use the two-level approximation, which takes into account only two energetic levels of atoms with the energies $E_1 - E_2 = \hbar\Omega$ [2]. The non-linear equation for the polarization vector, which generalizes the linear theory formulae [1], is obtained and used for computations in the papers [3-5].

In a gas with electron beam the Cerenkov instability builds up in similarity to the plasma beam one [6,7]. Yet, as different from plasma, the beam interacts with electrons, bound in atoms, the oscil- lators. Under the conditions of the anomalous Doppler effects the instability is realized for a rather dense beam at the resonance frequency $\omega = \Omega$, being accompanied by the transformation of the beam kinetic energy into atomic internal energy, with atoms simultaneously passing into the state with inverse population of energy levels for the time that is less than the reverse collision rate of electrons with atoms, and a system with a negative temperature comes into being [8,9].

The reverse effect of energy transformation of the three-level system into the plasma wave energy under the conditions of the normal Doppler effect is considered in the paper [10]. In the meanwhile, unstable becomes the medium with inverse level population, with the beam acting as a waveguide for the Langmuir wave. Conversion of polarization and Cerenkov losses of energy of the modulated beam in a medium with inverse level population [11] creates a principal possibility for charged particle beam acceleration in an unstable gaseous medium [12].

This Chapter studies the nonlinear Langmuir and electromagnetic oscillations in two-level and three-level resonance systems with electron beam. Consideration is given to the instability of a two-level atomic beam moving at the super-speed of light under the conditions, in which the collective pumping mechanism of laser system is realized [13].

### §6.1 Excitation of a two-level system by an electron beam.

The connection between the dipole moment $\mathbf{d}$ of an individual two-level atom and the electric field $\mathbf{E}$ are given by the equation [3-5]

$$\ddot{\mathbf{d}} + \Omega^2\mathbf{d} = -\frac{2d_0}{\hbar}\,w\mathbf{E}, \quad \dot{w} = \frac{2d_0}{\hbar}\,\mathbf{E}\dot{\mathbf{d}}, \tag{6.1.1}$$

where $d_0$ is a dipole-transition constant, $\Omega$ is the resonance frequency of the atom, and the dots denote derivatives with respect to time. We integrate equations (6.1.1) to find "internal" integral (6.1.1)

$$\dot{\mathbf{d}}^2 + \Omega^2 \mathbf{d}^2 + w^2 = \Omega^2 d_0^2 \qquad (6.1.2)$$

characterizing the properties of the atom [3,4] (see **Appendix 6.1**).

To describe the properties of the two-level medium, we introduce the polarization vector $\mathbf{P} = N\mathbf{d}$ ($N$ is the density of active atoms) and we rewrite Eqs. (6.1.1) and (6.1.2) in the form

$$\ddot{\mathbf{P}} + \Omega^2 \mathbf{P} = -\frac{2Nd_0^2\Omega}{\hbar} W\mathbf{E},$$
$$W = \frac{1}{(\Omega N d_0)^2} \sqrt{(\Omega N d_0)^2 - \dot{\mathbf{P}}^2 - \Omega^2 \mathbf{P}^2}. \qquad (6.1.3)$$

The difference of populations of the medium levels $W = w/\Omega d_0$ (which is proportional to the density of energy stored in active atoms) is determined by the second equation (6.1.3).

The equation (6.1.3) jointly with the Maxwell equations

$$\operatorname{curl} \mathbf{E} = -\frac{1}{c}\frac{\partial \mathbf{B}}{\partial t}, \quad \operatorname{curl} \mathbf{B} = \frac{1}{c}\frac{\partial \mathbf{E}}{\partial t} + \frac{4\pi}{c}\left(\frac{\partial \mathbf{P}}{\partial t} + en\mathbf{v}\right) \qquad (6.1.4)$$

and the system of hydrodynamic equations

$$\frac{\partial n}{\partial t} + \operatorname{div}(n\mathbf{v}) = 0, \quad \frac{\partial \mathbf{v}}{\partial t} + (\mathbf{v}\nabla)\mathbf{v} = \frac{e}{m}\mathbf{E} \qquad (6.1.5)$$

describe coupled nonstationary longitudinal-transverse oscillations in the system composed of the beam and the active medium.

In the linear approximation $\sim \exp(ikx - i\omega t)$, system (6.1.3)–(6.1.5) reduces to the following dispersion relation:

$$\left(\frac{c^2 k^2}{\omega^2} - \epsilon_\pm + \frac{\omega_b^2}{\omega^2}\right)\left[\epsilon_\pm - \frac{\omega_b^2}{(\omega - kv_0)^2}\right] = 0, \qquad (6.1.6)$$

$$\epsilon_\pm = 1 \pm \frac{\omega_g^2}{\Omega^2 - \omega^2}, \quad \omega_g^2 = \frac{8\pi N d_0^2 \Omega}{h}, \quad \omega_b^2 = \frac{4\pi e^2 n_b}{m}$$

corresponding to both of the independent branches of oscillation. Here the sign "+" relates to normal [$\operatorname{sign}(W_0) = -1$], and that for "−" relates to inverse [$\operatorname{sign}(W_0) = +1$] medium; $n_b$ and $v_0$ are the density and the velocity of the beam.

The electromagnetic perturbation is unstable if a majority of the atoms are in the upper level. Substituting

$$\omega = \Omega + \Delta\omega_\perp, \quad |\Delta\omega_\perp| \ll \Omega,$$

in the first cofactor in Eq. (6.1.6), we find that a small addition to the frequency

$$\Delta\omega_\perp^2 = \pm\frac{\omega_g^2}{4}, \quad \mathcal{K}^2 = \frac{\Omega^2 - \omega_b^2}{c^2}, \qquad (6.1.7)$$

becomes an imaginary only in the case $\operatorname{sign}(W_0) = +1$.

The electrostatic perturbation

$$\omega = \Omega + \Delta\omega_\parallel, \quad \omega = kv_0 - \omega_b + \Delta\omega_\parallel,$$
$$|\Delta\omega_\parallel| \ll \Omega, \quad |\Delta\omega_\parallel| \ll \omega_b$$

180

is unstable if $\Delta\omega_\parallel^2 < 0$,

$$\Delta\omega_\parallel^2 = -\frac{\omega_g^2\omega_b}{4\Omega}, \quad \Omega - kv_0 = \pm\omega_b. \tag{6.1.8}$$

The last requirement coincides with the condition for the normal Doppler effect [sign "+" in Eq. (6.1.8)] and an inverse medium or for the anomalous Doppler effect [sign "−" in Eq. (6.1.8)] and a normal medium.

The conditions for the applicability of solutions (6.1.7) and (6.1.8) are determined by the following inequalities:

$$\frac{\omega_g}{\Omega} \ll 1, \quad \frac{\omega_g}{\sqrt{\Omega\omega_b}} \ll 1. \tag{6.1.9}$$

The transverse electric field, $E_\perp$, satisfy the wave equation

$$\left(\frac{\partial^2}{\partial x^2} - \frac{1}{c^2}\frac{\partial^2}{\partial t^2}\right)E_\perp = \frac{4\pi}{c^2}\frac{\partial^2 P_\perp}{\partial t^2}, \tag{6.1.10}$$

and the longitudinal field, $E_\parallel$, satisfy the Maxwell equation in the electrostatic approximation ($\mathbf{B} = \operatorname{curl}\mathbf{E} = 0$) and also of the linearized equations of motion of the beam.

$$\left(\frac{\partial}{\partial t} + v\frac{\partial}{\partial x}\right)^2 (E_\parallel + 4\pi P_\parallel) + \omega_b^2 E_\parallel = 0. \tag{6.1.11}$$

Both types of the above oscillation are coupled in a nonlinear medium (6.1.3).

Since we are interested in nonstationary processes in the system composed of the beam and the active medium, we suppose that a majority of the atoms are in the lower level so that the instability develops under the conditions of the anomalous Doppler effect [8]. We can express $E_\parallel$ in terms of $P_\parallel$, from Eq. (6.1.11)

$$E_\parallel(t, x) = -4\pi P_\parallel(t, x) + 4\pi\omega_b \int_0^t \sin[\omega_b(t - t')]P_\parallel[t, x - v(t - t')]dt'.$$

On substituting the above result in Eq. (6.1.3), we will seek a solution of this nonlinear equation in the form of the wave with amplitude and phase varying slowly with time:

$$P_\parallel(t, x) = Nd_0 a_\parallel(t)\cos[kx - \omega t + \theta_\parallel(t)], \tag{6.1.12}$$

where the frequency $\omega$ and the wave number $k$ satisfy the condition (6.1.8).

As the result, we get the following first-order equations for $a_\parallel$ and $\theta_\parallel$:

$$\frac{da_\parallel}{d\tau} = -\operatorname{sign}(W_0)\pi q\omega_0\sqrt{1 - a_\parallel^2}\int_0^\tau a_\parallel(\tau')\,d\tau',$$

$$\frac{d\theta_\parallel}{d\tau} = -\operatorname{sign}(W_0)2\pi q\sqrt{1 - a_\parallel^2}, \tag{6.1.13}$$

where the dimensionless variables are

$$\tau = \Omega t, \quad q = 2Nd_0^2/\hbar\Omega, \quad \omega_0 = \omega_b/\Omega.$$

It is easy to see that when $\operatorname{sign}(W_0) = -1$, growth of amplitude of the longitudinal oscillations occurs. Then in the small amplitude range ($a_\parallel^2 \ll 1$), the growth is of exponential character, with an increment $\delta_\parallel = \sqrt{\pi q\omega_0}$ determined by the linear theory (6.1.8).

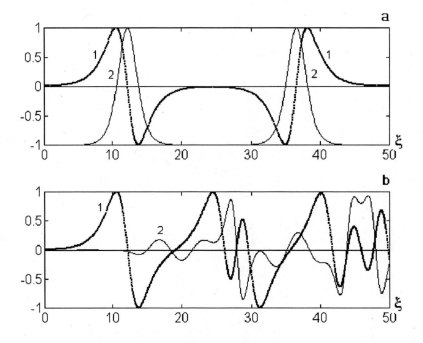

Fig. 6.1.1a. Nonlinear oscillations of (1) the amplitude of polarization vector $a_\|(\xi)$, $\xi=\sqrt{q}\tau$ and (2) the difference of populations $W$ for $\omega_0=0.3$;

Fig. 6.1.1b. The growth of (2) the amplitude of transverse polarization $a_\perp(\xi)$ and (1) the amplitude of unstable longitudinal polarization $a_\|(\xi)$.

For treatment of large amplitudes, it is convenient to introduce the substitution $a_\| = \sin\varphi$. Then Eq. (6.1.13) takes the following form:

$$d^2\varphi/d\tau^2 - \delta_\|^2 \sin\varphi = 0. \qquad (6.1.14)$$

This is known equation, which determines the oscillations of a nonlinear pendulum with the equilibrium position at the topmost point. Thus in the system under consideration, there takes place a periodically recurring transfer of the energy of the longitudinal oscillations of the field into internal energy of the matter.

An important fact is that the difference of populations of the levels (6.1.3)

$$W = -\cos\varphi(\tau), \qquad (6.1.15)$$

can as a result of this process assume positive values. The time that the system remains in the inverted state during each period of the oscillations is, in order of magnitude, $T \simeq \delta_\|^{-1} \gg \omega_0^{-1}$; the period of the oscillations increases logarithmically with decrease of the initial amplitude of the oscillations. The result of numerical integration is represented in Fig. 6.1.1a.

The amplitude of longitudinal field $\mathcal{E}_\| = E_\|/Nd_0$ is determined by the equation

$$d\mathcal{E}_\|/d\tau = -2\pi\omega_0 a_\| \qquad (6.1.16)$$

which follows from Eqs. (6.1.11) and (6.1.12). Comparing Eqs. (6.1.16) and (6.1.13) and using the energy integral of Eq. (6.1.14)

$$(d\varphi/d\tau)^2 = 2\delta_\|^2 (1 - \cos\varphi) = 0,$$

182

we find

$$\mathcal{E}_{\parallel} = 4 \sqrt{\frac{\omega_0}{q}} \sin \frac{\varphi}{2}. \tag{6.1.17}$$

We now consider the problem of the development, with time, of small fluctuations of the transverse field in the system under consideration (beam and active medium). For small transverse oscillations it may be supposed that the internal energy of the medium changes with time according to the law (6.1.14) determined by purely longitudinal oscillations.

As in the case of longitudinal oscillations, we will seek a solution for the field $E_{\perp}$ and the polarization $P_{\perp}$ in the form of a wave with slowly varying amplitude and phase:

$$P_{\perp}(t, x) = N d_0 a_{\perp}(t) \cos[\mathcal{K}x - \omega t + \theta_{\perp}(t)], \tag{6.1.18}$$

where the frequency $\omega$ and the wave number $\mathcal{K}$ satisfy the condition (6.1.7). Then the relation between $E_{\perp}$ and $P_{\perp}$, in accordance with (6.1.10), has the form

$$E_{\perp}(t, x) = -4\pi P_{\perp}(t, x) + 4\pi c \mathcal{K} \int_0^t \sin[c\mathcal{K}(t - t')] P_{\perp}(t', x) \, dt'. \tag{6.1.19}$$

On substituting into Eq. (6.1.3) the value of $P_{\perp}$ and $E_{\perp}$ from Eqs. (6.1.18) and (6.1.19), and $W$ from Eq. (6.1.15), we get the following equation for the amplitude of the transverse oscillations:

$$\frac{da_{\perp}}{d\tau} = -\pi q \cos \varphi(\tau) \int_0^\tau a_{\perp}(\tau') d\tau', \tag{6.1.20}$$

(we assume that $1 - a_{\parallel}^2 \gg a_{\perp}^2$).

From Eq. (6.1.20) it is clear that the amplitude of the transverse field begins to grow only upon transition of the system to the inverted state ($\cos \varphi(\tau) < 0$). Since $\varphi(\tau)$ changes appreciably over a time of order $(\sqrt{\pi} q \omega_0)^{-1/2} \gg (\sqrt{\pi} q)^{-1/2}$, a solution of Eq. (6.1.20) can be found by the WKB method:

$$a_{\perp}(\tau) = a_{\perp}(\tau_0) \sqrt[4]{\left| \frac{\cos \varphi(\tau)}{\cos \varphi(\tau_0)} \right|} \exp \left( \int_0^\tau \sqrt{\pi q \cos \varphi(\tau')} \right) d\tau'. \tag{6.1.21}$$

This expression is valid up to $\tau_0's$ for which $\cos \varphi(\tau_0) < 0$ and $\omega_0 \ll |\cos \varphi(\tau_0)|$. The increase of amplitude of the transverse oscillations will continue until the transverse wave begins to have an appreciable influence on the populations of the levels, that is until a time $\tau_m$ determined by the condition $|\cos \varphi(\tau_m)| \leq a_{\perp}(\tau_m)$.

In the general case, $a_{\perp} \approx a_{\parallel}$, coupled longitudinal–transverse oscillations satisfy the following system

$$\frac{d}{d\tau} \frac{\dot{a}_{\parallel}}{W} = \pi q \omega_0 a_{\parallel}, \quad \frac{d}{d\tau} \frac{\dot{a}_{\perp}}{W} = -\pi q a_{\perp}, \quad W = \sqrt{1 - a_{\parallel}^2 - a_{\perp}^2}, \tag{6.1.22}$$

at which the energy conservation law is

$$\frac{\dot{a}_{\parallel}^2}{\omega_0} - \dot{a}_{\perp}^2 + 2\pi q W^3 = C. \tag{6.1.23}$$

Study of the equations (6.1.22) shows that the system cannot be in a state in which the population of the levels or the direction of the polarization vector does not changed with

183

time. This conclusion is easily reached by studying the small oscillations in the system. By means of substitution

$$a_\perp = \sin \varphi \sin \vartheta, \quad a_\parallel = \sin \varphi \cos \vartheta,$$

$$\varphi = \varphi_0 + \varphi_1, \ |\varphi_1| \ll 1; \quad \vartheta = \vartheta_0 + \vartheta_1, \ |\vartheta_1| \ll 1,$$

we get the system of linearized equations:

$$\cos \varphi_0 \cos \vartheta_0 (\ddot{\varphi}_1 - \pi q \omega_0 \cos \varphi_0 \varphi_1) = \sin \varphi_0 \sin \vartheta_0 (\ddot{\vartheta}_1 - \pi q \omega_0 \cos \varphi_0 \vartheta_1),$$

$$\cos \varphi_0 \sin \vartheta_0 (\ddot{\varphi}_1 + \pi q \cos \varphi_0 \varphi_1) = - \sin \varphi_0 \cos \vartheta_0 (\ddot{\vartheta}_1 + \pi q \cos \varphi_0 \vartheta_1).$$

Representing the solution in the form $\exp(\lambda \tau)$, we obtain the following characteristic equation

$$(\lambda^2 - \pi q \omega_0 \cos \varphi_0)(\lambda^2 + \pi q \cos \varphi_0) = 0. \tag{6.1.24}$$

Because of $\mathrm{Re}\lambda > 0$, the small oscillations are always unstable [8].

Figure 6.1.1b illustrates a nonlinear transformation of exited by the beam longitudinal wave into an electromagnetic wave. The accuracy of calculations is monitored on the basis of integral (6.1.23).

A growth of longitudinal-transverse oscillations accompanies by the slowing of the beam. It follows from Eq. (6.1.5) that under the conditions of the anomalous Doppler effect, $\omega - kv_0 = -\omega_b$, a beam remains "linear", $v_0 - v \ll v_0$ if the inequality $eE_\parallel \ll mv_0\omega_b$ holds. Substituting the field amplitude from Eq. (6.1.17), we obtain the following restriction

$$\frac{n_b m v_0^2}{2} \gg \frac{\mathcal{E}_{\max}^2}{8\pi} = N\hbar\omega_b \tag{6.1.25}$$

which sets the applicability limits for the above approximation.

An alternative physical mechanism causing the effect of beam-gas instability under the conditions of the Cherenkov resonance is discussed in Ref. [6]:

$$\omega = kv_0 + \Delta\omega, \quad |\Delta\omega| \ll kv_0, \quad kv_0 = \omega_\parallel = \sqrt{\Omega^2 \pm \omega_g^2}.$$

A dispersion relation (6.1.6) has complex roots

$$\Delta\omega = -(1 \pm i\sqrt{3}) \left( \frac{\omega_g^2 \omega_b^2}{16\omega_\parallel} \right)^{1/3} \tag{6.1.26}$$

corresponding to an excitation of electrostatic oscillations in the gas, $\epsilon_\pm(\omega) = 0$.

As distinct from the anomalous Doppler instability, this effect accompanies by a deep modulation of the beam (by analogy with plasma wave has discussed in Chap. 1).

## §6.2. Population inversion in a three-level system.

The idea of the possibility of producing a three level maser [Basov and Prokhorov, ZhETF, **28**, 249, 1955; Bloembergen, Phys. Rev., **104**, 324, 1956] consists in subjecting a nonlinear active medium in the form of an aggregate of atoms with three energy levels $E_1 < E_2 < E_3$ to a high frequency electric field of frequency $\omega$ close to the greatest of the natural frequencies of the medium $\Omega_{31} = (E_3 - E_1)/\hbar$. If the field amplitude is large enough the population of levels $E_1$ and $E_3$ are equalized (although at the initial instant there are more atoms with energy $E_1$). Consequently, spontaneous transitions of atoms

184

from level $E_3$ to level $E_2$ occur, thereby giving rise to conditions such that the number atoms in level $E_2$ is greater than in $E_1$ and any small perturbations in an electric field of frequency $\Omega_{21} = (E_3 - E_1)/\hbar$ will be amplified.

This problem is considered in Ref. [9] on the assumption that the system of three-level atoms is situated in the field of a plane wave with a resonance frequency and a phase velocity equal to the velocity of light. We shall determine the conditions under which the level populations (the transition $E_2 \to E_1$) can be inverted and also the time required for the system to reach the inverted state.

### 1. Three-level atom equations.

In order to obtain the dipole moment $\mathbf{d}$ of one atom as a function of the amplitude of the electric field, let us consider the behavior of the atom in an external electric field $\mathbf{E}$. We use here Schrödinger's equation, assuming that the operator describing the interaction between the atom and the field is of the form $\hat{H}_1 = -\mathbf{d}\mathbf{E}$:

$$i\hbar \frac{\partial \psi}{\partial t} = (\hat{H}_0 - \mathbf{d}\mathbf{E})\psi \tag{6.2.1}$$

[where $\hat{H}_0$ describes the stationary state of the atom (in the absence of the field), and $\mathbf{d}$ is the operator of the dipole moment of the atom.

We seek a solution of (6.2.1) in the form

$$\psi(t, \mathbf{r}) = A(t)\psi_1(\mathbf{r}) + B(t)\psi_2(\mathbf{r}) + C(t)\psi_3(\mathbf{r}), \tag{6.2.2}$$

where $\psi_1(\mathbf{r})$, $\psi_2(\mathbf{r})$, and $\psi_3(\mathbf{r})$ are the wave functions corresponding to the energy levels $E_1 < E_2 < E_3$, and $A(t)$, $B(t)$, and $(t)$ are the complex functions of the time.

Substituting (6.2.2) in (6.2.1) and then multiplying equation (6.2.1) by $\psi_i(\mathbf{r})$ we get, after integrating with respect to $\mathbf{r}$, a system of equations for the coefficients $A(t)$, $B(t)$, and $(t)$:

$$\frac{dA}{dt} = -\frac{i}{\hbar}\left[E_1 A - (\mathbf{d}_{21}B + \mathbf{d}_{31}C)\mathbf{E}\right], \quad \frac{dB}{dt} = -\frac{i}{\hbar}\left[E_2 B - (\mathbf{d}_{21}A + \mathbf{d}_{32}C)\mathbf{E}\right],$$

$$\frac{dC}{dt} = -\frac{i}{\hbar}\left[E_3 C - (\mathbf{d}_{31}A + \mathbf{d}_{32}B)\mathbf{E}\right], \tag{6.2.3}$$

where

$$E_i = \langle \psi_i | \hat{H}_0 | \psi \rangle, \quad \mathbf{d}_{ik} = \langle \psi_i | \mathbf{d} | \psi_k \rangle$$

with $i \neq k$ and $d_{ii} = 0$, and the brackets denote averaging over $\mathbf{r}$.

The dipole moment of the atom then takes the form

$$\mathbf{d} = \mathbf{d}_{21}(AB^* + A^*B) + \mathbf{d}_{31}(AC^* + A^*C) + \mathbf{d}_{32}(BC^* + B^*C). \tag{6.2.4}$$

### 2. Self-consistent system of equations.

We shall simulate a nonlinear active medium by a system of $N$ three-level atoms with energy levels $E_1 < E_2 < E_3$, and we shall assume that the polarization vector $\mathbf{P}$ of the medium is determined by the sum of dipole moments of the individual atoms: $\mathbf{P} = N\mathbf{d}$. The system (6.2.3), together with Maxwell's equations for the fields, in which it is necessary to substitute the polarization current $\mathbf{J}_\mathbf{P} = \partial^2\mathbf{P}/\partial t^2$, is the self-consistent system of equations describing the interaction of the electromagnetic waves with the system of three-level atoms.

185

Accordingly to §6.1, the solution of Eqs. (6.1.10) and (6.2.3) can be represented in the form of the wave with amplitude slowly varying with time:

$$E(t, x) = \mathcal{E}_{21} \sin \Phi_{21} + \mathcal{E}_{31} \sin \Phi_{31} + \mathcal{E}_{32} \sin \Phi_{32},$$

$$A(t) = a(t) \exp\left(-i\frac{E_1}{\hbar}\xi\right), \quad B(t) = b(t) \exp\left(-i\frac{E_2}{\hbar}\xi\right), \quad C(t) = c(t) \exp\left(-i\frac{E_3}{\hbar}\xi\right),$$
$$(6.2.5)$$

where $a(t)$, $b(t)$ and $c(t)$ are real coefficients, and

$$\Omega_{mn} = (E_m - E_n)/\hbar, \quad \Phi_{mn} = \Omega_{mn}\xi, \quad \xi = t - x/c.$$

Averaging over the periods $T_{ik} = 2\pi/\Omega_{ik}$ of the high-frequency fields, we obtain a nonlinear set of equations for the coefficients $a(t)$, $b(t)$ and $c(t)$ and for the field amplitudes $\mathcal{E}_{31}$, $\mathcal{E}_{32}$, and $\mathcal{E}_{21}$:

$$\dot{a} = \frac{1}{2\hbar}\left(d_{21}\mathcal{E}_{21}b + d_{31}\mathcal{E}_{31}c\right), \quad \dot{\mathcal{E}}_{21} = 4\pi N d_{21}\Omega_{21}ab,$$

$$\dot{b} = \frac{1}{2\hbar}\left(-d_{21}\mathcal{E}_{21}a + d_{32}\mathcal{E}_{32}c\right), \quad \dot{\mathcal{E}}_{31} = 4\pi N d_{31}\Omega_{31}ac, \qquad (6.2.6)$$

$$\dot{c} = -\frac{1}{2\hbar}\left(d_{31}\mathcal{E}_{31}a + d_{32}\mathcal{E}_{32}b\right), \quad \dot{\mathcal{E}}_{32} = 4\pi N d_{32}\Omega_{32}bc,$$

In the derivation of (6.2.6), the higher-order terms of expansion in the small parameters

$$\frac{q_{mn}}{\Omega_{mn}} = \sqrt{\frac{2\pi N d_{mn}^2}{\hbar\Omega_{mn}}} \ll 1$$

are dropped.

Using (6.2.6) we obtain the integrals of motion

$$8\pi\hbar N a^2 - \frac{\mathcal{E}_{21}^2}{\Omega_{21}} - \frac{\mathcal{E}_{31}^2}{\Omega_{31}} = C_1, \quad 8\pi\hbar N b^2 + \frac{\mathcal{E}_{21}^2}{\Omega_{21}} - \frac{\mathcal{E}_{32}^2}{\Omega_{32}} = C_2,$$

$$8\pi\hbar N c^2 + \frac{\mathcal{E}_{31}^2}{\Omega_{31}} + \frac{\mathcal{E}_{32}^2}{\Omega_{32}} = C_3, \qquad (6.2.7)$$

($C_1$, $C_2$, and $C_3$ are constants which depend on the initial conditions), which reflect energy conservation in the medium-field system and the law of conservation of probability

$$a^2 + b^2 + c^2 = 1,$$

$$N(E_1 a^2 + E_2 b^2 + E_3 c^2) + \frac{1}{8\pi}(\mathcal{E}_{21}^2 + \mathcal{E}_{31}^2 + \mathcal{E}_{32}^2) = W_0. \qquad (6.2.8)$$

It is not hard to show that system (6.2.6) is equivalent to a set of three second-order nonlinear equations describing three coupled nonlinear oscillators with frequencies $q_{21}$, $q_{31}$, $q_{32}$, so that our problem concerning the natural oscillations in a three-level system can be reduced to the problem of three coupled two-level systems.

### 3. Three-level system in an external field.

Let us now consider the "illumination" of the system by an external transverse wave of frequency $\Omega_{31}$ (in this case we must assume $\mathcal{E}_0 \gg \mathcal{E}_{31}(t) \simeq \sqrt{8\pi N\hbar\Omega_{31}}$). Furthermore, we assume that the transition probability at the frequency $\Omega_{21}$ is small in comparison

186

with the transition probabilities at frequencies $\Omega_{31}$ and $\Omega_{32}$, i.e., $d_{21} \ll d_{31}$, $d_{32}$ so that we may neglect at first terms proportional to $d_{21}$. Under this conditions the set (6.2.6) takes the form

$$\dot{a} = \frac{d_{31}}{2\hbar}(\mathcal{E}_{31} + \mathcal{E}_0)c, \quad \dot{b} = \frac{d_{32}}{2\hbar}\mathcal{E}_{32}c, \quad \dot{c} = -\frac{1}{2\hbar}\left[d_{31}(\mathcal{E}_{31} + \mathcal{E}_0)a + d_{32}\mathcal{E}_{32}b\right],$$
$$\dot{\mathcal{E}}_{32} = 4\pi N d_{32}\Omega_{32}bc, \quad \dot{\mathcal{E}}_{31} = 4\pi N d_{31}\Omega_{31}ac. \tag{6.2.9}$$

Expressing $b$ and $c$ in terms of $a$ and $\dot{a}$ by means of the probability conservation law (6.2.8) and the first equation (6.2.9)

$$b = \sqrt{1 - a^2 - \dot{a}^2/\omega_0^2}, \quad c = \dot{a}/\omega_0, \quad \omega_0 = d_{31}\mathcal{E}_0/2\hbar,$$

and eliminating this functions from Eqs. (6.2.9), we obtain the following system of equations

$$\ddot{a} + a = -\sqrt{1 - a^2 - \dot{a}^2}\,\epsilon_{32}, \quad \dot{\epsilon}_{32} = \frac{q_{32}^2}{\omega_0^2}\sqrt{1 - a^2 - \dot{a}^2}\,\dot{a}, \tag{6.2.10}$$

where $\epsilon_{32} = d_{32}\mathcal{E}_{32}/d_{32}\mathcal{E}_0$, and the dot denotes differentiation with respect to dimensionless time $\tau = \omega_0 t$.

Substituting $\dot{\epsilon}_{32}$ from the second equation into the first, we obtain a third-order nonlinear equation for $a$:

$$\frac{d}{d\tau}\frac{\ddot{a} + a}{\sqrt{1 - a^2 - \dot{a}^2}} = -\frac{q_{32}^2}{\omega_0^2}\dot{a}\sqrt{1 - a^2 - \dot{a}^2},$$

which can be reduced to a second-order linear equation by means of the substitution $b = \sqrt{1 - a^2 - \dot{a}^2}$ and $da = \dot{a}\tau$:

$$\frac{d^2b}{da^2} - \frac{q_{32}^2}{\omega_0^2}b = 0. \tag{6.2.11}$$

The solution of (6.2.11) satisfying the initial conditions $b(0) = b_0$ and $\dot{b}(0) = 0$ is

$$b(a) = b_0 \cosh\left[\frac{q_{32}}{\omega_0}(a - a_0)\right]$$

and constitutes a first-order differential equation for $a$:

$$\dot{a} = \sqrt{1 - a^2 - b_0^2\cosh^2\left[\frac{q_{32}}{\omega_0}(a - a_0)\right]}. \tag{6.2.12}$$

We assume that at the initial instant $\tau = 0$ almost all atoms are situated in the lower level ($a_0 \simeq 1$, $b_0 \ll 1$, $c_0 = \dot{a}_0 = 0$), and we shall seek solution of (6.2.12) for which the population of level $E_2$ increases ($a \to 0$, $b \to 1$). The minimum value of $a_{\min}$ can be determined from the condition $\dot{a}(a_{\min}) = 0$. Putting $\dot{a} = 0$ in (6.2.12), we obtain the following algebraic equation for $a_{\min}$:

$$1 - a_{\min}^2 - b_0^2\cosh^2\left[\frac{q_{32}}{\omega_0}(a_{\min} - a_0)\right] = 0. \tag{6.2.13}$$

The solution of this equation depends on the relationship between the parameters $q_{32}$ and $\omega_0$. Carrying out the substitution

$$\cosh\left[\frac{q_{32}}{\omega_0}(a_{\min} - a_0)\right] \simeq 1,$$

187

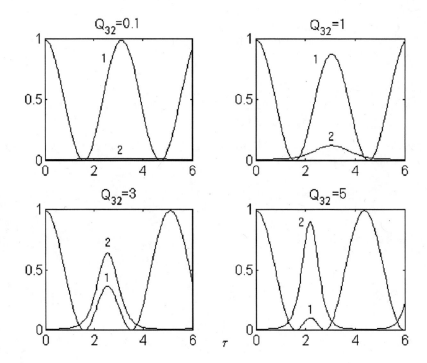

Fig. 6.2.1. Functions $a^2$ (curves 1) and $b^2$ (curves 2) for $Q_{31}=0$ and for different initial values of the parameter $Q_{32}$.

for the case $q_{32} \ll \omega_0$, we obtain

$$a_{\min} = \sqrt{1 - b_0^2} = -a_0, \quad a_{\max} = -a_{\min},$$

i.e., the system executes an oscillatory motion with turning points $\pm a_0$. The same result can be obtained directly from the equation of (6.2.10). Indeed, neglecting the term proportional to $q_{32}^2$ in the first equation of this set, we obtain $a = a_0 \cos \tau$ and $\dot{a} = a_0 \sin \tau$. This case corresponds to the problem of a two-level system in a strong electric field [2].

However, despite the fact that the population of level $E_1$ periodically reduces to zero, the population of level $E_2$ remains unchanged:

$$b^2 = 1 - a^2 - \dot{a}^2 = 1 - a_0^2 = b_0^2.$$

In the opposite limiting case when $q_{32} > \omega_0$, Eq. (6.2.13) can have solution $a_{\min} \ll a_0$. Neglecting $a_{\min}$ in comparison with $a_0$ in the argument of the hyperbolic cosine, we obtain

$$a_{\min}^2 = 1 - b_0^2 \cosh^2\left(\frac{q_{32}}{\omega_0} a_0\right) \ll 1. \tag{6.2.14}$$

It follows from (6.2.13) that $a_{\min}$ is smaller the greater the ratio $q_{32}/\omega_0$. However, for $q_{32}/\omega_0 \to \infty$ and $b_0 > 0$ Eq. (6.2.13) does not have a solution, so that the maximum value of $q_{32}/\omega_0$ (for given $b_0 > 0$) can be found from the condition $a_{\min} = 0$: $\cosh(a_0 q_{32}/\omega_0) = 1$.

The effect considered above admits of a simple physical interpretation. In a strong electric field of frequency close to $\Omega_{31}$, atoms periodically transfer from level $E_1$ to level $E_3$ with a frequency $\omega_0$, i.e., in a time $T_0 = 2\pi/\omega_0$ [2]. Spontaneous transitions $E_3 \to E_2$ thereupon become possible, the time for which is inversely proportional to the frequency

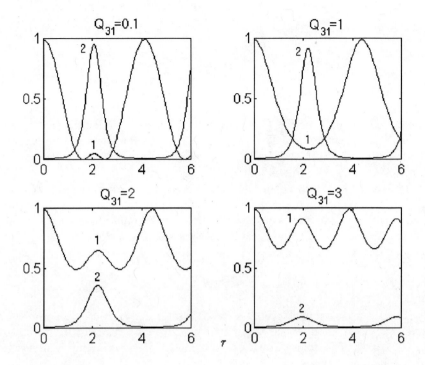

Fig. 6.2.2. Functions $a^2$ (curves 1) and $b^2$ (curves 2) for $Q_{32}=6$ and for different initial values of the parameter $Q_{31}$.

of the natural oscillations of the system, $T_{32} \sim 1/q_{32}$. However, if $T_0 \ll T_{32}$ the atoms will return to the level $E_1$ without having managed to reach level $E_2$. The population of level $E_2$ will thus begin to change only if $T_{32}$ is comparable with $T_0$. The efficiency of this method of population inversion is enhanced when $T_0 \gg T_{32}$, since in this case each atom arriving at level $E_3$ transfers to level $E_2$. In this manner, under the effect of an external transverse wave of frequency $\Omega_{31}$, the population of the level $E_2$ increases from its initial value of $b_{\min} \ll 1$ to a value $b_0 \simeq 1$, the time required to reach the inverted state being of order $T_0 \sim \omega_0^{-1}$.

As shown above the populations can be inverted if $q_{32} > \omega_0$. This condition [together with the condition that the natural oscillations of frequency $q_{31}$ ($q_{31} \ll \omega_0$) be negligible] leads to the following restriction on the field amplitude $\mathcal{E}_0$:

$$N\hbar\Omega_{31} \ll \frac{\mathcal{E}_0^2}{8\pi} < \frac{d_{32}^2}{d_{31}^2}\,N\hbar\Omega_{32}. \tag{6.2.15}$$

Clearly, condition (6.2.15) can only be satisfied provided $d_{31} < d_{32}$, since $\Omega_{31}$ is always greater than $\Omega_{32}$.

To take account the natural oscillations of frequency $\Omega_{31}$, we rewrite Eqs. (6.2.9) in the dimensionless variables:

$$\dot{a} = (1 + \epsilon_{31})\,c, \quad \dot{b} = \epsilon_{32}\,c, \quad \dot{c} = -(1 + \epsilon_{31})\,a - \epsilon_{32}\,b,$$
$$\dot{\epsilon}_{32} = Q_{32}\,bc, \quad \dot{\epsilon}_{31} = Q_{31}\,ac, \tag{6.2.16}$$

where

$$\epsilon_{31} = \mathcal{E}_{31}/\mathcal{E}_0, \quad Q_{31} = q_{31}^2/\omega_0^2, \quad Q_{32} = q_{32}^2/\omega_0^2.$$

189

Numerical solution of set (6.2.16) which satisfy the conditions

$$a_0^2 = 0.99, \quad b_0^2 = 0.1, \quad c_0 = \epsilon_{31}(0) = \epsilon_{32}(0) = 0$$

is represented in Fig. 6.2.1 and Fig. 6.2.2.

It follows from this figures that in the case $Q_{31} \ll 1$ a numerical solution is in agreement with an analytical solution of Eq. (6.2.12). A population inversion of level $E_2$ occurs for $Q_{32} > 1$ when the transition time $T_{32} \sim q_{32}^{-1}$ ($E_3 \to E_2$) is much less in comparison with $T_0$ ($E_3 \to E_1$) [see Fig. 6.2.1]. The efficiency of this method of population inversion is enhanced when $\epsilon_{31} \ll 1$ ($Q_{31} \to 0$) [see Fig. 6.2.2].

We note that the conditions under which the populations can be inverted can be altered greatly if allowance is made for the dissipation of the energy of the wave of frequency $\Omega_{31}$. Consequently, the approximation considered above is valid only provided the time constant $T_0$ is small in comparison with the decay time constant of the wave: $T_0 \ll \gamma_{31}^{-1}$, where $\gamma_{31}$ is the damping decrement of the wave.

### 4. Two-level approximation in a three-level system.

Once the population of level $E_2$ exceeds that of $E_1$ ($b^2 \gg a^2$), the problem becomes one of investigating the spontaneous transitions in a two level system with initial conditions $b(0) = 1$, $a(0) = 0$. The solution of this problem yields the maximum intensities of the fields emitted at frequency $\Omega_{32}$.

The required equations can be obtained by putting $d_{31} = d_{32} = 0$, $\mathcal{E}_{31} = \mathcal{E}_{32} = 0$, and $c = 0$ in (6.2.6):

$$\dot{a} = \frac{d_{21}}{2\hbar}\mathcal{E}_{21}\, b, \quad \dot{b} = -\frac{d_{21}}{2\hbar}\mathcal{E}_{21}\, a, \quad \dot{\mathcal{E}}_{21} = 4\pi N d_{21}\Omega_{21}\, ab. \tag{6.2.17}$$

Making in (6.2.17) the change of variable $a = \sin(\varphi/2)$, and $b = \cos(\varphi/2)$ (the normalization condition $a^2 + b^2 = 1$ is thereupon satisfied automatically), we arrive at two first-order equations

$$\dot{\varphi} = \frac{d_{21}}{\hbar}\mathcal{E}_{21}, \quad \dot{\mathcal{E}}_{21} = 2\pi N d_{21}\Omega_{21}\sin\varphi \tag{6.2.18}$$

which are equivalent to a second-order nonlinear equation for $\varphi(t)$:

$$\ddot{\varphi} - q_{21}^2 \sin\varphi = 0, \quad q_{21}^2 = \frac{2\pi N d_{21}^2 \Omega_{21}}{\hbar} \tag{6.2.19}$$

(see §6.1). The first integral of this equation has the form

$$\dot{\varphi}^2 = \dot{\varphi}_0^2 + 4q_{21}^2 \sin\frac{\varphi}{2}. \tag{6.2.20}$$

Putting $\varphi = \pi$ in (6.2.20) yields $\varphi_{\max}$ and thereby $(\mathcal{E}_{21})_{\max}$:

$$(\mathcal{E}_{21})_{\max} = \sqrt{\mathcal{E}_{21}^2(0) + 8\pi N \hbar \Omega_{21}}\,. \tag{6.2.21}$$

We note that in accordance with (6.2.18) the electric field at the beginning of the process ($\varphi \ll 1$) increases exponentially with a time constant $q_{21}$.

190

## §6.3. Transformation of the energy confined in a three-level system into the energy of a longitudinal wave in a plasma.

The production of longitudinal large amplitude waves in a plasma is of interest in connection with plasma heating, and also for acceleration in a plasma [15]. We consider in this section the direct transformation of the internal energy of an active medium used as the working medium in a laser into the energy of a longitudinal wave in a plasma, bypassing the stage of transverse generation [10].

We shall assume that henceforth that the active medium is chosen in such a way that the transitions between the levels $E_2$ and $E_1$ are completely forbidden ($\mathbf{d} = 0$). Then, if we assume that the initial instant of time the system is in the state such that all the atoms are at the levels $E_2$, Then such state of the medium is stable, since the transition to the levels $E_3$ is not favored from the energy point of view. However, if such a medium is placed in a high frequency field of a frequency closed to the frequency of the transition between the levels $E_2$ and $E_3$, namely $\Omega_{32} = (E_3 - E_2)/\hbar$, then the atoms of the medium can go over to the level $E_3$ after absorbing energy from the field. This opens up the possibility of the spontaneous transition $E_3 \to E_1$ with emission of the transverse wave of frequency $\Omega_{31} = (E_3 - E_1)/\hbar$. This is precisely the frequency-conversion mechanism which is used in three-level quantum amplifiers.

We are interested, however, in the case when the system considered above can radiate a longitudinal plasma wave, that is, a wave in which the projection of the electric field on the direction of wave propagation differs from zero. One of the main conditions for obtaining a longitudinal wave is in this case the creation of a waveguide in which such a wave could propagate. The simplest waveguide of this type is a plasma waveguide, which can be obtained by producing a plasma in the volume occupied by the active medium, for example, by passing an electron beam through a gas or through a channel in a solid.

We shall neglect below processes of energy dissipation in the active medium, and disregard likewise the thermal motion of the atoms of the medium. In this case we can assume that dipole moments of the atoms of the medium are parallel to the electric field, and the vector quantities are replaced by scalar quantities.

The initial system of equations of the problem in question consists of the equations of the system (6.2.3), in which we must put $d_{21} = 0$, and the Maxwell equation that describes the variation of the plasma wave field with frequency $\Omega_{31}$. In addition, we shall assume that a plane transverse wave of frequency $\Omega_{32}$ and of phase velocity $v_{\text{ph}}$ is incident on the system:

$$E_0 = \mathcal{E}_0 \sin \Omega_{32} \xi, \quad \xi = t - x/v_{\text{ph}}. \tag{6.3.1}$$

The Maxwell equation describing the variation of the longitudinal field $E_{\parallel}$ can be obtained by considering a system of equations consisting of linearized equations of motion and the continuity equation, and also the Poisson equation:

$$\left( \frac{\partial}{\partial t} + v_0 \frac{\partial}{\partial x} \right)^2 \left( E_{\parallel} + 4\pi P_{\parallel} \right) + \omega_b^2 E_{\parallel} = 0. \tag{6.3.2}$$

where $P_{\parallel} = N d_{31} \left( AC^* + A^*C \right)$, $v_0$ is the velocity of the beam, and $\omega_b^2 = 4\pi e^2 n_b/m$ is the plasma frequency of the beam. In derivation of (6.3.2) we have assumed that the velocity of the beam electrons can be represented in the form $v = v_0 + \tilde{v}$, where $\tilde{v}$ is the small addition to $v_0$ ($|\tilde{v}| \ll v_0$), and neglect the term of order $\tilde{v} \partial \tilde{v}/\partial z$. The condition for this approximation will be presented below.

In the case considered below, a nonlinear system of equation for the functions $a(t)$, $b(t)$, and $c(t)$ can be obtained from general equations (6.2.6):

$$\dot{a} = \frac{d_{31}}{2\hbar} \mathcal{E}_{31} c, \quad \dot{b} = \frac{d_{32}}{2\hbar} \mathcal{E}_0 c, \quad \dot{c} = -\frac{1}{2\hbar} \left[ d_{32} \mathcal{E}_0 a + d_{31} \mathcal{E}_{31} b \right]. \tag{6.3.3}$$

In order to obtain an equation for $\mathcal{E}_{31}(t)$, we substitute

$$E = \mathcal{E}_{31} \sin \Omega_{31} \xi, \quad P = \mathcal{P}_{31} \cos \Omega_{31} \xi \qquad (6.3.4)$$

in equation (6.3.2). In the same approximation as above, and also assuming that the relation

$$\Omega_{31} \left( 1 - \frac{v_0}{v_{\text{ph}}} \right) = \omega_b, \qquad (6.3.5)$$

is satisfied, we get

$$\dot{\mathcal{E}}_{31} = 4\pi N d_{31} \omega_b \, ac \qquad (6.3.6)$$

(under the condition that the natural oscillations of frequency $q_{32}$ be negligible, $\mathcal{E}_{32} \ll \mathcal{E}_0$).

Relation (6.3.5), which determines the dependence of $v_{\text{ph}}$ on $\Omega_{31}$, is the condition for the normal Doppler effect: the frequency of the wave emitted by the stationary atoms of the medium, recalculated in the reference frame connected with the beam, is equal to the frequency $\omega_b$.

The system (6.3.3) and (6.3.6) can be reduced to a first-order nonlinear equation (by analogy with §6.2)

$$\dot{b} = \omega_0 \sqrt{1 - b^2 - a_0^2 \cosh^2 \left[ \frac{q_{31}^b}{\omega_0} (b - b_0) \right]}, \qquad (6.3.7)$$

where

$$\omega_0 = \frac{d_{32} \mathcal{E}_0}{2\hbar}, \quad q_{31}^b = \sqrt{\frac{2\pi N d_{31} \omega_b}{\hbar}}.$$

Solving (6.3.7), we can find the dependence of $b$ on the time $t$. However, the solution of (6.3.7) cannot be expressed in terms of elementary functions, it is of interest to investigate it qualitatively.

We assume that at the initial instant of time all the atoms are at the level $E_2$: $b_0^2 \approx 1$ and $a_0^2 \approx 0$, and we consider the character of the solutions of (6.3.7) as functions of the relation between the parameters $q_{31}^b$ and $\omega_0$.

We assume first that the time of the spontaneous transition $E_3 \to E_1$ is large compared with the time of the simulated transition $E_2 \to E_3$: $1/q_{31}^b \gg 1/\omega_0$. In this case the parameter entering in the right side of (6.3.7) is small, $q_{31}^b/\omega_0 \ll 1$, so we can assume that $\cosh \left[ q_{31}^b/\omega_0 (b - b_0) \right] \approx 1$, and obtain a solution of (6.3.7) in explicit form: $b = b_0 \sin \omega_0 t$. Thus, the atoms of the medium execute simulated transition between the levels $E_2$ and $E_3$ with frequency $\omega_0$, and the population of level $E_1$ remains unchanged ($a \sim b_0 q_{31}^b/\omega_0$).

In the opposite limiting case when the parameter $q_{31}^b/\omega_0$ is not small ($q_{31}^b/\omega_0 > 1$), the character of the solutions of (6.3.7) can change, for in this case the last term of (6.3.7) is comparable to unity and the right side of (6.3.7) can vanish at small values of $b(t)$: $b = b_{\min} \ll 1$. The minimum value of the function $b_{\min}$ is obtained by equating $\dot{b}(b_{\min})$ to zero:

$$1 - b^2 - a_0^2 \, \text{ch}^2 \left[ \frac{q_b}{\omega_0} (b_{\min} - b_0) \right] = 0. \qquad (6.3.8)$$

Since $b_{\min}^2 \ll 1$ and $c(b_{\min}) = \dot{b}(b_{\min}) = 0$, we get $a(b_{\min}) \approx 1$, that is, practically all the atoms go over to the level $E_1$, emitting a longitudinal wave of frequency $\Omega_{31}$. According to (6.3.7) the transition time is of the order of $1/\omega_0$.

In order to determine the maximum energy of the generated longitudinal wave, we can use the energy conservation law which follows from (6.3.6)

$$\frac{\dot{b}^2}{\omega_0^2} + b^2 + \frac{\mathcal{E}_{31}^2}{8\pi N \hbar \omega_b} = 1. \qquad (6.3.9)$$

(We chose the constant on the right side of (6.3.9) from the condition $\dot{b}_0 = \mathcal{E}_{31}(0) = 0$ $b_0 = 1$). If we assume that all the atoms have gone over to the level $E_1$, namely $b_{\min} = \dot{b}_{\min} = 0$, then, according to (6.3.9), we have

$$\left(\frac{\mathcal{E}_{31}^2}{8\pi}\right)_{\max} = N\hbar\omega_b. \qquad (6.3.10)$$

Thus, the amplitude of the longitudinal wave can be increased by increasing the plasma density (however, the resonance condition (6.3.5) must not be violated). The maximum energy of the longitudinal wave cannot exceed in this case the value $N\hbar\omega_{31}$, that is, the maximum energy stored in the medium.

We now obtain the conditions under which it is possible to use the linear equations of motion of the beam. Using relations (6.3.5) and (6.3.10) we can show that the approximation considered above is valid if

$$\frac{n_b m v_0^2}{2} \gg \left(\frac{\mathcal{E}_{31}^2}{8\pi}\right)_{\max} = N\hbar\omega_b, \qquad (6.3.11)$$

that is, if the kinetic energy of the beam electrons is much higher than the wave energy. In the case when an inequality inverse to (6.3.11) is satisfied, the decisive nonlinear effects are the effects of nonlinear interaction of the beam electrons with the wave field. In such a case it is necessary to use the nonlinear equations of motion of the beam, and the equations of motion of the medium can be regarded as linear.

## §6.4. Cherenkov radiation in a medium with an inverted population level.

At the present time, Cherenkov radiation in equilibrium media has been studied extensively and used widely in various fields of physics [16]. On the other hand, it is of interest to investigate the features and potentialities, for practical use, of Cherenkov radiation in nonequilibrium media since the existence of nonequilibrium condition provides the possibility for conversion of the force of the Cherenkov field that acts on the radiating particle. This effect was first discussed and investigated by Tamm [17] for the case of a charge interacting with a moving dielectric. It was pointed out by Veksler that it would be possible to realize a moving dielectric by means of electron beam, and to use the conversion of the sign of the Cherenkov field for coherent acceleration of ions [18].

A similar effect obtains in a medium at rest in which the nonequilibrium is due to the inversion of the population of its energy levels [19]. The first suggestion for acceleration of charge particles in an active dielectric medium by means of direct conversion of energy of the medium into kinetic energy of the beam of accelerated particles was given by Fainberg [20]. This section is devoted to a theoretical analysis of the Cherenkov acceleration of a beam of charge particles in a transparent dielectric with an inverted population level, the inversion being due to the field induced by the beam. The field is comprised of longitudinal (polarization) oscillations and transverse waves (Cherenkov radiation) [11].

In the general case the accelerating field induced by the beam due to the interaction of this beam with the inverted level population of the medium is comparable with the retarding field that acts on the beam due to the noninverted levels. The efficiency of the beam with the inverted atoms of the medium can be enhanced by modulating the beam at an appropriate frequency. In this case the intensity of the emission of the inverted level population is enhanced by virtue of the coherent addition of the fields of individual bunches in the beam, as a consequence of which the amplitude of the accelerating field is increased substantially. In view of this circumstance we will consider the excitation of longitudinal and transverse waves in the medium by a modulated beam.

The complete system of equations which, in the nonlinear approximation, describes the collective interaction of a beam of charge particles with the dielectric medium consists of Maxwell's equations with a forcing term due to the current $\mathbf{j}$ together with equations of motion for the medium:

$$\mathrm{curl}\,\mathbf{E} = -\frac{1}{c}\frac{\partial \mathbf{B}}{\partial t}, \quad \mathrm{curl}\,\mathbf{B} = \frac{1}{c}\frac{\partial \mathbf{E}}{\partial t} + \frac{4\pi}{c}\left(\frac{\partial \mathbf{P}}{\partial t} + \mathbf{j}\right), \tag{6.4.1}$$

$$\frac{\partial^2 \mathbf{P}}{\partial t^2} + \Omega^2 \mathbf{P} = \pm \frac{2d_0}{\hbar}\sqrt{(\Omega N d_0)^2 - \dot{\mathbf{P}}^2 - \Omega^2 \mathbf{P}^2}\,\mathbf{E}. \tag{6.4.2}$$

Here, $\mathbf{P}$ indicates the dielectric polarization of the medium, $N$ is the density of active centers, $d_0$ is the electric dipole moment per center, $\Omega = (E_2 - E_1)/\hbar$ is the appropriate resonance frequency. In the general case the right side of Eq (6.4.2) is proportional to $N_1 - N_2$, the difference in the populations of the lower $(E_1)$ and upper $(E_2)$ energy levels; hence the upper sign in front of the radical in (6.4.2) corresponds to the normal condition while the lower sign corresponds to an inverted population level at the initial time $(t = \mathbf{E} = \mathbf{B} = \mathbf{P} = \dot{\mathbf{P}} = 0)$.

*1. Linear retarding medium.*

The two-level approximation (6.4.2) applies for the description interaction medium with the field when the characteristic frequencies of the electromagnetic fields are close to the resonance frequency of the medium $\Omega$. In the case of Cherenkov radiation and the excitation of polarization oscillations these conditions are satisfied for modest densities of active centers $(2\pi N d_0^2/\hbar\Omega \ll 1)$ and for the excitation of a medium by a modulated current if the modulation frequency $\omega_\mathrm{m}$ is reasonably close to the appropriate resonance frequency $(|\Omega - \omega_\mathrm{m}| \ll \Omega)$.

In the linear approximation $[\dot{\mathbf{P}}^2 + \Omega^2 \mathbf{P}^2 \ll (\Omega N d_0)^2]$ which corresponds to neglecting the change in the level of population in the medium under the effect of the radiation, using (6.4.2) we can find an effective dielectric constant for the medium:

$$\epsilon_\pm = 1 \pm \frac{\omega_g^2}{\Omega^2 - \omega_\mathrm{m}^2}, \quad \omega_g^2 = \frac{8\pi N d_0^2 \Omega}{\hbar}. \tag{6.4.3}$$

Thus, in an active dielectric the frequency region in which the Cherenkov condition is satisfied $(\beta^2 \epsilon_- > 1$ $(V = \beta c$ is the velocity of charge particle) is higher than the resonance frequency $\Omega$ while the refractive index $\sqrt{\epsilon_-}$ falls of with frequency in the transparency region (negative dispersion). Under these conditions the frequency polarization oscillations is smaller than the resonance frequency $\omega_\mathrm{p} \equiv (\Omega^2 - \omega_g^2)^{1/2}$.

Substituting (6.4.3) in the general expressions for the Cherenkov field and the polarization field [21] that act on the particle and taking account of change in the phase of the logarithm due to the change in the order of traversing the zeroes and poles of the argument of the logarithm in the active medium, we have

$$E_\mathrm{ch} = \mp\frac{e^2}{2c^2}\int\limits_{\beta^2\epsilon_\pm > 1}\left[1 - \frac{1}{\beta^2\epsilon_\pm(\omega)}\right]\omega\,d\omega, \quad E_\mathrm{p} = \mp\frac{e^2}{2V^2}\omega_g^2\ln\left(\frac{k_m^2 V^2}{\Omega^2 - \omega_g^2}\right), \tag{6.4.4}$$

where $e$ is the particle charge and $V = \beta c$ is the particle velocity.

The feasibility of practical use of this effect for the indicated purposes is determined by the amplitude of the accelerating field and the emission time. These are essentially nonlinear characteristics which only can be found from a nonlinear analysis, to which we now tern.

194

## 2. Nonlinear polarization oscillations.

We start with the polarization oscillations, which can be considered in the one-dimensional approximation. Assume that polarization oscillations in the dielectric are excited by a sequence of charged plane sheets with surface charge density $\sigma e$ which move in the medium with velocity $V$ in the direction of the normal to the plane surface. In this case the field in the medium is determined by the one-dimensional system of equation consisting Eq. (6.4.2) and Poisson's equation:

$$\frac{\partial}{\partial z}(E + 4\pi P) = 4\pi\sigma e \sum_{s=-\infty}^{\infty} \delta(z - Vt - s\ell),$$

$$\frac{\partial^2 P}{\partial t^2} + \Omega^2 P = \pm\frac{2d_0}{\hbar}\sqrt{(\Omega N d_0)^2 - \dot{P}^2 - \Omega^2 P^2}\, E,$$

(6.4.5)

where $\ell$ is the distance between charge sheets.

Assume that the modulation frequency $\omega = 2\pi V/\ell$ is equal to the frequency $\omega(0) = \sqrt{\Omega^2 - \omega_g^2}$ which corresponds to the linear polarization oscillations; we now seek the nonstationary solution of the system in (6.4.5) in the form:

$$E(t,z) = N d_0[A(t)\cos\Phi + B(t)\sin\Phi], \quad P(t,z) = N d_0[a(t)\cos\Phi + b(t)\sin\Phi],$$

$$\Phi = \omega - k_{\parallel}z, \quad k_{\parallel} = 2\pi/\ell.$$

(6.4.6)

The dimensionless functions $a(t)$ and $b(t)$, which vary slowly in time, are described by the following system of equations:

$$2\frac{da}{d\tau} - nb + n(b + I_p)\sqrt{1 - a^2 - b^2} = 0, \quad 2\frac{db}{d\tau} + na - na\sqrt{1 - a^2 - b^2} = 0,$$

(6.4.7)

where

$$A + iB = -4\pi(a + ib) - i\frac{4\sigma e}{N d_0},$$

$$n = \frac{8\pi N d_0^2}{\hbar\Omega} = \frac{\omega_g^2}{\Omega^2} \ll 1, \quad I_p = \frac{\sigma e}{\pi N d_0}, \quad \tau = \Omega t.$$

Making use the notation

$$a + ib = \sin\psi\exp i\varphi, \quad \varphi(0) = \pi, \quad \psi(0) = 0,$$

we have from equation (6.4.7)

$$\frac{d\psi}{d\tau} = \frac{n}{2} I_p \left[1 - \frac{1}{I_p^2}\sin^4\frac{\psi}{2}\,\mathrm{tg}^2\frac{\psi}{2}\right]^{1/2},$$

$$\sin\varphi = \frac{1}{I_p}\sin^2\frac{\psi}{2}\,\mathrm{tg}\,\frac{\psi}{2}.$$

(6.4.8)

It follows from Eq. (6.4.8) that the excitation of polarization oscillations alternates periodically with the absorption of the field of the oscillations in the medium. For small values of the current ($I_p \ll 1$) the period $T_p$ in units of $\Omega^{-1}$ is $4(nI_p^{2/3})^{-1}$. The peak amplitude of the accelerating field and is determined from the condition $\dot{A} = 0$ and for $I_p \ll 1$ is found to be

$$E_\mathrm{m} = 2\sqrt{3}\left(\pi\sigma e N^2 d_0^2\right)^{1/3}.$$

(6.4.9)

195

It is evident that in the cases being considered $\sigma e/Nd_0 \ll 1$ this field is large compared with the surface charge field $2\pi\sigma e$. The increment of kinetic energy per unit surface of the layer referred to the density of energy stored in the medium (efficiency) is found to be proportional to $I_p^{-2/3}$.

Physically, the saturation of the amplitude of the field arises as follows. Excitation of polarization oscillations occurs by virtue of the energy stored in the atoms in the media so that as the field amplitude increases there is a reduction in the difference of populations in the levels. The frequency of the longitudinal oscillations $\omega_p$ increases and violates the resonance between the frequency of the exciting force $\omega_m$ and the frequency of the characteristic oscillations of the medium. When $\tau \simeq T$ the phase of the field changes sign and the acceleration becomes a retardation, even though the difference in the population levels ($W = \cos\psi$) does change sign.

### 3. Nonlinear Cherenkov oscillations.

We now consider the problem of Cherenkov radiation due to a modulated current in a dielectric at a negative temperature. In this case, in order to estimate the peak amplitude of the Cherenkov field and the corresponds lifetime of the pulse we consider a model problem in which the Cherenkov radiation of the inverted medium is excited by a current with a density

$$\mathbf{j} = \mathbf{n} j_0 \cos\Psi, \quad \Psi = \omega t - k_{\|}z - k_{\perp}x,$$

where $\mathbf{n}$ is the unit vector along the $z$-axis while the wave numbers $k_{\|}$ and $k_{\perp}$ satisfy the dispersion equation obtained from the linear theory

$$k_{\|}^2 + k_{\perp}^2 = \frac{\omega^2}{c^2}\epsilon_-(\omega), \quad k_{\|} = \frac{\omega}{V}. \tag{6.4.10}$$

Substituting the current (6.4.10) in the Maxwell's equations (6.4.1), we seek a solution of these equations together with the equation of state of the medium (6.4.2), again using the above method:

$$\mathbf{E}(t, \mathbf{r}) = Nd_0[\mathbf{A}(t)\cos\Phi + \mathbf{B}(t)\sin\Phi], \quad \mathbf{P}(t, \mathbf{r}) = Nd_0[\mathbf{a}(t)\cos\Phi + \mathbf{b}(t)\sin\Phi],$$

$$\Phi = \omega t - k_{\|}z, \quad k_{\|} = 2\pi/\ell.$$

We then obtain the following system of equations in total derivatives for the slowly varying amplitudes $a_\perp$ and $b_\perp$ of the transverse component of the polarization vector, assuming

$$2\Delta \frac{d}{d\tau}\left(\frac{u}{\sqrt{1 - |u|^2}}\right) - \frac{2n}{\Delta}\frac{1}{\sqrt{1 - |u|^2}}\frac{du}{d\tau} - inu\left(\frac{1}{\sqrt{1 - |u|^2}} - 1\right) = -iK_\perp nI, \tag{6.4.11}$$

where

$$n = \frac{8\pi Nd_0^2}{\hbar\Omega}, \quad I = \frac{j_0}{\Omega Nd_0}, \quad \tau = \Omega t,$$

$$\Delta = 1 - \omega^2/\Omega^2, \quad K_\perp = k_\perp c/\Omega, \quad u = a_\perp + ib_\perp.$$

In this case the amplitude of the longitudinal field $E_{\|}$, which determines the effectiveness of the interaction of the beam with medium, is given by the relation

$$A_{\|} = -K_\perp A_\perp = 4\pi\frac{K_\perp}{n}\frac{2\dot{b}_\perp + \Delta a_\perp}{\sqrt{1 - |u|^2}}. \tag{6.4.12}$$

196

An investigation of linear solutions of the system (6.4.11) shows that in the nonstationary regime acceleration of the beam by radiation from the medium can occur only when the condition $1 - n/\Delta^2 < 0$ is satisfied. It is evident from (6.4.10a), that this condition is equivalent to the requirement that the group velocity of the excited wave be negative. Assuming that this inequality is satisfied with margin, making use the substitution $u = \sin\psi\exp(i\varphi)$, we have

$$\frac{d\psi}{d\tau} = \frac{\psi_{\mathrm{m}}}{T}\left[1 - \left(\frac{\psi}{\psi_{\mathrm{m}}}\right)^3\right]^{1/2}, \quad \psi_{\mathrm{m}} = 2(K_\perp I)^{-2/3},$$

$$\sin\varphi = \left(\frac{\psi}{\psi_{\mathrm{m}}}\right)^3, \quad T = \frac{4\Delta}{n}\left(1 - \frac{n}{\Delta^2}\right)(K_\perp I)^{2/3}.$$

(6.4.13)

It will be evident from this equation that if the relaxation effects are neglected the process of emission of the Cherenkov field in the medium and the acceleration of the beam alternate periodically with the inverse process of absorption in the medium and retardation of the beam. The characteristic period for the radiation is of order $T$ (in units of $\Omega^{-1}$) while the peak amplitude of the longitudinal component of the field is given by the expression

$$(A_{\|})_{\max} = -K_\perp A_\perp = 4\pi\frac{K_\perp}{n}|\Delta|(K_\perp I)^{1/3}.$$

(6.4.14)

Strictly speaking, this expression applies when $K_\perp^2 \simeq n/|\Delta| \ll 1$ $(n > \Delta^2)$. However, for the purpose of making estimates we can take $K_\perp \sim 1$. Thus we find

$$(E_{\|})_{\max} = 4\pi N d_0\left(\frac{2j_0}{\Omega N d_0}\right)^{1/3}.$$

(6.4.15)

Physically, the stabilization in the growth of the field amplitude is explained by the dependence of the characteristics frequency of the field in the medium on its amplitude: as the amplitude increases the synchronism between the wave and beam is disturbed and as a result the field radiation and the particle acceleration are converted to field absorption and beam retardation. When $\psi_{\mathrm{m}} \ll 1$, in a time corresponding to one cycle only a small fraction of the energy stored in the medium $(\simeq \psi_{\mathrm{m}}^2)$ is converted into radiation. The energy of the field can be increased by programming a change in the frequency of modulation of the beam.

Above, in analyzing the Cherenkov interaction of a beam of charged particles with an active dielectric we have assumed that beam current is specified. Inasmuch as it is difficult to provide modulation of a beam by external sources in the shortwave region, one could make use of the Cherenkov instability of an unmodulated beam [see (6.1.26)].

### §6.5. Acceleration of a modulated beam of charged particles in a medium with inverted population levels.

As shown in §6.5, a sign reversal of the Cherenkov field occurs when a beam of charged particles moves through a medium with inverted population levels, so that the retardation of the beam, which takes place in a passive medium, is replaced by acceleration. The approximation employed in §6.5, in which the beam resembles a succession of "infinitely heavy" charge bunches, is to neglect the phase motion of the bunches. In this case, the fundamental nonlinear effect that limits the amplitude of the accelerating field is the dependence of the longitudinal wave frequency $\omega(t) = \sqrt{\Omega^2 - \omega_g^2\cos\psi}$ on the field amplitude $|E| = 4\pi N d_0\sin\psi$; here $\omega_g^2 = 8\pi N d_0^2/\hbar\Omega$, $\Omega$ is the resonance frequency, and $N$

is the density of active centers in the medium. If there is no field at time zero, $\psi(0) = 0$, and the beam modulation frequency is equal to the frequency of longitudinal oscillations of the medium, $\omega_m = 2\pi v_0/\ell = \omega(0)$ ($v_0$ is the beam velocity and $\ell$ is the distance between bunches, and $\omega(0) = \sqrt{\Omega^2 - \omega_g^2}$ corresponds to a medium with inverted population levels); then $\omega(t)$ increases with the growth of the field amplitude [$\psi(t) > 0$], and the phase velocity of the waves $v_{\rm ph}(t) = (\ell/2\pi)\omega(t)$ becomes larger than the initial beam velocity. As a result, the wave and the beam are no longer in resonance and only part of the energy stored in the medium is converted to field energy $\psi_{\rm max} \ll 1$.

The object of the present work is to extend the previous treatment to take account of phase motion of the accelerating bunches. One significant result is that the beam velocity increases in proportion to increases in the phase velocity of the waves, corresponding to acceleration by the wave field. As shown below, it is possible for the resonance between the beam and wave to be maintained during the entire acceleration process so that the energy stored in the medium is completely converted to beam energy.

Assume that a train of charged planes with surface charge density $\sigma e$ moves through the medium with velocity $v_0$ in the direction normal to the planes. The system of equations describing the interaction of such a beam with the medium consists the equation of state of the medium and Poisson's equation [see Eqs. (6.4.5)]; and the equations of motion of the bunches [12]

$$\frac{\partial}{\partial x}(E + 4\pi P) = 4\pi\sigma e \sum_{s=-\infty}^{\infty} \delta\left[z - s\ell - z_s(t)\right],$$

$$\frac{\partial^2 P}{\partial t^2} + \Omega^2 P = \pm\frac{2d_0}{\hbar}\sqrt{(\Omega N d_0)^2 - \dot{P}^2 - \Omega^2 P^2}\, E, \tag{6.5.1}$$

$$\frac{d}{dt}(\dot{z}_s \gamma_s) = \frac{e}{m}\, E\left[t, z_s(t)\right],$$

where $z_s(t)$ is the coordinate of the bunch denoted by index $s$, $\dot{z}_s(t)$ is the velocity of the bunch, and $\gamma_s = \left(1 - \dot{z}_s^2/c^2\right)^{-1/2}$. The dot means the derivative with respect to time.

We write the solution of (6.5.1) in the form of a traveling wave with slowly varying amplitude and phase:

$$E(t, z) = \operatorname{Re}\mathcal{E}\exp i\Phi, \quad P(t, z) = \operatorname{Re}a\exp i\Phi,$$
$$\Phi = kz - \omega t, \quad \dot{\mathcal{E}} \ll \omega\mathcal{E}, \quad \dot{a} \ll \omega a. \tag{6.5.2}$$

Substituting (6.5.2) in the field equation and averaging over the spatial period of the system, $\ell$, we find the relation between the complex amplitudes $\mathcal{E}$ and $a$:

$$\mathcal{E}(t) = -4\pi a(t) - i4\sigma e\exp[-i\Phi(t)], \quad \Phi(t) = kz(t) - \omega t. \tag{6.5.3}$$

Putting $a = N d_0 \sin\psi \exp(i\vartheta)$ in (6.5.3), and then substituting (6.5.2) and (6.5.3) in (6.5.1), we find

$$\frac{d\psi}{dt} = -\frac{4\sigma e d_0}{\hbar}\cos(\vartheta + \Phi),$$

$$\frac{d\vartheta}{dt} = -\frac{4\pi N d_0^2}{\hbar}(1 - \cos\psi) + \frac{4\sigma e d_0}{\hbar}\operatorname{ctg}\psi\sin(\vartheta + \Phi), \tag{6.5.4}$$

$$\frac{d}{dt}\gamma\dot{z} = -\frac{4\pi N d_0}{m}\sin\psi\cos(\vartheta + \Phi).$$

198

We have verified that this result (6.5.4) satisfied the equality $\omega^2(0) = \Omega^2 - \omega_g^2$, and the following inequalities hold:

$$\frac{8\pi N d_0^2}{\hbar \Omega} = \frac{\omega_g^2}{\Omega^2} \ll 1, \quad \frac{4\sigma e d_0}{\hbar \Omega} \ll 1. \tag{6.5.5}$$

From the first and third equations in (6.5.4) we have momentum conservation:

$$n_b \dot{z} \gamma = n_b v_0 \gamma_0 + \frac{\pi N \hbar}{\ell} (1 - \cos \psi), \tag{6.5.6}$$

where $n_b = \sigma/\ell$ and $v_0 = c\beta_0$ are the density and initial velocity of the beam, and

$$\psi(0) = 0, \quad k = 2\pi/\ell \simeq \Omega/v_0, \quad \gamma_0 = (1 - \beta_0^2)^{1/2}.$$

According to (6.5.6), the momentum density of the accelerated beam is proportional to the difference in populations energy levels in the medium, $\cos \psi = \sqrt{1 - (Nd_0)^{-2}|a|^2}$, and increases with "de-excitation" of the medium ($\cos \psi < 1$).

Transformation of the system of equations (6.5.4) to the new variable $X = \vartheta + \Phi$, after expressing the quantity $\dot{\Phi} = (2\pi/\ell)\dot{z} - \omega$ in terms of the function $\psi$ from relation (6.5.6), yields the result

$$\frac{dX}{dt} = \frac{2\pi c}{\ell} \frac{\mathcal{B}}{\sqrt{1 + \mathcal{B}^2}} - \omega - \frac{4\pi N d_0^2}{\hbar}(1 - \cos \psi) + \frac{4\sigma e d_0}{\hbar} \mathrm{ctg}\psi \sin X, \tag{6.5.7}$$

where $\mathcal{B} = \beta_0 \gamma_0 + (\pi N \hbar/m\sigma)(1 - \cos \psi)$.

We consider first a nonrelativistic beam, $\beta_0 \ll 1$ and $\gamma_0 \simeq 1$. Neglecting $\mathcal{B}^2$ in comparison to unity in the radical in Eq. (6.5.7) and integrating (6.5.7) together with the first equation in (6.5.4), we find

$$\sin X = I^{-1} \sin^2 \frac{\psi}{2} \mathrm{tg} \frac{\psi}{2}, \tag{6.5.8}$$

where

$$I = \frac{\sigma e}{\pi N d_0} \left(1 - \frac{N\hbar\Omega}{n_b m v_0^2} \frac{\Omega^2}{\omega_g^2}\right)^{-1}$$

takes into account a possibility of synchronous acceleration of the beam by the wave.

Expressing $\cos X$ on the right-hand side of the first equation (6.5.4) in terms of $\psi$, we obtain the nonlinear equation

$$\frac{d\psi}{dt} = \frac{4\sigma e d_0}{\hbar} \left(1 - I^{-2} \sin^4 \frac{\psi}{2} \mathrm{tg}^2 \frac{\psi}{2}\right)^{1/2} \tag{6.5.9}$$

which corresponds to Eq. (6.4.8) in the limiting case

$$n_b m v_0^2 \gg (\Omega^2/\omega_g^2) N\hbar\Omega.$$

The maximum value of the function $\psi_m$ is determined by the relation $\dot{\psi}(\psi_m) = 0$. Since $|I| \ll 1$, it follows that $\psi_m = 2|I|^{1/3}$. In the opposite limit, the phase motion dominates.

If the beam energy density approximates the resonance value,

$$\left| n_b m v_0^2 - (\Omega^2/\omega_g^2) N\hbar\Omega \right| \to 0, \tag{6.5.10}$$

it follows from Eq. (6.5.9) that the function $\psi$ changes linearly with time $\psi \simeq= -4\sigma e d_0 t / \hbar$; consequently, the dielectric reaches a stable state $\cos \psi = -1$ in a time $T \simeq \hbar \pi / 4 \sigma e d_0$, for which the change in the beam momentum density is, from (6.5.6),

$$n_b m (v_{\max} - v_0) = N \hbar \Omega / v_0, \qquad (6.5.11)$$

where $v_{\max}$ is the maximum velocity of the beam.

Thus, if the condition (6.5.10) for resonant acceleration of the beam is satisfied, then coherence between the beam and wave is maintained during the entire acceleration process can begin and the beam energy can be converted in the internal energy of the medium.

In the relativistic case integration of Eq. (6.5.7) together with the first equation of (6.5.4) yields the relation

$$
\begin{aligned}
2 \sin X \sin \psi = & \frac{\pi N d_0}{\sigma e} (1 - \cos \psi)^2 + \frac{\hbar \Omega}{2 \sigma e d_0} (1 - \cos \psi) \\
& - \frac{mc}{2 d_0 e l N} \left( \sqrt{1 + \mathcal{B}^2} - c \gamma_0 \right).
\end{aligned}
\qquad (6.5.12)
$$

Substituting $\psi = \pi$ in (6.5.12), we determine the resonance condition for acceleration of a relativistic beam:

$$\frac{n_b m v_0^2 \gamma^3}{N \hbar \Omega} = \frac{\Omega^2}{\omega_g^2} - \beta_0^2 \gamma^2. \qquad (6.5.13)$$

It is evident that Eq. (6.5.13) represents a generalization of Eq. (6.5.10).

In conclusion, we observe that representation of a beam by a succession of bunches is physically justified if a wave field with increasing phase stability, as in the case for linearly accelerated charged particles, resulting in the partial subdivision of a modulated beam into bunches.

## §6.6. Self-excitation of a molecular beam moving through a slow wave system.

It is well known that the energy of translational motion of an oscillator moving through a slow-wave system at a velocity greater than the velocity of light is transforms into oscillation energy if the longitudinal velocity of the oscillator satisfies the condition for the anomalous Doppler effect [22]. This effect can be used for producing inverse populated levels in a molecular beam that constitutes an aggregate of moving two-level oscillators [13].

### 1. Electromagnetic instability.

The initial system of equations consists of the equation for the polarization vector **P** of the molecular beam [3]

$$\frac{d^2 \mathbf{P}}{dt^2} + \Omega^2 \mathbf{P} = -\frac{2 d_0}{\hbar} \sqrt{(\Omega N d_0)^2 - \Omega^2 \mathbf{P}^2 - \left( \frac{d \mathbf{P}}{dt} \right)^2} \ \mathbf{E}, \qquad (6.6.1a)$$

$[d/dt = \partial/\partial t + V \partial/\partial z;$ $N$ and $V$ are the beam density and velocity, $\Omega$ is the resonance frequency, and $d_0$ is the atom dipole moment], and the Maxwell equation for the field **E** in a medium with effective refractive index $n_0$

$$\frac{\partial^2 \mathbf{E}}{\partial x^2} - \frac{1}{c^2} \frac{\partial^2}{\partial t^2} \left( n_0^2 \mathbf{E} + 4 \pi \mathbf{P} \right) = 0. \qquad (6.6.1b)$$

In order to investigate the stability of the system, we shall first consider the problem in the linear approximation,

$$\dot{P}^2 + \Omega^2 P^2 \ll (\Omega N d_0)^2.$$

By linearizing Eqs. (6.6.1) and assuming that all the quantities depend on the time $t$ and the longitudinal coordinate $z$ only through the factor $\exp[i(\omega t - kz)]$ (plane wave propagating along the beam velocity $V$), we obtain the dispersion equation, which describes the relationship between the frequency $\omega$ and the wave number $k$:

$$\left(\frac{ck}{\omega}\right)^2 = n_0^2 + \frac{\omega_g^2}{\Omega^2 - (\omega - kV)^2} = 0, \qquad (6.6.2)$$

where $\omega_g^2 = 8\pi N d_0^2 \Omega / \hbar$.

We seek the solution of Eq. (6.6.2) in the following form:

$$\omega = ck/n_0 + i\delta, \quad \omega = kV - \Omega + i\delta, \quad \delta \ll \omega. \qquad (6.6.3)$$

(the anomalous Doppler effect). Substituting (6.6.3) in (6.6.2), we find

$$\omega_0 = \frac{\Omega}{\beta n_0 - 1}, \quad \delta = \sqrt{\frac{\pi q}{\beta n_0 - 1}}, \qquad (6.6.4)$$

where $\beta = V/c$ and $q = \omega_g^2/4\pi n_0^2$.

The theory above describes only the beginning of self-excitation of the beam, when the amplitude op the polarization vector is still rather small $|P| \ll N d_0$, so that it cannot be used for analyzing the changes in the difference between the populated levels of the beam. In order to estimate this effect we consider the problem in the nonlinear approximation.

We shall seek the nonlinear solution of the self-consistent system of equation (6.6.1) in the form of

$$E_x = \mathcal{E}(t) \sin(\omega_0 t - kz), \quad P_x = a(t) \cos(\omega_0 t - kz), \qquad (6.6.5)$$

where $\omega$ and $k$ satisfy the conditions (6.6.3). Then, assuming that the amplitudes $a(t)$ and $\mathcal{E}$ are slowly varying functions ($\dot{a} \ll a$, $\dot{\mathcal{E}} \ll \mathcal{E}$), we obtain

$$\dot{a} = -\text{sign} W_0 \frac{d_0}{\hbar} \sqrt{N^2 d_0^2 - a^2}, \quad \dot{\mathcal{E}} = \frac{2\pi}{n_0^2} \omega_0 a. \qquad (6.6.6)$$

By substituting the variable $a = N d_0 \sin \varphi$, we can reduce the system in (6.6.6) to the nonlinear pendulum equation

$$\ddot{\varphi} + \delta^2 \text{sign} W_0 \sin \varphi = 0, \qquad (6.6.7)$$

where $\delta$ is determined by (6.6.4).

According to (6.6.7), for $\text{sign} W_0 = -1$ (when all the beam particles are at the energy level), the amplitude of beam polarization increases exponentially at the beginning of the process $\varphi \ll 1$. For large amplitudes, $\varphi \simeq 1$, the solution of (6.6.7) is periodic with the period $T \sim \delta^{-1}$. In this, the difference between the populated levels of the beam,

$$W(t) = -\frac{1}{\Omega N d_0} \sqrt{(\Omega N d_0)^2 - \Omega^2 \mathbf{P}^2 - \left(\frac{d\mathbf{P}}{dt}\right)^2} = -\cos\varphi(t),$$

$[\varphi(0) = 0]$ can periodically assume positive values [for $\cos\varphi(t) < 0$], so that the system periodically passes into the inverted state.

## 2. Polarization instability.

The effect occurs only if $V > c/n_0$, i.e., there is a limit to the beam velocity for any slow-wave system with fixed $n_0$. In this connection the use of an electron plasma as a slow-wave structure, where the instability develops as a longitudinal wave, while the beam is polarized in the direction of motion, is of special interest.

In this case the field $E_z$ in the plasma with a molecular beam is determined by the one-dimensional system of equation consisting Eq. (6.6.1a) and Poisson's equation:

$$\frac{\partial}{\partial x}\left(E_z + 4\pi P_z\right) = 4\pi e\left(n_e - n_i\right) \tag{6.6.8}$$

($n_i$ is the density of background ions).

For small polarization oscillations in a cold plasma, we can use the following system of linear equations

$$\frac{\partial \tilde{v}}{\partial t} = \frac{e}{m}E_z, \quad \frac{\partial \tilde{n}}{\partial t} + \frac{\partial \tilde{v}}{\partial z} = 0. \tag{6.6.9}$$

to eliminate $\tilde{n} = n_e - n_i$ ($|\tilde{n}| \ll n_i$) from Eq. (6.6.8). As a result, we get

$$\frac{\partial^2}{\partial t^2}\left(E_z + 4\pi P_z\right) + \omega_p^2\, E_z = 0, \tag{6.6.10}$$

where $\omega_p = \sqrt{4\pi e^2 n_i/m}$ is the Langmuir density of the plasma.

By linearizing Eq. (6.6.1a), and using this equation together with Eq.(6.6.10), we obtain the dispersion relation

$$1 - \frac{\omega_p^2}{\omega^2} + \frac{4\pi q}{\Omega^2 - \left(\omega - kV\right)^2} = 0. \tag{6.6.11}$$

By analogy with (6.6.3), the solution of algebraic can be presented in the form

$$\omega = \omega_p + i\delta_p, \quad \omega = kV - \Omega + i\delta_p, \quad \delta_p \ll \omega, \tag{6.6.12}$$

where

$$\delta_p = \sqrt{\pi q\,\frac{\omega_p}{\Omega}}, \quad k = \frac{\omega_p + \Omega}{V}. \tag{6.6.13}$$

Then, the phase velocity of the disturbance is

$$v_{\mathrm{ph}} = \frac{\omega_p}{k} = \frac{\omega_p}{\omega_p + \Omega_p}\, V, \tag{6.6.14}$$

i.e., it is always lower than the beam velocity $V$. Thus, if the disturbance velocity is fixed and equal to $c/n_0$ in the case of a transverse-wave instability; the beam in the plasma "selects" its own wave with the required phase velocity $v_{\mathrm{ph}} < V$. Consequently, an instability occurs for any beam velocity.

The final equation of the nonlinear theory coincides with (6.6.7) if we put $\delta = \delta_p$.

We can show that the approximation of linear plasma considered above is valid if the inequality

$$\frac{|\tilde{n}|}{n_i} \approx \frac{eE}{m\,\omega_p v_{\mathrm{ph}}} \ll 1 \tag{6.6.17}$$

is violated.

It should be mentioned that above approximation holds for $t \ll T_{\mathrm{rel}}$, since we have neglected relaxation processes, thereby assuming $T_{\mathrm{rel}} \to \infty$. In the opposite limiting case, the relaxation processes can modify the character of the solution.

### Appendix 6.1

According to Ref. [3], we consider an interaction of the atom with the field $\mathbf{E}$ when the characteristic frequency of the electromagnetic wave is close to the resonance frequency of the atom $\Omega = (E_2 - E_1)/\hbar$.

Under conditions of two-level approximation, we can seek the solution of the Schrödinger's equation

$$i\hbar\dot{\psi}(t) = \left(\hat{H}_0 + \hat{H}'\right)\psi \qquad (A6.1.1)$$

in the form

$$\psi(t) = a(t)\psi_1 + b(t)\psi_2. \qquad (A6.1.2)$$

Here the eigenfunctions $\psi_1$ and $\psi_2$ correspond to the energy levels $E_1$ and $E_2$, and the Hamiltonian satisfy the conditions

$$< \psi_m|H_0|\psi_{m'} >= E_m \delta_{mm'},$$

$$< \psi_m|H'|\psi_{m'} >=< \psi_m| - \hat{\mathbf{d}}\mathbf{E}|\psi_{m'} >= -\mathbf{d}\mathbf{E}\left(1 - \delta_{mm'}\right), \qquad (A6.1.3)$$

$$\mathbf{d}_{12} =< \psi_1|\hat{\mathbf{d}}|\psi_2 >, \quad m, m' = 1, 2.$$

Substituting (A6.1.2) in (A6.1.1) and using (A6.1.3), we get the following set of equation

$$i\hbar\dot{a} = E_1 a - \mathbf{d}_{12}\mathbf{E}\,b, \quad i\hbar\dot{b} = E_2 b - \mathbf{d}_{12}\mathbf{E}\,a. \qquad (A6.1.4)$$

Taking into account that the dipole moment of the atom is equal to

$$\mathbf{d} =< \psi|\hat{\mathbf{d}}|\psi >= \mathbf{d}_{12}(ab^* + a^*b), \qquad (A6.1.5)$$

we can rewrite the above equations in the form

$$\ddot{\mathbf{d}} + \Omega^2\mathbf{d} = -k^2 U\mathbf{E}, \quad \dot{U} = \dot{\mathbf{d}}\mathbf{E}, \qquad (A6.1.6)$$

where $k^2 = 4d_{12}^2/\hbar^2$.

The function

$$U(t) =< \psi|H_0|\psi > -\frac{1}{2}\left(E_1 + E_2\right) \quad = \frac{\hbar\Omega}{2}\left(bb^* - aa^*\right) \qquad (A6.1.7)$$

is an averaged internal energy of the atom which zero point is determined by $(1/2)\left(E_1 + E_2\right)$.

Putting $w = kU$ and $d_0 = d_{12}$, we obtain the system (1.6.1).

# REFERENCES

## To Introduction

1. Akhiezer A.I., Fainberg Ya.B. *Interaction of Charge-particle Beams with Plasma.* Dokl. Akad. Nauk. SSSR, **64**, 555, (1949); *On the high-frequency oscillations in an electron plasma.* Zh. Eksp. Teor. Fiz., **21**, 1262, (1951).

2. Bohm D., Gross E.P. *Theory of plasma oscillations. B. Excitation and damping of oscillations.* Phys. Rev., **75**, 1864, (1949).

3. Collection of Works: *Plasma Electronics* (ed. by V.I.Kurilko), Kiev, Naukova Dumka, 1989.

4. Kuzelev M.V., Rukhadze A.A. *Electrodynamics of High-Density Electron Beams in a Plasma* [in Russian], Nauka, Moscow, 1990.

5. Fainberg Ya.B. *Interaction of Charge-particle Beams with Plasma.* Atomnaya Energiya (Atomic Energy), **11**, 313, (1961).

6. Silin V.P., Rukhadze A.A. *Electromagnetic Properties of Plasma and Plasma-like Media*, Atomizdat, Moscow, 1961.

7. Akhiezer A.I., Akhiezer I.A., Polovin R.V., Sitenko A.G., Stepanov K.N. *Plasma Electrodynamics* [in Russian], Nauka, Moscow, 1974.

8. Mikhailovskii A.B.. *Theory of Plasma Instabilities*, v. 1 and 2 Plenum, New York, 1974.

9. Zavoiskii E.K., Rudakov L.I. *Plasma Physics. (Collective Processes in Plasma and Turbulent Heating)* [in Russian], Znanie, Moscow, 1967.

10. Vedenov A.A., Velikhov E.P., Sagdeev R.Z. *Nonlinear oscillations of rarefied plasma.* Nuclear fusion, **1**, 82, (1961); Nuclear fusion, Appendix. **2**, 465, (1962).

11. Romanov Yu.A., Filippov G.F. *Interaction of electron fluxes with longitudinal plasma waves.* Zh. Eksp. Teor. Fiz., **40**, 123, (1961) [Sov. Phys. JETP, **13**, , (1961)].

12. Drummond V.E., Pines D. *Nonlinear saturation of plasma oscillations.* Nuclear fusion, Appendix. **2**, 1049, (1962).

13. Ivanov A.A., Rudakov L.I. *Dynamics of quasilinear relaxation in collisionless plasma.* Zh. Eksp. Teor. Fiz., **51**, 1522, (1966) [Sov. Phys. JETP, **24**, 1027, (1967)].

14. Ivanov A.A. *Physics of Highly Nonequilibrium Plasmas* [in Russian], Atomizdat, Moscow, 1977.

15. Fainberg Ya.B. *The use of plasma wave guides as accelerating structure.* Proc. Symp. CERN, **1**, 84, (1956).

16. Fainberg Ya.B. *Charge particles acceleration in a plasma.* Usp. Fiz. Nauk., **93**, 617, (1967).

17. Briggs R.S. *Collective accelerator for electrons*. Phys. Rev. Lett., **54**, 2588, (1985).

18. Katsouleas T. *Physical mechanisms in the plasma wave field accelerator*. Phys. Rev., **33A**, 2056, (1986).

19. Krasovitskii V.B., Kurilko V.I. *Propagation and excitation of electromagnetic waves in a nonlinear medium*. Zh. Eksp. Teor. Fiz., **48**, 353, (1965) [Sov. Phys. JETP, **21**, 232, (1965)]; *Contribution to the nonlinear theory of two-stream instability in the presence of the anomalous Doppler effect*. Zh. Eksp. Teor. Fiz., **49**, 1832, (1965) [Sov. Phys. JETP, **22**, 1252, (1965)].

20. Onischenko I.N., Linetskii A.R., Matsiborko N.G., Shapiro V.D., Shevchenko V.I. *On nonlinear theory of excitation of monochromatic plasma wave by an electron beam*. Pis'ma Zh. Eksp. Teor. Fiz., **12**, 407, (1970); Matsiborko N.G., Onischenko I.N., Shapiro V.D., Shevchenko V.I. *On nonlinear theory of instability of a monoenergetic electron beam in plasma*. Plasma Physics, **14**, 591, (1972).

21. Thode L.E., Sudan R.N. *Two-stream instability heating of plasmas by relativistic electron beams*. Phys. Rev. Letters, **30**, 732, (1973); *Plasma heating by relativistic electron beams*. I. *Two-stream instability*, II. *Return current interaction*. Phys. Fluids, **18**, 1552, 1564 (1975).

22. Krasovitskii V.B., Kurilko V.I., Strzhemechnii M.A. *Nonlinear theory of interaction between modulated beam and plasma*. Atomnaya Energiya (Atomic Energy), **24**, 545, (1968).

23. Krasovitskii V.B. *The nonlinear theory of the interaction of a modulated relativistic electron beam with a plasma*. Izv. Vyssh. Uchebn. Zaved., Radiofiz., **13**, 1902 (1970) [Radiophys. Qu. Electr., **13**, 1468, (1970)]; *Excitation of a regular plasma wave by a modulated beam with high energy density*. Pis'ma Zh. Eksp. Teor. Fiz., **15**, 346, (1972) [JETP Lett. **15**, 244, (1972)]; *Nonlinear theory of interaction between a bounded relativistic beam and a plasma*. Zh. Eksp. Teor. Fiz., **62**, 995, (1972) [Sov. Phys. JETP, **35**, 525, (1972)]. *Excitation of a regular plasma wave by a charged particle beam*. Zh. Eksp. Teor. Fiz., **66**, 154, (1974) [Sov. Phys. JETP, **39**, 71, (1974)].

24. Chen P., Dawson I.M., Huff R.W., Katsouleas T. *Acceleration of electrons by the interaction of a bunched electron beam with a plasma*. Phys. Rev. Lett., **54**, 693, (1985).

25. Bliokh Yu.P., Balakirev V.A, Mukhin V.V., Onischenko I.N., Fainberg Ya.B. *Charge particles acceleration in a plasma by the charge density waves excited by electron beams and laser radiation*. In: *Plasma Electronics* (ed. by V.I.Kurilko), Kiev, Naukova Dumka, 1989.

26. Bogolyubov N.N., Mitropol'skii Ya.A. *Asymptotic Methods in Theory of Nonlinear Oscillations*, Gordon and Breach, 1965.

27. Krasovitskii V.B., Kurilko V.I. *Theory of amplification of longitudinal waves by a charged particle beam in a nonlinear plasma*. Zh. Eksp. Teor. Fiz., **51**, 445, (1966) [Sov. Phys. JETP, **24**, 300, (1966)].

28. Shapiro V.D., Shevchenko V.I. *The excitation of a monochromatic wave during steady injection of an electron beam into a plasma*. Nucl. Fusion, **12**, 133, (1972).


**To Chapter 1**


1. Onischenko I.N., Linetskii A.R., Matsiborko N.G., Shapiro V.D., Shevchenko V.I. *On nonlinear theory of excitation of monochromatic plasma wave by an electron beam*. Pis'ma Zh. Eksp. Teor. Fiz., **12**, 407, (1970).

2. Drummond W.E., Malberg J.H., O'Neil T.M., Thompson J.R. Phys. Fluids, **13**, 2422, (1970).

3. Ivanov A.A., Parail V.V., Soboleva T.K. *Nonlinear theory of interaction of monoenergetictic beam with a dense plasma*. Zh. Eksp. Teor. Fiz., **63**, 1678, (1972).

**4.** Ivanov A.A. *Physics of Highly Nonequilibrium Plasmas* [in Russian], Atomizdat, Moscow, 1977.

**5.** Fainberg Ya.B., Shapiro V.D., Shevchenko V.I. *To nonlinear theory of interaction of monochromatic relativistic beam with a plasma.* Zh. Eksp. Teor. Fiz., **57**, 966, (1969). [Sov. Phys. JETP, **30**, 528, (1970)].

**6.** Kovtun R.I., Rukhadze A.A. *To theory of nonlinear interaction of a low-density relativistic beam with a plasma.* Zh. Eksp. Teor. Fiz., **58**, 1709, (1970) [Sov. Phys. JETP, **31**, 915, (1970)].

**7.** Matsiborko N.G., Onischenko I.N., Shapiro V.D., Shevchenko V.I. *On nonlinear theory of instability of a monoenergetic electron beam in plasma.* Plasma Physics, **14**, 591, (1972).

**8.** M.Lampe, P.Sprangle. *Saturation of the relativistic two-stream instability by electron trapping.* Physics of fluids, **18**, 475, (1975).

**9.** Thode L.E., Sudan R.N. *Two-stream instability heating of plasmas by relativistic electron beams.* Phys. Rev. Letters, **30**, 732, (1973); *Plasma heating by relativistic electron beams. I. Two-stream instability, II. Return current interaction.* Phys. Fluids, **18**, 1552, 1564 (1975).

**10.** Sudan R.N. *Collective interaction between a beam and a plasma.* In: *Basic Plasma Physics: Supplement to the Second Volume* (ed. by Galeev A.A. and Sudan R.N.), Energoatomizdat, Moscov, 1984.

**11.** Kuzelev M.V., Rukhadze A.A. *Electrodynamics of High-Density Electron Beams in a Plasma* [in Russian], Nauka, Moscow, 1990.

**12.** Krasovitskii V.B. *Excitation of a regular plasma wave by a charged particle beam.* Zh. Eksp. Teor. Fiz., **66**, 154, (1974) [Sov. Phys. JETP, **39**, 71, (1974)].

**13.** Krasovitskii V.B. *Relaxation oscillations in a plasma with an ultrarelativistic electron beam.* Zh. Eksp. Teor. Fiz., **83**, 1324, (1982) [Sov. Phys. JETP, **56**, 760, (1982)].

**14.** Krasovitskii V.B. *Kinetic features of the instability of an ultrarelativistic electron beam in a dense plasma.* Fiz. Plazmy, **22**, 728, (1996) [Plasma Physics Report, **22**, 559, (1996)].

**15.** Dorofeenko V.G., Krasovitskii V.B., Fomin G.V. *Control of ultrarelativistic electron beam instability in a nonlinear plasma.* Zh. Eksp. Teor. Fiz., **98**, 419, (1990) [Sov. Phys. JETP, **71**, 234, (1990)].

**16.** Krasovitskii V.B., Fomin G.V. *Evolution of symmetric solitons in a plasma with a modulated kinetic beam.* Fiz. Plazmy, **24**, 904, (1998) [Plasma Physics Report, **24**, 841, (1998)].

**17.** Volkov Yu.A., Krasovitskii V.B. *Numerical model of a plasma with a monoenergetictic relativistic electron beam.* Fiz. Plazmy, **26**, 73, (2000) [Plasma Physics Report, **26**, 70, (2000)].

**18.** Volkov Yu.A., Krasovitskii V.B. *Nonlinear dispersion in a plasma with a relativistic electron beam.* Fiz. Plazmy, **26**, 1110, (2000) [Plasma Physics Report, **26**, 1039, (2000)].

**19.** Krasovitskii V.B. *The nonlinear theory of the interaction of a modulated relativistic electron beam with a plasma.* Izv. Vyssh. Uchebn. Zaved., Radiofiz., **13**, 1902 (1970) [Radiophys. Qu. Electr., **13**, 1468, (1970)];

**20.** Bliokh Yu.P., Balakirev V.A, Mukhin V.V., Onischenko I.N., Fainberg Ya.B. *Charge particles acceleration in a plasma by the charge density waves excited by electron beams and laser radiation.* In: *Plasma Electronics* (ed. by V.I.Kurilko), Kiev, Naukova Dumka, 1989.

**21.** Bogolyubov N.N., Mitropol'skii Ya.A. *Asymptotic Methods in Theory of Nonlinear Oscillations*, Gordon and Breach, 1965.

**22.** Abramowitz, M. and Stegun, I. A. (Editors) *Handbook of Mathematical Functions with Formulas, Graphs, and Mathematical Tables*, Dover, New York, 1972.

**23.** Nayfeh. *Perturbation Methods*, Wiley, New York, 1973.

**24.** Krall N.A., Trivelpiece A.W. *Principles of Plasma Physics*, McGraw-Hill, New York, 1972.

**25.** Kadomtsev B.B. *Collective Processes in Plasmas*, Pergamon, Oxford, 1976.

**26.** Krasovitskii V.B. , Mitin L.A. *Amplification of a regular wave by an electron beam at the nonlinear phase resonance in a plasma.* Fiz. Plazmy, **23**, 230, (1997) [Plasma Physics Report, **23**, 209, (1997)].

**27.** Shapiro V.D., Shevchnko V.I. *Wave-particle interaction in nonequilibrium media.* Izv. Vyssh. Uchebn. Zaved., Radiofiz., **19**, No. 5-6, 767, (1976).

**28.** Mikhailovskii A.B.. *Theory of Plasma Instabilities*, v. 1, Plenum, New York, 1974.

**29.** Akhiezer A.I., Akhiezer I.A., Polovin R.V., Sitenko A.G., Stepanov K.N. *Plasma Electrodynamics* [in Russian], Nauka, Moscow, 1974.

## To Chapter 2

**1.** Akhiezer A.I., Fainberg Ya.B. *Interaction of charge-particle beams with a plasma.* Dokl. Akad. Nauk. SSSR, **64**, 555, (1949);

**2.** Bohm D., Gross E.P. *Theory of plasma oscillations. B. Excitation and damping of oscillations.* Phys. Rev., **75**, 1864, (1949).

**3.** Shapiro V.D. *On nonlinear theory of interaction between a monoenergetic beam and a plasma.* Zh. Eksp. Teor. Fiz., **44**, 613, (1963) [Sov. Phys. JETP, **17**, 416, (1963)].

**4.** Bogdanov E.V., Kislov V.Ya., Chernov Z.S. *Interaction of electron flux with a plasma; Interaction of slow plasma waves with an electron flux.* Radiotehknika i elektronika (Radio Engineering and Elecrtronics), **5**, 229; 1974 (1960).

**5.** Frieman E.A., Goldberger M.L., Watson K.M., Weinberg S., Rosenbluth M.N. *Two-stream instability in finite beams.* Phys.Fluids, **5**, 196, (1962).

**6.** Gorbatenko M.F. *Interaction of an electron flux with a plasma.* In: *Plasma Physics and the Problem of Controlled Thermonuclear Fusion* (ed. by K.D.Sinel'nikov), Nauchnaya mysl', 1963.

**7.** Krasovitskii V.B. *Waveguide properties of plasma-gas system.* Ukr. Fiz. Zh., **7**, 692, (1964).

**8.** Trivelpiece A.W. *Slow wave propagation in plasma waveguides.* San Francisco, 1967.

**9.** Briggs R.J. *Two-stream instability.* In: *Advances in Plasma Physics*, Vols. 3 and 4 (ed. by A.Simon and W.B.Thompson), Interscience Publishers, New York, 1969.

**10.** Davidson R.C. *Theory of Nonneutral Plasmas*, W.A.Benjamin, Inc., Advanced Book Program Reading, Massachusets, 1974.

**11.** Rukhadze A.A., Bogdankevich L.S., Rosinskii S.E., Rukhlin V.G. *Physics of High-Current Relativistic Electron Beams*, Atomizdat, Moscow, 1980.

**12.** Ivanov A.A., Popkov N.G. *Kinetic theory of two-stream instability of bounded electron beams.* In: *Plasma Electronics* (ed. by V.I.Kurilko), Kiev, Naukova Dumka, 1989.

**13.** Kondratenko A.N. *Cherenkov instability of electron beam in a plasma.* In: *Plasma Electronics* (ed. by V.I.Kurilko), Kiev, Naukova Dumka, 1989.

**14.** Krasovitskii V.B. *Radial self-focusing of an electron beam due to the two-stream instability in a plasma.* Zh. Eksp. Teor. Fiz., **56**, 1253, (1969) [Sov. Phys. JETP, **29**, 674, (1969)];

**15.** Winterberg F. *Electrostatic self-focusing of intense relativistic electron beams in dense plasma.* Bull. Amer. Phys. Soc., 11, 1453, (1970); Atomkernenergie, **22**, 142, (1973).

**16.** Krasovitskii V.B. *On the nonlinear theory of the beam-plasma instability.* Zh. Eksp. Teor. Fiz., **64**, 1597, (1973) [Sov. Phys. JETP, **37**, 809, (1973)];

17. Krasovitskii V.B. *Radiation pressure at the surface of a beam of coherently radiating electrons.* Zh. Tekh. Fiz., **47**, 10, (1977) [Sov. Phys. Tekh. Phys., **22**, 4, (1977)].

18. Krasovitskii V.B., Razdorskii V.G. *Instability of a relativistic monoenergetic beam of electrons with curved orbits propagating in a plasma.* Fiz. Plazmy,, **4**, 785, (1978) [Sov. Plazma Phys, **4**, 441, (1978)].

19. Landau L.D, Lifshits. *Electrodynamics of Continuous Media*, Pergamon, Oxford, 1984.

20. Miller M.A. *Motion of charge particles in a high-frequency electromagnetic field.* Izv. Vyssh. Uchebn. Zaved., Radiofiz., **1**, 110, (1958); Gaponov A.V., Miller M.A. Zh. Eksp. Teor. Fiz., **34**, 242, (1958) [Sov. Phys. JETP, **7**, 168, (1958)].

21. Fainberg Ya.B., Shapiro V.D., Shevchenko V.I. *On the nonlinear theory of the interaction of a monochromatic relativistic electron beam with a plasma.* Zh. Eksp. Teor. Fiz., **57**, 966, (1969) [Sov. Phys. JETP, **30**, 528, (1970)].

22. Krasovitskii V.B., Osmolovskii S.I. *Dynamics of the formation of relativistic electron clumps in a plasma.* Fiz. Plazmy, **19**, 1385, (1993) [Plasma Physics Report, **19**, 726, (1993)].

23. Krasovitskii V.B. *Self-focusing of relativistic electron ring beam exposed to cogerent radiation.* Zh. Tekh. Fiz., **43**, 2599, (1973) [Sov. Phys. Tech. Phys., **18**, 1632, (1973)].

24. Kamke E. *DifferentialGleichungen*, Leipzig, 1959.

25. Gradshteyn I.S., Ryshik I.M. *Table of Integrals, Series, and Products*, Academic Press, New York, 1965.

26. Winterberg F. *Production of dense thermonuclear plasma by means of intense relativistic electron beams.* Phys. Rev., **174**, 212, (1968); *Physics of High Energy Density*, Proceedings of the International School of Physics (Enrico Fermi), Academic Press, New York and London, 1971, p.421.

## To Chapter 3

1. Fainberg Ya.B., Shapiro V.D., Shevchenko V.I. *On the nonlinear theory of the interaction of a monochromatic relativistic electron beam with a plasma.* Zh. Eksp. Teor. Fiz., **57**, 966, (1969) [Sov. Phys. JETP, **30**, 528, (1970)].

2. Kuzelev M.V., Rukhadze A.A. *Electrodynamics of High-Density Electron Beams in a Plasma* [in Russian], Nauka, Moscow, 1990.

3. Krasovitskii V.B. *Nonlinear theory of interaction between a bounded relativistic beam and a plasma.* Zh. Eksp. Teor. Fiz., **62**, 995, (1972) [Sov. Phys. JETP, **35**, 525, (1972)].

4. Lawson J.D. *The Physics of Charge-Particle Beam.* Clarendon Press, Oxford, 1977.

5. Fainberg Ya.B. *Charge-particles acceleration in a plasma.* Usp. Fiz. Nauk., **93**, 617, (1967); *Interaction of charge-particle beams with a plasma.* A survey of phenomena in ionized gases. Vienna, IAEA, 1968.

6. Krasovitskii V.B. *Excitation of regular transverse plasma waves of finite amplitude by a bounded relativistic electron beam.* Fiz. Plazmy, **3**, 105, (1977) [Sov. Plazma Phys, **3**, 60, (1977)].

7. Zinchenko V.P., Krasovitskii V.B. *Nonlinear stabilization of transverse Cherenkov mode in a plasma by an external magnetic field.* Izv. Vyssh. Uchebn. Zaved., Radiofiz., **22**, 51 (1979). [Radiophys. Qu. Electr., **22**, 33, (1979)].

8. Krasovitskii V.B. *Excitation of regular quasitransverse plasma waves by a bounded relativistic electron beam with an anomalous Doppler effect.* Fiz. Plazmy, **5**, 271, (1979) [Sov. Plazma Phys, **5**, 152, (1977)].

9. Dorofeenko V.G., Krasovitskii V.B. *Nonlinear stabilization of the collisionless radial defocusing of a relativistic electron beam.* Fiz. Plazmy, **9**, 357, (1983) [Sov. Plazma Phys,

**9**, 208, (1983)].

10. Krasovitskii V.B., Osmolovskii S.I. *Dynamics of the formation of relativistic electron clumps in a plasma*. Fiz. Plazmy, **19**, 1385, (1993) [Plasma Physics Report, **19**, 726, (1993)].

11. Dorofeenko V.G., Krasovitskii V.B., Osmolovskii S.I. *Nonlinear radial oscillations in a plasma with a bounded relativistic electron beam*. Fiz. Plazmy, **19**, 1371, (1993) [Plasma Physics Report, **19**, 720, (1993)].

12. Dorofeenko V.G., Krasovitskii V.B., Osmolovskii S.I. *Nonlinear dynamics of a thin electron-ion beam in a plasma*. Fiz. Plazmy, **21**, 407, (1995) [Plasma Physics Report, **21**, 385, (1995)].

13. Fedoryuk M.V. Asymptotics. Integrals and Series [in Russian], Moscow, Nauka, 1978.


## To Chapter 4

1. Krasovitskii V.B., Kurilko V.I. *Theory of amplification of longitudinal waves by a charged particle beam in a nonlinear plasma*. Zh. Eksp. Teor. Fiz., **51**, 445, (1966) [Sov. Phys. JETP, **24**, 300, (1966)].

2. Shapiro V.D., Shevchenko V.I. *The excitation of a monochromatic wave during steady injection of an electron beam into a plasma*. Nucl. Fusion, **12**, 133, (1972).

3. Kuzelev M.V., Rukhadze A.A. *Electrodynamics of High-Density Electron Beams in a Plasma* [in Russian], Nauka, Moscow, 1990.

4. Krasovitskii V.B., Mitin L.A. *Amplification of a regular wave by an electron beam at the nonlinear phase resonanse in a plasma*. Fiz. Plazmy, **23**, 230, (1997) [Plasma Physics Report, **23**, 230, (1997)].

5. Dorofeenko V.G., Krasovitskii V.B. *Steady injection of a bounded modulated relativistic electron beam in a plasma*. Izv. Vyssh. Uchebn. Zaved., Radiofiz., **32**, 1535 (1989) [Radiophys. Qu. Electr., **32**, 1140, (1989)].

6. Mitin L.A. *Nonlinear theory of interaction of an electron beam with hybrid waves in a plasma-resonators retarding system*. Fiz. Plazmy, **19**, 445 (1993).

7. Talanov V.I. *Self-focusing of electromagnetic waves in a nonlinear media*. Izv. Vyssh. Uchebn. Zaved., Radiofiz., **7**, 564, (1964).

8. Ivanov A.A. *Physics of Highly Nonequilibrium Plasmas* [in Russian], Atomizdat, Moscow, 1977.

9. Kamke E. *DifferentialGleichungen*, Leipzig, 1959.

10. Abramowitz, M. and Stegun, I. A. (Editors) *Handbook of Mathematical Functions with Formulas, Graphs, and Mathematical Tables*, Dover, New York, 1972.

11. Shapiro V.D., Shevchenko V.I. *Wave-particle interaction in nonequilibrium media*. Izv. Vyssh. Uchebn. Zaved., Radiofiz., **19**, 767, (1976).

12. Filimonov G.F. Radiotehknika i elektronika (Radio Engineering and Elecrtronics), **6**, 1508 (1961).


## To Chapter 5

1. Ginzburg V.L. *On the theory of radiation of super-light charge-particle motion in a retarding medium*. Usp. Fiz. Nauk., **69**, 537, (1959) [Sov. Phys. Usp., **2**, 874, (1969)];

2. Zheleznyakov V.V. *Magnetic bremsstrahlung and instability of charge-particle system in a plasma*. Izv. Vyssh. Uchebn. Zaved., Radiofiz., **2**, 14, (1959); **3**, 57, (1960).

3. Nezlin M.V. *Waves with negative mass and the anomalous Doppler effect*. Usp. Fiz. Nauk., **120**, 481, (1976) [Sov. Phys. Usp., , , (1976)];

4. Mikhailovskii A.B.. *Theory of Plasma Instabilities*, v. 1, Plenum, New York, 1974.

5. Krasovitskii V.B., Kurilko V.I. *Contribution to the nonlinear theory of two-stream instability in the presence of the anomalous Doppler effect.* Zh. Eksp. Teor. Fiz., **49**, 1831, (1965) [Sov. Phys. JETP, **22**, 1252, (1965)];

6. Krasovitskii V.B. *Excitation of a regular plasma wave by a charged particle beam.* Zh. Eksp. Teor. Fiz., **66**, 154, (1974) [Sov. Phys. JETP, **39**, 71, (1974)].

7. Kuzelev M.V., Rukhadze A.A. *Electrodynamics of High-Density Electron Beams in a Plasma* [in Russian], Nauka, Moscow, 1990.

8. Krasovitskii V.B. *Collective slowing of a relativistic electron beam of oscillators in a nonlinear dielectric medium.* Zh. Eksp. Teor. Fiz., **71**, 1358, (1976) [Sov. Phys. JETP, **44**, 710, (1976)].

9. Bachin I.V., Krasovitskii V.B. *Cogerent electromagnetic radiation of an electron-ion beam.* Ukr. Fiz. Zh., **28**, 371, (1983).

10. Bachin I.V., Korchagin V.I., Krasovitskii V.B. *Excitation of an nonlinear low-frequency electromagnetic wave by a relativistic electron beam.* Fiz. Plazmy, 4, 443, (1978) [Sov. Plazma Phys, 4, 248, (1978)].

11. Krasovitskii V.B. *Charge-particle acceleration by the field of a plane wave with alternating phase velocity.* Atomnaya Energiya (Atomic Energy), **20**, 347, (1966).

12. Fainberg Ya.B., Kurilko V.I. *On theory of acceleration by a pressure of light.* Zh. Tekh. Fiz., **29**, 939, (1959).

13. Kolomenskii A.A., Lebedev A.N. *Autoresonant motion of a charge-particle in a plane electromagnetic wave.* Dokl. Akad. Nauk. SSSR, **145**, 1259, (1962) [Sov. Phys.-Dokl., **7**, 745, (1962)]; *Resonance phenomena of a charge-particle motion in a plane electromagnetic wave.* Zh. Eksp. Teor. Fiz., **44**, 261, (1963) [Sov. Phys. JETP, **17**, 179, (1963)].

14. Davydovskii V.Ya. *About possibility of a resonance charge-particle acceleration by electromagnetic waves in a constant magnetic field.* Zh. Eksp. Teor. Fiz., **43**, 886, (1962) [Sov. Phys. JETP, **16**, 629, (1963)].

15. Roberts C.S., Buchsbaum S.J. *Motion of a charge particle in a constant magnetic field and a transverse electromagnetic wave propagating along the field.* Phys. Rev., **A135**, 381, (1964).

16. Jory H.R., Trivelpiece A.W. *Charge particle motion in a large amplitude electromagnetic fields.* J. Appl. Phys., **39**, 3053, (1968).

17. Krasovitskii V.B., Kurilko V.I. *Influence of radiation on resonance charge-particle acceleration in a field of plane wave.* Izv. Vyssh. Uchebn. Zaved., Radiofiz., **7**, 1193, (1964).

18. Krasovitskii V.B., Razdorskii V.G. *Influence of Coulomb collisions on autoresonant acceleration.* Zh. Tekh. Fiz., **54**, 700, (1984) [Sov. Phys. Tech. Phys., **29**, 414, (1984)].

19. Krasovitskii V.B. *Resonant acceleration of a beam of oscillators in a medium with population inversion.* Zh. Eksp. Teor. Fiz., **57**, 1760, (1969) [Sov. Phys. JETP, **30**, 951, (1970)].

20. Krasovitskii V.B. *Resonant acceleration of a beam of oscillators in a field of plane wave.* Atomnaya Energiya (Atomic Energy), 28, 434, (1970).

21. Bliokh Yu.P., Lyubarskii M.P., Onischenko I.N. *Interaction of a relativistic electron beam with an electromagnetic wave under conditions of autoresonance.* Fiz. Plazmy, **6**, 114, (1980) [Sov. Plazma Phys, **6**, , (1980)].

22. Landau L.D, Lifshits. *Electrodynamics of Continuous Media*, Pergamon, Oxford, 1984.

23. Krasovitskii V.B., Kurilko V.I. *Propagation and excitation of electromagnetic waves in a nonlinear medium.* Zh. Eksp. Teor. Fiz., **48**, 353, (1965) [Sov. Phys. JETP, **21**, 232, (1965)].

24. Krasovitskii V.B., Kurilko V.I. *Interaction of electromagnetic waves with a two-level system*. Zh. Tekh. Fiz., **36**, 401, (1966) [Sov. Phys. Tech. Phys., **11**, 414, (1966)].

25. Askaryan G.A. *Influence of field gradient of intense electromagnetic beam on electrons and atoms*. Zh. Eksp. Teor. Fiz., **42**, 1567, (1962) [Sov. Phys. JETP, **15**, 1088, (1962)].

26. Akhiezer A.I., Akhiezer I.A., Polovin R.V., Sitenko A.G., Stepanov K.N. *Plasma Electrodynamics* [in Russian], Nauka, Moscow, 1974.

27. Bachin I.V., Krasovitskii V.B. *Amplitude-modulation electromagnetic waves in a non-equilibrium magnetized plasma*. Ukr. Fiz. Zh., **27**, 139, (1982).

28. Rowlands J., Krasovitskii V.B., Kurilko V.I. *Concerning the stability of phase oscillators*. Pis'ma Zh. Eksp. Teor. Fiz., **2**, 511, (1965) [JETP Lett. **2**, 319, (1965)].

29. Dzhavakhishvili D.I., Tsintsadze N.L. *Transport phenomena in an ionized ultra-relativistic plasma* Zh. Eksp. Teor. Fiz., **64**, 1314, (1973) [Sov. Phys. JETP, **37**, 666, (1974)].

30. Connor J.W., Hastie R.J. *Relativistic limitation on runaway elrctrons*. Nucl. fusion, **15**, 415, (1975).

31. Abramowitz, M. and Stegun, I. A. (Editors) *Handbook of Mathematical Functions with Formulas, Graphs, and Mathematical Tables*, Dover, New York, 1972.

32. Fainberg Ya.B. *Dynamics of charge-particles in linear accelerator with a traveing wave*. In: *Linear Accelerator Theory and Design* [in Russian], Gosatomizdat, Moscow, 1962.

33. Landau L.D, Lifshits E.M. *The Classical Theory of Fields*, Pergamon, New York, 1976.

### To Chapter 6

1. Feynman R., Leighton R., Sands M. *Feynman Lectures on Physics*, v. 2, Addison-Wesley, Reading, Massachusets, 1964.

2. Landau L.D, Lifshits E.M. *Quantum Mechanics*, Addison-Wesley, 1958.

3. Davis L.W. *Semiclassical theory of optical generators*, Proceedings of the IEEE, **51**, 112, (1963).

4. Allen L., Eberly J.H. *Optical Resonance and Two-level Atoms*, Wiley-Interscience, New York, 1975.

5. Fain V.M., Khanin Ya.I. *Quantum Radiophysics* [in Russian], Soviet Radio Press, 1965.

6. Neufeld J. Radiation produced by an electron beam passing through a dielectric medium. Phys.Rev., **116**, 785, (1959).

7. Tsytovich V.N., Shapiro V.D. *Interaction of electron beam with an optical active medium*. Zh. Tekh. Fiz., **34**, 764, 1964 [Sov. Phys. Tech. Phys., **9**, 583, (1964)].

8. Krasovitskii V.B., Kurilko V.I. *Propagation and excitation of electromagnetic waves in a nonlinear medium*. Zh. Eksp. Teor. Fiz., **48**, 353, (1965) [Sov. Phys. JETP, **21**, 232, (1965)].

9. Krasovitskii V.B., Linetskii A.R. *Population inversion in a three-level system*. Izv. Vyssh. Uchebn. Zaved., Radiofiz., **11**, 224, (1968) [Izvestia Vuz. Radiofizika, , 126, (1968)].

10. Krasovitskii V.B. *Transformation of the energy accumulated in an active medium into the energy of a longitudinal wave in a plasma*. Zh. Eksp. Teor. Fiz., **53**, 573, (1967) [Sov. Phys. JETP, **26**, 371, (1968)].

11. Krasovitskii V.B., Kurilko V.I. *Cherenkov radiation in a medium with an inverted population level*. Zh. Eksp. Teor. Fiz., **57**, 863, (1969) [Sov. Phys. JETP, **30**, 473, (1970)].

12. Krasovitskii V.B. *Acceleration of a modulated beam of charge particle in a medium with inverted population level*. Zh. Tekh. Fiz., **61**, 1093, (1971) [Sov. Phys. Tech. Phys., **16**, 864, (1971)].

**13.** Krasovitskii V.B. *Self-excitation of a molecular beam moving through a slow-wave system.* Zh. Tekh. Fiz., **40**, 1328, (1970) [Sov. Phys. Tech. Phys., **15**, 1028, (1970)].

**14.** Siegman A.E. *Microwave Solid State Masers*, McGraw-Hill, New York, 1964.

**15.** Fainberg Ya.B. *Dynamics of charge-particles in linear accelerator with a traveing wave.* In: *Linear Accelerator Theory and Design* [in Russian], Gosatomizdat, Moscow, 1962.

**16.** Jelly J.V. *Cherenkov radiation*, Pergamon Press, New York, 1958.

**17.** Tamm I.E. *Radiation of charge-particle which is at rest in a moving with super speed-of-light medium.* J. USSR, **1**, 439, 1939.

**18.** Veksler V.I. *Coherent principle of acceleration of charged particles.* Proc. Symp. CERN, **1**, 80, (1956); Usp. Fiz. Nauk., **66**, 99, (1958) [Sov. Phys. Usp., **1**, 54, (1958)].

**19.** Ginzburg V.L., Eidman V.Ya. *Responce of radiation in the case of media with negative damping.* Zh. Eksp. Teor. Fiz., **36**, 1823, (1959) [Sov. Phys. JETP, **9**, 1300, (1959)].

**20.** Fainberg Ya.B. *Charge-particles acceleration by a pressure of light.* In: *Plasma Physics and the Problem of Controlled Thermonuclear Fusion* (edit by K.D.Sinel'nikov), Nauchnaya mysl', 1963.

**21.** Landau L.D, Lifshits. *Electrodynamics of Continuous Media*, Pergamon, Oxford, 1984.

**22.** Ginzburg V.L. *On the theory of radiation of super-light charge-particle motion in a retarding medium.* Usp. Fiz. Nauk., **69**, 537, (1959) [Sov. Phys. Usp., **2**, 874, (1969)].

# INDEX

## #

4G, 178

## A

Aβ, 31, 74, 109, 115, 120, 124, 138, 140, 141, 185, 186, 195, 196
accelerator, 205, 211, 212
accuracy, 135, 137, 155, 184
active centers, 194, 198
adiabatic, 27, 30, 33, 124
alternative, 3, 99, 112, 127, 153, 184
amplitude, 3, 4, 5, 8, 9, 10, 11, 13, 14, 16, 19, 20, 21, 22, 23, 24, 25, 26, 27, 28, 30, 32, 33, 34, 35, 36, 39, 40, 41, 42, 43, 44, 46, 47, 50, 51, 52, 53, 55, 56, 58, 60, 66, 72, 73, 74, 75, 79, 83, 93, 99, 100, 102, 103, 104, 105, 109, 110, 111, 112, 113, 115, 116, 117, 118, 119, 121, 122, 123, 124, 125, 127, 128, 130, 131, 132, 133, 134, 135, 136, 137, 138, 139, 140, 141, 142, 144, 145, 147, 149, 150, 151, 152, 155, 156, 157, 158, 160, 161, 162, 163, 168, 172, 174, 176, 181, 182, 183, 184, 185, 186, 189, 191, 193, 194, 195, 196, 197, 198, 201, 208, 210
angmuir, 5, 179
angular velocity, 98
anisotropic, 64, 66
anisotropy, 156
anomalous, 1, 3, 5, 93, 106, 113, 114, 115, 116, 117, 119, 147, 148, 149, 151, 152, 153, 155, 156, 158, 162, 179, 181, 184, 200, 201, 205, 208, 209, 210
argument, 24, 35, 44, 74, 123, 188, 194
assumptions, 64, 98
asymmetry, 37, 43

asymptotic(s), 8, 11, 14, 19, 21, 24, 28, 30, 32, 33, 34, 43, 48, 52, 61, 71, 82, 84, 95, 101, 104, 117, 123, 135, 141, 144, 152, 160, 170, 171, 209
asymptotically, 19, 37, 163
asynchronous, 4, 8, 37, 39, 43, 59
atoms, 179, 180, 181, 184, 185, 187, 188, 189, 191, 192, 193, 196, 211
attention, 3, 4
autoresonance, 5, 147, 210
averaging, 10, 23, 45, 54, 59, 79, 100, 105, 106, 109, 185, 198

## B

beam density, 5, 11, 18, 19, 21, 23, 24, 25, 28, 30, 33, 39, 45, 66, 70, 75, 80, 81, 82, 84, 85, 87, 90, 105, 140, 142, 148, 161, 200
beams, 3, 4, 53, 55, 57, 59, 63, 64, 66, 90, 93, 96, 120, 137, 205, 206, 207, 208
behavior, 19, 20, 28, 32, 33, 34, 35, 56, 61, 74, 80, 83, 101, 110, 112, 119, 121, 168, 185
bell, 33
Bessel, 71, 74, 77, 80, 85, 92, 99, 169
bleaching, 36
blocks, 140, 141
boundary conditions, 44, 69, 88, 94, 142
bounds, 106
bremsstrahlung, 164, 209

## C

cell, 4, 8, 43, 55
CERN, 204, 212
charge density, 4, 8, 105, 195, 198, 205, 206
charged particle, 3, 4, 5, 6, 147, 156, 164, 169, 172, 177, 179, 197, 200, 205, 206, 209, 210, 212

## D

## E

electron beam(s), 1, 3, 4, 5, 7, 8, 14, 15, 17, 23, 24, 25, 26, 36, 40, 41, 42, 43, 53, 63, 66, 74, 75, 81, 85, 87, 94, 112, 119, 120, 121, 125, 127, 128, 129, 135, 137, 141, 145, 146, 147, 153, 156, 162, 179, 191, 193, 205, 206, 207, 208, 209, 210, 211

electron density, 45, 129, 145

electrostatic, 3, 54, 64, 66, 83, 87, 93, 96, 97, 98, 129, 145, 147, 167, 168, 180, 181, 184

elongation, 39

emission, 72, 93, 191, 193, 194, 197

energy, 3, 4, 5, 6, 7, 8, 9, 11, 13, 15, 16, 20, 21, 23, 24, 25, 26, 28, 34, 35, 36, 37, 38, 39, 43, 44, 45, 46, 47, 48, 50, 51, 52, 53, 55, 56, 58, 59, 60, 63, 66, 68, 70, 72, 74, 81, 85, 90, 93, 106, 109, 114, 116, 128, 135, 137, 147, 149, 151, 152, 153, 157, 158, 160, 162, 164, 165, 166, 167, 168, 169, 171, 172, 174, 176, 177, 179, 180, 182, 183, 184, 185, 186, 190, 191, 192, 193, 194, 196, 197, 198, 199, 200, 201, 203, 205, 211

energy density, 7, 15, 23, 28, 45, 51, 52, 55, 56, 58, 59, 66, 74, 109, 116, 153, 157, 158, 160, 166, 199, 205

energy transfer, 16, 51, 81

equality, 199

equating, 95, 192

equilibrium, 27, 65, 74, 83, 87, 99, 104, 120, 121, 124, 129, 144, 145, 182, 193, 211

Euler, 75

evolution, 5, 15, 19, 27, 40, 42, 45, 49, 52, 55, 56, 57, 58, 80, 94, 104, 120, 121, 122, 124

excitation, 3, 4, 5, 36, 58, 64, 87, 93, 99, 109, 110, 112, 125, 147, 155, 168, 184, 194, 195, 200, 201, 205, 209, 210, 211, 212

expansions, 144, 170

experts, 1

exponential, 3, 4, 8, 23, 38, 41, 43, 44, 52, 101, 108, 140, 181

F

Fermi, 208

Feynman, 211

flow, 5, 48, 63

fluctuations, 116, 183

focusing, 1, 5, 64, 66, 74, 75, 78, 81, 93, 99, 103, 104, 106, 110, 112, 119, 120, 121, 127, 141, 144, 145, 207, 208, 209

Fourier, 67, 75, 77, 84, 123

fusion, 3, 204, 211

G

gases, 208

generalization, 65, 200

generation, 191

generators, 211

graduate students, 6

graph, 145

groups, 43, 50

growth, 3, 4, 5, 7, 8, 14, 15, 23, 25, 26, 27, 28, 30, 36, 37, 41, 42, 43, 44, 45, 50, 51, 52, 53, 55, 56, 57, 58, 61, 66, 71, 73, 77, 78, 79, 81, 86, 87, 89, 90, 93, 94, 99, 101, 102, 106, 107, 108, 109, 110, 111, 112, 113, 114, 116, 117, 118, 119, 120, 121, 124, 127, 141, 147, 148, 149, 151, 153, 154, 155, 156, 159, 161, 172, 174, 181, 182, 184, 197, 198

growth rate, 5, 7, 8, 15, 23, 25, 26, 27, 30, 36, 37, 41, 42, 44, 45, 50, 51, 52, 53, 56, 57, 58, 61, 66, 71, 73, 77, 79, 81, 86, 87, 89, 90, 101, 102, 106, 107, 108, 109, 110, 111, 112, 114, 116, 117, 118, 119, 121, 124, 148, 149, 154, 155, 156, 159, 161

growth time, 99

gyrofrequency, 97, 98, 106, 109, 116, 117, 153, 158, 162, 164, 169

H

Hamiltonian, 27, 203

harmonics, 4, 10, 11, 72, 74, 93, 99, 102, 113, 123, 137, 139, 142

heat, 169

heating, 3, 4, 8, 16, 37, 43, 47, 50, 169, 191, 205, 206

height, 79

hybrid, 56, 106, 113, 117, 209

hydrodynamic, 4, 8, 17, 18, 21, 23, 24, 37, 38, 39, 40, 43, 44, 45, 46, 47, 48, 52, 53, 56, 57, 58, 67, 74, 78, 81, 93, 96, 136, 153, 169, 180

hydrodynamics, 8, 23, 75

hyperbolic, 188

I

IAEA, 208

identity, 69

inclusion, 169

inequality, 19, 23, 37, 63, 78, 80, 81, 84, 91, 96, 100, 105, 112, 114, 143, 144, 151, 155, 160, 161, 173, 177, 184, 193, 197, 202

infinite, 63, 79, 83, 91, 92, 98, 102, 142

inhomogeneity, 4, 63

instability, 1, 3, 4, 5, 7, 8, 14, 15, 16, 17, 19, 20, 21, 23, 26, 27, 28, 30, 35, 36, 43, 44, 47, 50, 51, 52, 53, 54, 55, 56, 58, 59, 63, 64, 66, 70, 71, 73, 74, 76, 78, 80, 85, 86, 88, 89, 90, 91, 93, 94, 99, 100, 101, 102, 103, 106, 108, 109, 110, 112, 113, 114, 116, 117, 118, 119, 120, 121, 122, 123, 124, 125, 127, 128, 138, 139, 140, 141, 145, 146, 147, 148, 152, 153, 155, 156, 162, 178, 179, 181, 184, 197, 200, 202, 205, 206, 207, 209, 210

integration, 10, 14, 15, 18, 30, 31, 33, 35, 37, 43, 45, 50, 52, 56, 58, 63, 67, 74, 78, 80, 81, 82, 89, 92, 97, 98, 101, 102, 104, 105, 106, 110, 116, 117, 121, 135, 140, 141, 155, 165, 167, 182, 200

intensity, 158, 193

interaction, 1, 3, 4, 5, 7, 15, 16, 23, 36, 43, 52, 55, 59, 64, 75, 77, 78, 80, 93, 127, 137, 145, 146, 147, 153, 185, 193, 194, 196, 197, 198, 203, 204, 205, 206, 207, 208, 209, 210, 211

interface, 66, 67

interpretation, 4, 12, 68, 188

interval, 15, 39, 45, 51, 99, 120, 121

inversion, 184, 189, 190, 193, 210, 211

ion beam, 120, 121, 124, 147, 153, 155, 158, 159, 209

ions, 8, 43, 45, 50, 120, 121, 124, 147, 153, 155, 156, 158, 159, 162, 165, 167, 174, 193, 202

---

## K

kinetic energy, 48, 63, 128, 168, 179, 193, 196

kinetic equations, 23, 52

kinetic model, 37, 39, 45, 47, 48

---

## L

Landau damping, 3, 4, 8, 43, 51

Langmuir oscillations, 5, 7, 14, 15, 19, 45, 46, 52, 53, 55, 58, 59, 109, 127, 137, 147

Langmuir wave, 5, 43, 50, 52, 58, 59, 63, 179

Langmuir wave packets, 50

laser, 5, 179, 191, 205, 206

laser radiation, 205, 206

law, 24, 60, 80, 163, 165, 176, 183, 186, 187, 192

lead, 7, 24, 64, 90, 168

lifetime, 196

limitation, 211

linear, 3, 4, 10, 14, 15, 16, 17, 19, 23, 25, 26, 28, 36, 38, 41, 44, 52, 53, 54, 55, 56, 58, 60, 63, 64, 74, 75, 77, 94, 99, 100, 102, 103, 107, 109, 110, 112, 115, 116, 118, 119, 121, 127, 128, 133, 134, 135, 136, 137, 138, 139, 140, 141, 144, 145, 147, 149, 152, 153, 155, 158, 159, 160, 161, 162, 179,

180, 181, 187, 193, 194, 195, 196, 197, 201, 202, 211, 212

London, 208

low temperatures, 64

---

## M

magnetic, 4, 5, 64, 66, 69, 70, 73, 81, 87, 88, 93, 94, 95, 97, 106, 107, 108, 110, 113, 116, 117, 119, 120, 127, 128, 137, 138, 140, 141, 142, 143, 144, 145, 146, 147, 148, 152, 156, 159, 161, 163, 164, 168, 173, 174, 175, 208, 210

magnetic field, 5, 64, 69, 70, 73, 81, 88, 93, 94, 95, 97, 106, 107, 108, 110, 113, 116, 117, 119, 120, 127, 128, 137, 138, 140, 141, 142, 143, 144, 146, 147, 148, 152, 208, 210

magnetic field effect, 64

matrix, 148

Maxwell equations, 52, 54, 69, 75, 94, 98, 158, 180, 185, 194, 196

mechanical, 28, 34

media, 5, 6, 193, 196, 207, 209, 212

metric, 37

microwave radiation, 1

mixing, 26, 39, 141

models, 4, 43, 46, 47, 87, 105, 174

modulation, 3, 4, 5, 8, 19, 21, 25, 26, 28, 30, 35, 36, 37, 39, 40, 42, 44, 50, 51, 52, 55, 56, 58, 93, 94, 99, 101, 102, 104, 107, 110, 119, 120, 127, 128, 129, 130, 131, 135, 136, 138, 140, 141, 142, 144, 145, 168, 184, 194, 195, 197, 198, 211

molecular beam, 200, 202, 212

momentum, 8, 9, 11, 13, 25, 38, 39, 40, 42, 45, 59, 60, 67, 68, 148, 151, 163, 169, 170, 171, 172, 174, 176, 177, 199, 200

monoenergetic, 4, 7, 8, 15, 43, 44, 45, 52, 53, 64, 76, 77, 78, 87, 93, 99, 205, 206, 207, 208

monograph, 4, 6

Moscow, 204, 206, 207, 208, 209, 210, 211, 212

motion, 7, 8, 11, 23, 24, 25, 34, 43, 44, 52, 58, 67, 68, 73, 75, 79, 80, 87, 90, 93, 94, 96, 98, 100, 120, 121, 123, 128, 129, 136, 140, 141, 147, 152, 153, 158, 160, 163, 164, 169, 173, 175, 177, 181, 186, 188, 191, 193, 194, 197, 198, 199, 200, 202, 209, 210, 212

---

## N

natural, 3, 35, 184, 186, 189, 192

Nd, 180, 181, 182, 183, 185, 191, 195, 196, 198, 199, 201

neglect, 60, 67, 121, 153, 170, 187, 191, 197

New York, 204, 206, 207, 208, 209, 210, 211, 212

nonequilibrium, 193, 207, 209

nonlinear, 1, 3, 4, 5, 7, 8, 10, 13, 14, 15, 16, 17, 19, 23, 24, 25, 26, 27, 28, 30, 33, 34, 35, 36, 37, 38, 43, 44, 47, 48, 49, 50, 52, 54, 55, 56, 58, 59, 73, 74, 93, 94, 96, 97, 99, 100, 101, 102, 104, 106, 110, 112, 116, 117, 119, 120, 121, 127, 128, 130, 131, 133, 134, 135, 136, 137, 138, 139, 141, 147, 149, 152, 153, 154, 156, 157, 158, 159, 160, 161, 162, 163, 164, 166, 167, 168, 177, 179, 181, 182, 184, 185, 186, 190, 191, 192, 193, 194, 197, 199, 201, 202, 205, 206, 207, 208, 209, 210, 211

nonlinear dynamics, 120

nonlinearities, 4, 8, 128, 135

## O

omentum, 169

operator, 26, 83, 85, 87, 174, 185

optical, 211

orbit, 91

ordinary differential equations, 44, 106

orthogonality, 99

oscillation, 3, 4, 5, 14, 16, 19, 27, 33, 34, 36, 45, 46, 50, 56, 58, 59, 63, 77, 78, 81, 116, 127, 138, 149, 151, 168, 177, 180, 181, 200

oscillator, 34, 116, 147, 200

## P

parabolic, 43, 50

parameter, 7, 11, 21, 22, 25, 26, 27, 28, 29, 31, 36, 38, 39, 41, 42, 70, 71, 72, 75, 83, 86, 89, 90, 100, 101, 102, 103, 106, 108, 109, 110, 117, 119, 121, 124, 128, 131, 132, 135, 136, 138, 140, 143, 144, 145, 155, 161, 165, 167, 172, 188, 189, 192

particles, 1, 3, 44, 45, 47, 48, 50, 63, 78, 79, 81, 87, 93, 103, 105, 110, 112, 121, 128, 136, 137, 149, 151, 156, 158, 160, 164, 168, 169, 170, 172, 174, 193, 194, 197, 200, 201, 204, 205, 206, 208, 211, 212

passive, 197

pendulum, 182, 201

periodic, 44, 48, 53, 54, 123, 128, 150, 201

permit, 4

permittivity, 25, 37, 53, 63, 82, 148

perturbation(s), 3, 4, 5, 8, 17, 18, 19, 23, 28, 30, 35, 36, 38, 39, 40, 42, 43, 44, 45, 46, 52, 53, 63, 66, 81, 82, 87, 88, 94, 107, 112, 116, 118, 119, 122, 124, 125, 127, 137, 138, 139, 147, 148, 150, 173, 180, 185

phase diagram, 53, 58

phase space, 15, 17, 128, 136

phonon, 64

physics, 1, 3, 193

plasma, 1, 3, 4, 5, 6, 7, 8, 9, 10, 11, 14, 15, 16, 21, 23, 24, 25, 26, 27, 28, 30, 35, 36, 37, 38, 39, 40, 41, 42, 43, 44, 45, 46, 47, 48, 49, 50, 51, 52, 53, 54, 55, 56, 57, 58, 59, 60, 63, 64, 65, 66, 68, 69, 70, 71, 72, 73, 74, 75, 77, 78, 79, 80, 81, 82, 83, 84, 85, 86, 87, 90, 93, 94, 96, 97, 98, 99, 100, 101, 104, 105, 106, 108, 110, 112, 113, 114, 115, 116, 119, 120, 122, 125, 127, 128, 129, 130, 131, 133, 134, 135, 137, 138, 139, 140, 141, 142, 144, 145, 146, 147, 152, 153, 156, 158, 159, 160, 161, 162, 164, 167, 168, 169, 174, 179, 184, 191, 193, 202, 204, 205, 206, 207, 208, 209, 210, 211

plasma current, 54

Pointing vector, 5, 63

Poisson, 8, 9, 23, 24, 43, 87, 97, 98, 129, 153, 191

Poisson equation, 8, 9, 23, 24, 87, 97, 129, 153, 191

polarization, 5, 86, 153, 158, 179, 180, 182, 183, 185, 193, 194, 195, 196, 200, 201, 202

polarized, 5, 93, 148, 162, 164, 202

population, 5, 6, 179, 183, 184, 187, 188, 189, 190, 192, 193, 194, 196, 197, 198, 210, 211

Population inversion, 184, 211

potential energy, 34

power(s), 10, 24, 27, 48, 66, 109, 114, 124, 135, 170

pressure, 5, 44, 63, 64, 65, 66, 69, 75, 81, 85, 90, 100, 121, 127, 128, 142, 147, 208, 210, 212

probability, 186, 187

production, 3, 191

programming, 197

propagation, 71, 72, 94, 96, 97, 156, 158, 191, 207

property, 3

pulse(s), 1, 5, 8, 15, 30, 31, 33, 34, 37, , 39, 41 63, 196

## Q

quantum mechanics, 26

quasilinear, 75, 78, 79, 80, 204

## R

radiation, 1, 5, 63, 64, 66, 68, 69, 72, 73, 74, 84, 93, 96, 147, 153, 156, 177, 193, 194, 196, 197, 205, 206, 208, 209, 210, 211, 212

radical, 194, 199

radius, 4, 5, 63, 65, 66, 72, 73, 74, 75, 78, 80, 81, 82, 87, 89, 93, 94, 96, 97, 99, 100, 103, 104, 105, 106, 111, 112, 113, 114, 117, 118, 119, 121, 122,

123, 124, 130, 137, 138, 140, 142, 143, 144, 145, 177

random, 76

range, 8, 14, 15, 19, 20, 21, 23, 27, 33, 37, 39, 44, 45, 53, 55, 58, 74, 84, 86, 99, 102, 104, 109, 119, 120, 128, 148, 156, 158, 172, 181

reasoning, 65

recalling, 171

rectilinear, 5, 66, 89

redistribution, 58

reduction, 23, 81, 141, 196

reference frame, 192

refining, 53

reflection, 64, 65

refractive index, 149, 151, 153, 155, 156, 159, 160, 161, 162, 194, 200

relationship(s), 135, 138, 140, 141, 143, 144, 145, 150, 187, 201

relaxation, 3, 4, 8, 36, 43, 48, 52, 55, 59, 197, 203, 204

relaxation effect, 197

relaxation processes, 203

retardation, 152, 153, 156, 196, 197

returns, 27, 30, 34

## S

satellite, 58

saturation, 4, 8, 14, 15, 16, 35, 36, 43, 55, 56, 58, 59, 93, 114, 119, 128, 137, 163, 167, 177, 196, 204

scalar, 191

scattering, 14, 141, 169

search, 93

separation, 55, 105, 129, 156

series, 67, 123, 157, 161

shape, 52, 140

sign, 26, 36, 72, 75, 80, 82, 142, 148, 178, 180, 181, 193, 194, 196, 197

similarity, 127, 147, 179

simulation, 4, 6, 26, 110

sine, 171

singular, 32, 82, 84

solid state, 66

soliton(s), 28, 34, 36, 37, 40, 150, 152, 156, 206

species, 44, 45, 48, 50, 159

spectra, 93

spectrum, 3, 4, 34, 35, 37, 43, 50, 59, 63, 64, 77, 78, 79, 81, 83, 84, 85, 87, 93, 99

speed, 5, 7, 147, 163, 164, 168, 179, 212

speed of light, 5, 7, 147, 163, 164, 168, 179

stability, 26, 28, 109, 116, 127, 139, 141, 144, 164, 200, 201, 211

stabilization, 28, 104, 106, 109, 110, 116, 118, 152, 155, 197, 208

stages, 4, 8, 19, 43, 52, 122

stars, 85, 131, 132, 133

stress, 72

students, 1, 6

substitution, 62, 70, 74, 77, 83, 88, 92, 114, 146, 152, 163, 164, 182, 184, 187, 197

successive approximations, 75

Sudan, 205, 206

suppression, 23, 110, 141

surface component, 68, 73

surface layer, 68, 70

symmetry, 76

synchronization, 55, 104

synchronous, 153, 173, 174, 199

systems, 1, 179, 186

## T

tangential electric field, 99

temperature, 44, 48, 49, 52, 76, 78, 83, 85, 86, 87, 91, 96, 97, 100, 121, 129, 156, 169, 171, 172, 179, 196

temporal, 9, 121, 124, 125, 151, 172

theoretical, 3, 4, 6, 43, 193

theory, 1, 3, 4, 5, 6, 8, 15, 19, 23, 24, 25, 35, 53, 54, 55, 56, 58, 64, 66, 78, 81, 94, 99, 105, 115, 121, 127, 139, 140, 141, 144, 147, 158, 160, 162, 177, 179, 181, 196, 201, 202, 205, 206, 207, 208, 209, 210, 211, 212

thermonuclear, 5, 66, 208

threshold, 26, 27, 28, 30, 36, 45, 56, 109, 110, 117, 119, 121, 139, 140, 141, 144

time, 4, 5, 7, 8, 10, 11, 15, 23, 26, 27, 30, 34, 35, 36, 39, 43, 45, 47, 48, 49, 50, 51, 52, 53, 54, 55, 56, 58, 59, 60, 66, 72, 74, 76, 78, 79, 81, 88, 99, 100, 102, 106, 110, 115, 116, 117, 119, 120, 121, 124, 127, 128, 129, 130, 131, 135, 138, 141, 149, 152, 153, 157, 162, 163, 168, 171, 172, 174, 179, 180, 181, 182, 183, 184, 185, 186, 187, 188, 189, 190, 191, 192, 193, 194, 195, 197, 198, 200, 201

trajectory, 27, 29, 173

transformation(s), 4, 5, 27, 32, 63, 142, 152, 168, 179, 184, 191

transition(s), 3, 14, 17, 23, 32, 33, 36, 67, 110, 123, 156, 180, 183, 184, 185, 186, 187, 188, 190, 191, 192

transparency, 194

transport, 36, 102, 110

transportation, 137

traveling waves, 115

turbulent, 3